Low Carbon Energy Supply Technologies and Systems

Low Carbon Energy Supply Technologies and Systems

Edited by
Atul Sharma
Amritanshu Shukla
Renu Singh

CRC Press
Taylor & Francis Group
Boca Raton London New York

CRC Press is an imprint of the
Taylor & Francis Group, an **informa** business

CRC Press
Taylor & Francis Group
52 Vanderbilt Avenue,
New York, NY 10017

Contents

Foreword

Energy is the backbone of human activity in the whole world. However, energy availability is a real concern for everyone. Population and income growth are the two most powerful driving forces behind the demand for energy. Societies have been dependent on different types of energy in the past and have been forced to change from one type of energy to another, for example, from wood to fossil fuels. Fossil fuels have been the fuels of choice in the twentieth century, and the first quarter of the twenty-first century is a transition period from fossil fuels to renewable energy. The transition requires a concerted effort for capacity building in the area of the renewable sector. There is also a needed growth of electric power with a reduction of CO_2 discharge. A broad spectrum of energy systems and novel technologies are required to address the challenges of expanding energy resources and sustainable development. The majority of the world's man-made carbon emissions are released by burning fossil fuels to create electricity, heat, or motion. This means that it will almost certainly be impossible to reduce greenhouse gas concentrations to sustainable levels unless large quantities of low-carbon energy can be brought on-stream to substitute for fossil fuels.

The main sources of low-carbon energy are renewables, which include categories such as solar, wind, hydro, biomass, marine energy, and nuclear. Efficient energy management and effective conservation procedures have also been very important considerations for our society for many years. Increasing the amount of energy from renewable and low-carbon technologies will help us to make sure the world has a secure energy supply, reduce greenhouse gas emissions to slow down climate change, and stimulate investment in new jobs and businesses. However, planning has an important role in the delivery of new renewable and low-carbon energy infrastructure in locations where the local environmental impact is acceptable.

Some renewable energy technologies including solar and wind are growing too quickly. However, the proportion of total energy that comes from low-carbon sources in the coming years and decades will depend not just on the roll-out of renewables and nuclear but also on total energy consumption, which without regulation at the global level could continue to grow, cancelling out the emissions savings of low-carbon sources. There are many options for lowering the current levels of carbon emissions. Some options, such as wind power and solar power, produce low quantities of total life-cycle carbon emissions, using renewable sources. The term "low-carbon power" can also include power that continues to utilize the world's natural resources, such as natural gas and coal, but only when they employ techniques that reduce carbon dioxide emissions from these sources when burning them for fuel, such as, as of 2012, the pilot plants performing carbon capture and storage.

This book contains chapters on solar, bio, fuel cell, and alternate fuels. This book should be of interest to a wide variety of individuals. Some of these include vocational-technical schools, teachers, policy makers, and students. No doubt that the editors from the Rajiv Gandhi Institute of Petroleum Technology (RGIPT) and IARI, New Delhi, had made enough efforts for the last several years to develop this type of book, which discusses the low-carbon energy supply technology and systems with full details. I feel that this book is a good study on low-carbon energy-based systems.

During the visit of the RGIPT Professors to SGV University, Jaipur India, later on, we had regularly discussed the energy scenarios in the days ahead. During such discussions, the need to publish *Low Carbon Energy Supply Technologies and Systems* was identified. I congratulate the editors for having come out with the book, which will be of immense help to students, practicing managers, and policymakers alike.

Professor D. Buddhi
Ex-Vice Chancellor
SGVU Jaipur, India

Preface

Climate change is the long-term change in average weather patterns throughout the world. Since the mid-1800s, humans have contributed to the release of carbon dioxide and other greenhouse gases (GHG) into the atmosphere. The Industrial Revolution began in the mid-1800s when humans began to burn fossil fuels, such as coal, oil, and gas for fuel. This causes global temperatures to rise, resulting in long-term changes to the climate. The same process was followed continuously until today and will be followed in the near future, which is why the GHG emissions are released into the atmosphere at a high concentration rate. Burning fossil fuels produce electricity, heat, or motion but also release GHGs, such as carbon dioxide, methane, and nitrous monoxide into the atmosphere. Overall, the climate is completely changed due to our lifestyle and energy requirements. Therefore, because of climate change, the whole world is suffering from temperature rise, drought, unseasonal rains, and floods. As per the current trends to produce electricity, it's not possible to reduce the GHG concentrations in the upper atmosphere until, or unless, the whole world shifts toward the low carbon energy resources to substitute the use of primary fossil fuels. Today, the energy demand with sustainable economic growth, including concerns about global warming and climate change, are very important issues in the entire world. These include environmental and social impacts and the question of the power generation and fuel production processes, as well as associated emissions. The earth's environment has been harmfully impacted during the production, distribution, and consumption of fossil-fuel-based energy recourses, which produce a lot of GHGs to the atmosphere.

Technologies and systems based on low carbon energy supplies have been promoted for several years into the front line of energy policies, which also proved how the world's societies address their energy needs on a more sustainable basis. Right now, the entire world is consuming fossil-fuel-based energy recourses for social upliftment, economic development, the advancement of knowledge and improvement of the human condition. The energy produced from any resources is also consumed in the domestic, commercial, industrial, and transport sectors in the whole world. Meeting the needs of the present energy demand without compromising the needs of the future, humanity has to pay attention to the technologies and systems based on low carbon energy supplies, which harness nonexhaustible, environmentally friendly renewable energy sources (RESs). The world's total energy demand is also expected to increase, and the share of energy supplied in the form of electricity is likely to rise rapidly. However, simultaneously many countries in the world, such as the USA, China, Germany, Japan, India, and Australia, are also increasing their energy capacities by RESs.

The main RESs include solar, wind, hydro, geothermal, biomass, biogas, and biofuels, which are available almost everywhere in the whole world and distributed worldwide to meet the expanding energy demands without adverse environmental impacts. However, overall, at present, the current renewable energy (RE) share is small in the worldwide energy production, and the RE share should have been much higher than that of current RE share and also in favor of the environment, which is also the main concern globally. The use of RE also increases income, improves trade balances, contributes to industrial development, and creates many local jobs. Almost every country pays much attention to RES, due to the ongoing environmental challenges, which may be solved only by the usage of renewables. Many countries are making significant efforts to increase the RE share as much as they can within economic and political constraints. Several reports mentioned that there are global trends and developments for RE in the electricity, heat, and transport sectors.

The aim of this book is to share the latest developments and advances in technologies and systems based on low carbon energy supplies involved in energy generation, transmission, distribution, and storage. The chapters were contributed by researchers in the energy and materials field, using original materials. This book may be used as a reference for college/university/training institute/professionals. It may also be referred to clean energy/green energy-related laboratories, industries, and academic libraries, as well as used as a reference book for alternative energy sources, RES, climate change, energy sustainability, and energy systems for undergraduate and graduate students. This book presents a timely combination of research and practice explained in a simple and comprehensive manner.

Atul Sharma
Jais, Amethi, U.P., India

Amritanshu Shukla
Jais, Amethi, U.P., India

Renu Singh
IARI, New Delhi

Acknowledgments

Before the start of this project in May 2019, we were assured that to develop such a book, we would need at least one year. We thought of bringing an edited book to share knowledge, development, and scientific advancement in the field of low carbon energy supply technologies and systems with a large group of interested readers. Therefore, we thought that we could simply get connected with the researchers around the world and request for them to provide their recent leading-edge research in the form of book chapters for this particular edited volume. As in the past, we had already worked on such types of projects, and we had very good experience in it. It felt so simple at the beginning; however, we were wrong. After the completion of this project, now we understood it in a better way, since the authors who contributed to this book are from the scientific and academic community. Throughout, however, we received remarkable encouragement and enthusiasm from our authors and reviewers. We are really honored to have all the contributors who have been very supportive, dedicated, and responsive throughout our interactions. All of us are highly thankful to all our passionate authors and reviewers!

We would like to express our gratitude to the many persons who saw us through this book; to all those who provided support, talked things over, read, wrote, offered comments, and allowed us to quote their remarks. This book would never have taken off without the generous support of the authors of the book chapters. It would never have been completed without the cooperation and administrative and editorial assistance of Dr. Gagandeep Singh and Ms. Mouli Sharma of CRC press. Our heartfelt thanks to all these dedicated and cooperative individuals!

This book would not have been possible without the support and encouragement of our families. Words cannot express our gratitude to them for their kind support and encouragement. There is no doubt that we are very excited, enthusiastic, and gratified about the final outcome. We also feel sorry about unwittingly neglecting them on many occasions, especially during weekends. For all their patience and moral support, we dedicate this book to our families.

Atul Sharma, Amritanshu Shukla, and Renu Singh
Rajiv Gandhi Institute of Petroleum Technology,
Jais, Amethi, U.P. and IARI, New Delhi

Editors

Dr. Atul Sharma completed his MPhil in energy and environment (August 1998) and PhD on the topic *Effect on Thermophysical Properties of PCMs due to Thermal Cycles and Their Utilization for Solar Thermal Energy Storage Systems* (June 2003) from the School of Energy and Environmental Studies, Devi Ahilya University, Indore (MP), India.

He has worked as a scientific officer at the Regional Testing Centre Cum Technical Backup Unit for Solar Thermal Devices, at the School of Energy and Environmental Studies, Devi Ahilya University, Indore, funded by the Ministry of Non-Conventional Energy Sources of the government of India. He also worked as a research assistant at the Solar Thermal Research Center, New and Renewable Energy Research Department at Korea Institute of Energy Research, Daejon, South Korea (April 1, 2004–May 31, 2005), and as a visiting professor at the Department of Mechanical Engineering, Kun Shan University, Tainan, Taiwan, R.O.C (August 1, 2005–June 30, 2009).

Dr. Sharma is currently associate professor at Rajiv Gandhi Institute of Petroleum Technology (RGIPT), which has been set up by the Ministry of Petroleum and Natural Gas, government of India, through an Act of Parliament (RGIPT Act 54/2007) along the lines of the IITs with an "Institute of National Importance" tag. The institute is co-promoted as an energy domain-specific institute by six leading PSUs—ONGC Ltd., IOCL, OIL, GAIL, BPCL, and HPCL in association with OIDB. The mission of the institute is to work toward promoting energy self-sufficiency in the country through its teaching and R&D efforts.

Recently, Dr. Sharma published four edited books: *Low Carbon Energy Supply— Trends, Technology, Management*, Springer (Pvt.) limited India, New Delhi, ISBN 978-981-10-7326-7 (2018); *Sustainability through Energy-Efficient Buildings*, CRC Press, Taylor & Francis Group, ISBN 978-113-80-6675-5 (2018); *Energy Security and Sustainability*, CRC Press, Taylor & Francis Group, ISBN 9781498754439 (2016); and *Energy Sustainability Through Green Energy*, Springer (Pvt.) limited India, New Delhi. ISBN 978-81-322-2337-5 (2015). Dr. Sharma has published several research papers in various international journals and conferences. He also published several patents related to the phase change materials (PCMs) technology in the Taiwan region. He is working on the development and applications of PCMs, green buildings, solar water heating systems, solar air heating systems, solar drying systems, etc. Dr. Sharma is conducting research at the Non-Conventional Energy Laboratory (NCEL), RGIPT, and is currently engaged with the Council of Science & Technology, U.P. sponsored project at his lab. Further, he served as an editorial board member and as a reviewer for many national and international journals, project reports, and book chapters.

Dr. Amritanshu Shukla received his master's degree in physics and completed his PhD from IIT Kharagpur in January 2005. He completed his postdoctoral research work from some of the premier institutes across the globe, namely at the Institute of Physics, Bhubaneswar (Department of Atomic Energy, Government of India); University of North Carolina Chapel Hill USA; University of Rome/ Gran Sasso National Laboratory, Italy; and Physical Research Laboratory Ahmedabad (Department of Space, Govt. of India). He is currently working as an associate professor in physics at Rajiv Gandhi Institute of Petroleum Technology (RGIPT) (set up through an Act of Parliament by the Ministry of Petroleum & Natural Gas, as an "Institute of National Importance" on the lines of IITs).

Recently, Dr. Shukla published edited books, *Low Carbon Energy Supply—Trends, Technology, Management*, Springer (Pvt.) limited India, New Delhi, ISBN 978-981-10-7326-7 (2018); *Sustainability through Energy-Efficient Buildings*, CRC Press, Taylor & Francis Group, ISBN 978-113-80-6675-5 (2018); and *Energy Security and Sustainability*, CRC Press, Taylor & Francis Group, ISBN 9781498754439 (2016). Dr. Shukla's research interests include nuclear physics and physics of renewable energy resources. He has published more than hundred research papers in various international journals and in various international and national conferences. He has delivered invited talks at various national and international institutes. Currently, he is participating in several national and international projects and active research collaborations from India and abroad on the topics of his research interests.

Dr. Renu Singh is a senior scientist at the Centre for Environment Science and Climate Resilient Agriculture, Indian Agricultural Research Institute, New Delhi. She holds a master's degree in environmental sciences, an MTech in environmental sciences and engineering, and a PhD degree in environmental impact assessment from the Indian Institute of Technology, Delhi. She has more than 17 years of research experience in the fields of environmental impact assessment, biofuels, ecology, and environmental microbiology. Dr. Singh has worked on various projects related to climate change and biofuels funded by different agencies. She also has exposure to remote sensing and dispersion modeling. She has published research papers in journals and at conferences in the field of environmental sciences. Currently, she has several national and international projects and active research collaborations from India and abroad on the topics of her research interests.

Contributors

Avlokita Agrawal
Department of Architecture and
    Planning
Indian Institute of Technology
Roorkee, India

Asim Ahmad
Department of Mechanical Engineering
Birla Institute of Technology
Ranchi, India

Abhishek Anand
Non-Conventional Energy Laboratory
Rajiv Gandhi Institute of Petroleum
    Technology
Jais, India

Ciril Arkar
Laboratory for Sustainable
    Technologies in Buildings
Faculty of Mechanical Engineering
University of Ljubljana
Ljubljana, Slovenia

Maitiniyazi Bake
Centre for Research in Built and Natural
    Environment
Coventry University
Coventry, United Kingdom

Tulika Banerjee
Department of Electrical & Electronics
School of Engineering
University of Petroleum & Energy
    Studies
Mumbai, India

Prashant Baredar
Energy Centre
Maulana Azad National Institute of
    Technology
Bhopal, India

Debajyoti Bose
Department of Electrical, Power &
    Energy
University of Petroleum & Energy
    Studies
Dehradun, India

and

Faculty of Applied Sciences &
    Biotechnology
Shoolini University of Biotechnology &
    Management Sciences
Solan, India

Salwa Bouadila
Thermal Processes Laboratory
Research and Technology Center
    of Energy
Tunis, Tunisia

Suzana Domjan
Faculty of Mechanical Engineering
University of Ljubljana
Ljubljana, Slovenia

Aymen El Khadraoui
Thermal Processes Laboratory
Research and Technology Center
    of Energy
Tunis, Tunisia

Shiva Gorijan
Biosystems Engineering Department
Faculty of Agriculture
Tarbiat Modares University
Tehran, Iran

Kashif Irshad
Center of Research Excellence in
    Renewable Energy (CoRERE)
King Fahd University of Petroleum &
    Minerals (KFUPM)
Dhahran, Saudi Arabia

Karunesh Kant
Department of Mechanical Engineering
Eindhoven University of Technology
Eindhoven, the Netherlands

Mária Kozlovská
Laboratory of Construction Technology
    and Management
Department of Construction Technology
    and Management
Faculty of Civil Engineering
Technical University of Košice
Košice, Slovak Republic

Anil Kumar
Department of Mechanical Engineering
Delhi Technological University
New Delhi, India

Shuli Liu
Centre for Research in Built and
    Natural Environment
School of Energy, Construction and
    Environment
Coventry University
Coventry, United Kingdom

Sumit Lonkar
Energy Centre
Maulana Azad National Institute of
    Technology
Bhopal, India

Abdur Rehman Mazhar
School of Energy, Construction and
    Environment
Coventry University
Coventry, United Kingdom

Sašo Medved
Faculty of Mechanical Engineering
University of Ljubljana
Ljubljana, Slovenia

Chaitali Morey
Energy Centre
Maulana Azad National Institute
    of Technology
Bhopal, India

Anukul Pandey
Department of Electronics and
    Communication Engineering
Dumka Engineering College
Dumka, India

Terézia Pošiváková
Department of the Environment
    Veterinary Legislation and Economy
University of Veterinary Medicine and
    Pharmacy of Košice
Košice, Slovakia

Om Prakash
Department of Mechanical Engineering
Birla Institute of Technology
Ranchi, India

Safa Skouri
Thermal Processes Laboratory
Research and Technology·Center
    of Energy
Tunis, Tunisia

Camilo C. M. Rindt
Department of Mechanical
    Engineering
Eindhoven University of Technology
Eindhoven, the Netherlands

Atul Sharma
Non-Conventional Energy Laboratory
Rajiv Gandhi Institute of Petroleum
    Technology
Jais, India

Madhu Sharma
Department of Electrical &
  Electronics
School of Engineering
University of Petroleum & Energy
  Studies
Mumbai, India

Amritanshu Shukla
Department of Basic Science and
  Humanities
Non-Conventional Energy Laboratory
Rajiv Gandhi Institute of Petroleum
  Technology
Jais, India

Ashish Shukla
Centre for Research in Built and
  Natural Environment
Coventry University
Coventry, United Kingdom

Anand Singh
Electrical & Electronics Department
Lakshmi Narain College of
  Technology
Bhopal, India

Guruwendra Singh
Energy Centre
Maulana Azad National Institute
  of Technology
Bhopal, India

Ramkishore Singh
Solar Energy Division
Sardar Patel Renewable Energy
  Research Institute
Anand, India

Renu Singh
Centre for Environment Science and
  Climate Resilient Agriculture, Indian
  Agricultural Research Institute
New Delhi, India

Appukuttan Sreekumar
Department of Green Energy
  Technology
Pondicherry University (A Central
  University)
Pondicherry, India

Jozef Švajlenka
Laboratory of Construction Technology
  and Management
Department of Construction Technology
  and Management
Faculty of Civil Engineering
Technical University of Košice
Košice, Slovak Republic

Chinnasamy Veerakumar
Department of Green Energy
  Technology
Pondicherry University (A Central
  University)
Pondicherry, India

1 A Review of Solar Air Heater

Asim Ahmad, Om Prakash,
Anil Kumar, and Anukul Pandey

CONTENTS

1.1 INTRODUCTION

Solar energy is an unlimited source of energy. The threat to the extinction of conventional sources of energy has led mankind to focus on alternative sources of energy. Radiation from the sun, which includes light and heat, is the most predominant source of energy [1].

A solar air heater (SAH) is a device that is used to heat air by consuming the sun's energy. A SAH device has multiple applications, for example, space heating and drying of agricultural products, such as wheat and corn. Heat transfer in an SAH is attained through simultaneous radiation, convection, and conduction. A conventional SAH chiefly comprises of an absorber plate, a glass or a plastic cover that is kept above the absorber plate, and thermal insulation, which is provided at the bottom and sides. Fabrication and maintenance of SAHs are quite simple. Another major advantage an SAH has over its water heating counterpart is that the effect of corrosion on the air heating system is less severe [2].

Despite its obvious desirable features, a major concern regarding a solar air heating system is that its thermal efficiency is poor because of the bad heat transfer between the flowing air and the heated solar absorbing plate. This is due to poor thermophysical properties of air and a sticky sublayer that appears near the absorber, which is resistant to the heat transfer [3].

A couple of common techniques that have been implemented in order to improve the thermal efficiency are:

- Increasing the heat transfer area with the use of extended surfaces or corrugated ones.
- Enhancing the heat transfer coefficient by introducing surface roughness on the absorber plate.

SAH can be classified into three types according to air passes:

- Through air flowing between the absorber plate and the shield coating.
- Through air flow between the absorber plate and the rear board.
- Through two air channels overhead and underneath the absorber plate [4].

Storage of energy is highly imperative in order to see the energy requirements during the night and also in daytime periods of cloud cover. The radiant solar energy can be converted into thermal, chemical, or kinetic energy. Thermal energy storage can be categorized according to either as workable heat or as latent heat [5].

1.2 TYPES OF SOLAR AIR HEATER

There are basically two types of SAH, namely, active SAH and passive SAH [6].

1.2.1 ACTIVE SOLAR AIR HEATING SYSTEM

The active system uses a fan in order to heat up the collector system. The complete system comprises a fan, a diverging section, glass cover, fixed steel matrix, etc., all of which have different functions [7]. The site selection is a very important parameter. The site selected should have a temperature of more than 40°C in the month of August and approximately 20°C during the winters. The daily average solar radiation should be around 180–190 kW/m² [8].

1.2.1.1 Working of an Active Solar Air Heating System

In the beginning, the absorber sheet is going to absorb the solar radiation. The fan is there to blow the air inside the system through the inlet duct at the speed of 1 m/sec [9].

An active solar heating system has a collector plate and an air moving equipment (fan) to flow the air that has been heated by the system inside a building for space heating or for drying of agricultural crops [10]. Absence of corrosion in the solar air heating system is one of its major advantages [11]. The flat plate collector has one or two translucent sheets in order to reduce heat losses due to convection, a dark absorbing surface plate, and insulation on the back of the collector. The absorber having a black coating, the coating should be of two types:

non-selective and selective coating [12]. The selective coating and second transparent cover are used to minimize the heat losses. The outlet temperature should be around 45°C to 55°C [13].

In a liquid heating system, the heat generated by the collector is sent to the storage section. Then the stored hot air is either used for space heating as well as being sent to a heat exchanger for preheating the cold water to use in the domestic hot water system [14].

The cold water, which is flowing inside the pipes, uses the heat from the hot air to increase its temperature. The collectors are connected with a pressure relief valve, and the rate of flow is reduced to increase the heat gain [15]. Whereas for space heating of buildings, the hot air is transferred from the thermal storage to the building through an insulating pipe [16].

There are two types of connections, namely, series and parallel connections. The setup of the series connection is easy to install, and it is highly beneficial because it increases the temperature of the water and air at a rapid rate. Therefore, the temperature of water extracted from the thermal storage is of a higher value than the temperature of the water entering the storage system. Whereas the parallel connection is used to obtain the desired temperature in the room [17].

An energy and exergy analysis of an SAH was done. The collector's length is 2 m, the volume is 0.28 m³, and the cover's thickness is about 0.004 m. Polythene insulation is used to reduce thermal losses [18].

Conclusion of the analysis is:

- There is no heat loss in the SAH at the time of discharge.
- The outlet temperature at night is 20°C.
- The energy efficiency is around 40%.
- The exergy efficiency is 22% [19].

There are two types of collectors, glazed and unglazed, according to the cover provided. The unglazed solar collector lacks a glass cover and thus is not costly when compared to the glazed collectors [20].

Figure 1.1 shows a glazed solar heater. The glazed cover is used basically to prevent heat losses. There are some colors of absorber plate that can be used, such as blue, black, red, and brown. It was found practically that blue, red, and brown have efficiencies near the black. Figure 1.2 shows the two different glazed SAHs [21].

The analysis concludes:

- The SAH's efficiency will rise by using a perforated glazed SAH.
- Thermal efficiency of unglazed is minimum compared to glazed.
- The efficiency of dark-colored glazed heaters is more compared to a bright color [22].

The analysis has been done to figure out the procedure for improving the heat transfer coefficient of the SAH [23]. There is a cubic or rectangular duct shown in Figure 1.2. The size of the duct is 2395 mm, area is 300 × 25 mm², and the inlet and outlet sizes would be 740 and 555 mm, respectively.

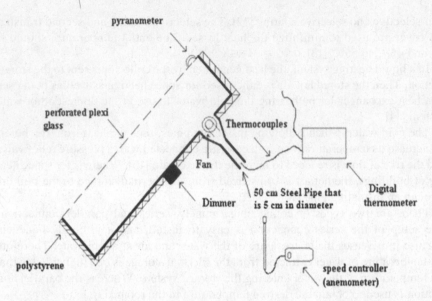

FIGURE 1.1 Experimental setup for perforated glazed solar air heater. (From Vaziri, R. et al., *Sol. Energy*, 119, 251–260, 2015.)

FIGURE 1.2 Two different perforated glazed solar air heaters. (From Vaziri, R. et al., *Sol. Energy*, 119, 251–260, 2015.)

The duct is protected with 50 mm broad polystyrene. The air is supplied in the pipe through a 2 HP blower. The inlet and outlet temperatures of the air were taken, and mean temperatures were measured by placing various thermocouples throughout the system.

The analysis concludes:

- Maximizing the distance between the three leads results in the reduction in the heat transfer rate.

1.2.2 Passive Solar Air Heating System

Generally, there is a lot of focus on the performance of different types of SAHs, but not much attention has been provided for the innovation on its structure. Therefore, in order to increase the efficiency of an SAH, it is combined with flat micro-heat pipe arrays (FMHPAs), aluminum fins, or vacuum tubes. An FMHPA comprises of a flat aluminum plate with several non-partisan micro-heat pipes. An FMHPA can transport heat efficiently, whereas vacuum tubes are used to reduce the heat losses [24].

1.2.2.1 Working of a Passive System

In this type of system, the coating film of the vacuum tube is used to absorb the solar radiation energy received from the sun. The air is then heated inside that tube due to radiation. Then the heat is transferred to the wall of the evaporator section of the FMHPA where the working fluid (generally, acetone with 20% liquid filling ratio) present in the evaporator unit traps the heat and starts to evaporate. The vapor then moves up to the condenser unit where it condenses into liquid after energy is removed from the fins, and then it is transferred to the evaporator section due to gravitational and capillary force [25].

In general, the heat is transmitted to the fins end (condenser section) via phase change through the flat micro-heat pipe arrays. Finally, the air is heated when it flows through the fins due to convection and radiation. Heat transfer of an SAH can be improved by the use of an FMHPA that has excellent performance [26].

The thermal efficiency of the system is low because of poor heat transfer between air and the collector. However, it increases as the solar radiation and ambient temperature rise [27].

The arrangement of a traditional SAH can generally be improved by the means of various heat transfer techniques with a focus on different types of surface enhancements. The use of obstructions on the absorber surface is one of the several methods used in the design of an SAH to improve the heat transfer because it creates a disturbance in flow and thus improves the thermal exchange by convection and improves the thermal proficiency of the system simultaneously [28].

Once the fluid collides with these obstacles, quite a bit of the fluid will flow back, affecting the flow to get separated near the obstacles. As the flow passes these obstructions/fins, some of the fluid will be circulated back, causing a minor flow to occur [29]. The fluid will get reattached again after some flow length next to the obstacle, and therefore it allows more fluid to stay in contact with the absorber plate; thus, the heat transfer to the fluid from the absorber plate is increased [30].

High heat transfer with low-pressure drop is a major criterion for designing. So it is imperative to select an obstacle geometry that should not only enhance the heat transfer rate but also keep pressure drop at its minimum possible level. Selection of the optimum values of roughness parameters involves a comparison between the enhancement of friction losses as a result of using these obstacles in the collector system with respect to that of the system without any such obstacles [31].

Figure 1.3 shows another method to improve the thermal proficiency of an SAH, by introducing mock surface unevenness on the solar absorber dish. The enhancement of heat transfer is found to be greater in the packed-bed-type SAHs than the conventional type that operates in similar circumstances and has similar working dimensions. There is a huge increase in the heat transfer coefficient that depends upon the permeability and the type of absorbent material used in the SAH. This is due to the fact that voids exist within the porous packed bed material, and the flowing air increases the heat transfer coefficient between the packed bed and air [32].

In recent years, a lot of methods have been proposed for improving the thermal efficiency of solar air collectors. The use of absorbent packing material in the double pass collector is one of the methods. Heat and mass transfer rates of double-pass flat plate solar air collectors are increased by increasing the fluid velocity with the same flow rate [33].

However, it has two demerits: first, the effect of increasing the convective heat transfer rate, and second, the result of extra inlet fluid getting warm by mixing it with hot air leads to a temperature dynamic force decrement, which further increases the heat loss from top glass covers [34].

Absorber plate material, copper, has the highest thermal conductivity amongst all the conductors after gold and silver and thus, it is the most suitable material to be used for the absorber plate [35].

Glass wool can be used for insulation at the bottom of the plate. It is produced by a centrifugal procedure, where molten glass is thrown from a huge number of tiny holes on the crosswise wall of a spinner by centrifugal strength due to rotation of the spinner and are formed into discontinuous fibers. Glass wool is an insulating material that traps numerous small pockets of air among the glass, and these small air pockets result in great thermal insulation of things [36].

FIGURE 1.3 Photographic view of the absorber plate with obstacles mounted for e/H = 0.50 at an angle of attack 60 (downside up). (From Bekele, A. et al., *Energy Convers. Manag.*, 85, 603–611, 2014.)

1.3 MODELING

1.3.1 ANALYTICAL MODELING

The efficiency calculation is performed along with the heat loss and the air temperature. The mean air temperature of the air is also determined [37]. Two solar air heating systems are used for analytical modeling in order to analyze thermal strategies. Considering the sectional energy balance in every area of both types of collectors, the numerical expression for the efficiency and heat loss is obtained, and air temperature is also calculated. Integrating the sharing among the inlet and outlet of the collector, a numerical expression of the mean temperature of the air is obtained. All these variables are very essential for calculating the heat transfer and mean temperature [38].

1.3.2 MULTI-OBJECTIVE OPTIMIZATION OF SOLAR AIR HEATER WITH OBSTACLES

A multi-objective growth of SAH on an observer sheet is basically used to improve heat transfer and reduce the pressure loss. It is taken with the three-dimensional Reynolds-averaged Navier–Stokes method [39].

Figure 1.4 shows the procedure of multi-objective optimization of a SAH. In which after the problem formulation is done, then numerical analysis by computational fluid dynamics (CFD) is performed, MATLAB optimization is also used there, and at last finally, the solution is represented in the functional space [40].

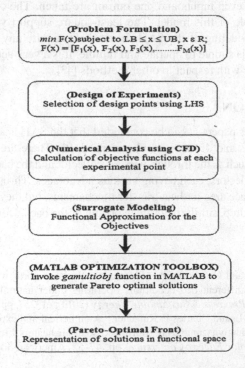

(Problem Formulation)
min F(x)subject to LB $\leq$ x $\leq$ UB, x ε R;
F(x) = [F_1(x), F_2(x), F_3(x),.........F_M(x)]

(Design of Experiments)
Selection of design points using LHS

(Numerical Analysis using CFD)
Calculation of objective functions at each experimental point

(Surrogate Modeling)
Functional Approximation for the Objectives

(MATLAB OPTIMIZATION TOOLBOX)
Invoke *gamultiobj* function in MATLAB to generate Pareto optimal solutions

(Pareto-Optimal Front)
Representation of solutions in functional space

FIGURE 1.4 Procedure for multi-objective optimization. (From Kulkarni, K. et al., *Sol. Energy*, 114, 364–377, 2015.)

The results of these experiments show that the trade-off amongst the two objective rates and the Pareto solution gives the good heat transfer in an SAH.

1.3.3 MASS FLOW, PRESSURE DROP, AND LEAKAGE-DEPENDENT MODELING

The thermal efficiency of the air collector totally depends on mass flow, whereas for liquid collectors it is not the same. This work justifies the relation between thermal, hydraulic act, and the output primary energy. The efficient mass flow relying on models for thermal efficiency is shown [41].

This model is only for simple mass flow relaying thermal efficiency, that follows simulation and infinitely changing mass flow. Moreover, a drop in air collector is established and can be put into the simulation purpose. The modified model is TRNSYS 832 [42].

1.3.4 MODELING OF SAH BY LEAST-SQUARES SUPPORT VECTOR MACHINES

In this modeling, a least-squares support vector machine is used to calculate the efficiency of the SAH. To study the process, take an absorber plate—this is made up of aluminum cans—and insert this plate into the double-pass channel in the SAH. The SAH is a multi-variable device, that is, it is difficult to model through nonrenewable methods. In this method, consider various mass flow rates and different types of collectors used for gaining optimum parameters [43].

In this method, seven inputs and one output are taken. The collector efficiency is the output variable of this model. The least-square support vector technique is used for efficient modeling of the SAH. It does not require any pre-knowledge of the device and thus is simple to operate and handle. It is a very accurate and reliable method and is faster with respect to other methods [43].

1.4 CONCLUSION

By reviewing various papers, it can be concluded that the SAH is a useful technology for heating buildings and drying agricultural products. There are different types of solar air collectors, such as the through-pass air collector and the back, front, and combination pass-air collectors, each having various advantages. Through-pass collectors have maximum surface area for heat transfer mechanisms, but they need fan power to reduce the pressure drop, similarly for the working of the back, front collector.

REFERENCES

1. Ahmad, Asim, and Om Prakash. "Thermal analysis of north wall insulated greenhouse dryer at different bed conditions operating under natural convection mode." *Environmental Progress & Sustainable Energy* (2019); e13257, pp. 1–12.
2. Gawlik, Keith M., and Charles F. Kutscher. "A numerical and experimental investigation of low-conductivity unglazed, transpired solar air heaters." In *ASME Solar 2002: International Solar Energy Conference*, pp. 47–55. American Society of Mechanical Engineers, 2002.
3. Othman, Mohd Yusof Hj, Baharudin Yatim, and Mohd Hafidz Ruslan. "Preliminary results of a V-groove back-pass solar collector." *Renewable Energy* 9, nos. 1–4 (1996): 622–625.

4. Bisht, Vijay Singh, Anil Kumar Patil, and Anirudh Gupta. "Review and performance evaluation of roughened solar air heaters." *Renewable and Sustainable Energy Reviews* 81 (2018): 954–977.

5. Ali, Mohammed Hadi. "Studying the performance of back–pass plain plate solar air heating un-glazed collector." *Journal of Engineering and Sustainable Development* 15, no. 4 (2011): 135–154.

6. Wazed, Md. Abdul, Yusoff bin Nukman, and Md. Tazul Islam. "Design and fabrication of a cost effective solar air heater for Bangladesh." *Applied Energy* 87, no. 10 (2010): 3030–3036.

7. Vaziri, Roozbeh, Mustafa İlkan, and F. Egelioglu. "Experimental performance of perforated glazed solar air heaters and unglazed transpired solar air heater." *Solar Energy* 119 (2015): 251–260.

8. Esen, Hikmet. "Experimental energy and exergy analysis of a double-flow solar air heater having different obstacles on absorber plates." *Building and Environment* 43, no. 6 (2008): 1046–1054.

9. Karsli, Suleyman. "Performance analysis of new-design solar air collectors for drying applications." *Renewable Energy* 32, no. 10 (2007): 1645–1660.

10. El Ouderni, Ahmed Ridha, Taher Maatallah, Souheil El Alimi, and Sassi Ben Nassrallah. "Experimental assessment of the solar energy potential in the gulf of Tunis, Tunisia." *Renewable and Sustainable Energy Reviews* 20 (2013): 155–168.

11. Lodhi, M. Arfin. K. "Collection and storage of solar energy." *International Journal of Hydrogen Energy* 14, no. 6 (1989): 379–411.

12. Tsatsaronis, George. "Definitions and nomenclature in exergy analysis and exergoeconomics." *Energy* 32, no. 4 (2007): 249–253.

13. Anderson, Timothy Nicholas, Mike Duke, and James K. Carson. "The effect of colour on the thermal performance of building integrated solar collectors." *Solar Energy Materials and Solar Cells* 94, no. 2 (2010): 350–354.

14. Prasad, S Bhushan, Joginder S. Saini, and Krishna Mohan Singh. "Investigation of heat transfer and friction characteristics of packed bed solar air heater using wire mesh as packing material." *Solar Energy* 83, no. 5 (2009): 773–783.

15. Pottler, Klaus, Carl Martell Sippel, Andreas Beck, and Jochen Fricke. "Optimized finned absorber geometries for solar air heating collectors." *Solar Energy* 67, nos. 1–3 (1999): 35–52.

16. Sopian, Kamaruzzaman, M. A. Alghoul, Ebrahim M. Alfegi, M. Y. Sulaiman, and E. A. Musa. "Evaluation of thermal efficiency of double-pass solar collector with porous–nonporous media." *Renewable Energy* 34, no. 3 (2009): 640–645.

17. Thakur, N. S., J. S. Saini, and S. C. Solanki. "Heat transfer and friction factor correlations for packed bed solar air heater for a low porosity system." *Solar Energy* 74, no. 4 (2003): 319–329.

18. Kurtbas, Irfan, and Aydın Durmuş. "Efficiency and exergy analysis of a new solar air heater." *Renewable Energy* 29, no. 9 (2004): 1489–1501.

19. Öztürk, Hasan Hüseyin. "Experimental determination of energy and exergy efficiency of the solar parabolic-cooker." *Solar Energy* 77, no. 1 (2004): 67–71.

20. Naphon, Paisarn. "On the performance and entropy generation of the double-pass solar air heater with longitudinal fins." *Renewable Energy* 30, no. 9 (2005): 1345–1357.

21. Benli, Hüseyin, and Aydın Durmuş. "Performance analysis of a latent heat storage system with phase change material for new designed solar collectors in greenhouse heating." *Solar Energy* 83, no. 12 (2009): 2109–2119.

22. Karwa, Rajendra. "Experimental studies of augmented heat transfer and friction in asymmetrically heated rectangular ducts with ribs on the heated wall in transverse, inclined, v-continuous and v-discrete pattern." *International Communications in Heat and Mass Transfer* 30, no. 2 (2003): 241–250.

23. Sahu, M. M., and J. L. Bhagoria. "Augmentation of heat transfer coefficient by using 90 broken transverse ribs on absorber plate of solar air heater." *Renewable Energy* 30, no. 13 (2005): 2057–2073.

24. Gupta, Dhananjay, S. C. Solanki, and J. S. Saini. "Thermohydraulic performance of solar air heaters with roughened absorber plates." *Solar Energy* 61, no. 1 (1997): 33–42.

25. Saini, S. K., and R. P. Saini. "Development of correlations for Nusselt number and friction factor for solar air heater with roughened duct having arc-shaped wire as artificial roughness." *Solar Energy* 82, no. 12 (2008): 1118–1130.

26. Singh, Sukhmeet, Subhash Chander, and J. S. Saini. "Heat transfer and friction factor correlations of solar air heater ducts artificially roughened with discrete V-down ribs." *Energy* 36, no. 8 (2011): 5053–5064.

27. Gill, R. S., Sukhmeet Singh, and Parm Pal Singh. "Low cost solar air heater." *Energy Conversion and Management* 57 (2012): 131–142.

28. Bekele, Adisu, Manish Mishra, and Sushanta Dutta. "Performance characteristics of solar air heater with surface mounted obstacles." *Energy Conversion and Management* 85 (2014): 603–611.

29. Choudhury, C., and H. P. Garg. "Design analysis of corrugated and flat plate solar air heaters." *Renewable Energy* 1, no. 5–6 (1991): 595–607.

30. Esen, Hikmet, Filiz Ozgen, Mehmet Esen, and Abdulkadir Sengur. "Modelling of a new solar air heater through least-squares support vector machines." *Expert Systems with Applications* 36, no. 7 (2009): 10673–10682.

31. Prasad, B. N., and J. S. Saini. "Effect of artificial roughness on heat transfer and friction factor in a solar air heater." *Solar Energy* 41, no. 6 (1988): 555–560.

32. Mittal, M. K., and L. Varshney. "Optimal thermohydraulic performance of a wire mesh packed solar air heater." *Solar Energy* 80, no. 9 (2006): 1112–1120.

33. Ramadan, M. R. I., A. A. El-Sebaii, S. Aboul-Enein, and E. El-Bialy. "Thermal performance of a packed bed double-pass solar air heater." *Energy* 32, no. 8 (2007): 1524–1535.

34. Yeh, Ho-Ming, and Chii-Dong Ho. "Solar air heaters with external recycle." *Applied Thermal Engineering* 29, no. 8–9 (2009): 1694–1701.

35. Ramani, B. M., Akhilesh Gupta, and Ravi Kumar. "Performance of a double pass solar air collector." *Solar Energy* 84, no. 11 (2010): 1929–1937.

36. Yadav, Avadhesh, and V. K. Bajpai. "An experimental study on evacuated tube solar collector for heating of air in India." *World Academy of Science, Engineering and Technology* 79, no. 2011 (2011): 81–86.

37. Hernandez, Alejandro L., and José E. Quiñonez. "Analytical models of thermal performance of solar air heaters of double-parallel flow and double-pass counter flow." *Renewable Energy* 55 (2013): 380–391.

38. Mohamad, A. A. "High efficiency solar air heater." *Solar Energy* 60, no. 2 (1997): 71–76.

39. Kulkarni, Kishor, Arshad Afzal, and Kwang-Yong Kim. "Multi-objective optimization of solar air heater with obstacles on absorber plate." *Solar Energy* 114 (2015): 364–377.

40. Husain, Afzal, and Kwang-Yong Kim. "Multiobjective optimization of a microchannel heat sink using evolutionary algorithm." *Journal of Heat Transfer* 130, no. 11 (2008): 114505-1–114505-3.

41. Welz, Christian, Christoph Maurer, Paolo Di Lauro, Gerhard Stryi-Hipp, and Michael Hermann. "Mass flow, pressure drop, and leakage dependent modeling and characterization of solar air collectors." *Energy Procedia* 48 (2014): 250–263.

42. Perers, Bengt. "An improved dynamic solar collector test method for determination of non-linear optical and thermal characteristics with multiple regression." *Solar Energy* 59, no. 4–6 (1997): 163–178.

43. Alvarez, G., J. Arce, L. Lira, and M. R. Heras. "Thermal performance of an air solar collector with an absorber plate made of recyclable aluminum cans." *Solar Energy* 77, no. 1 (2004): 107–113.

2 Development of Different Sun-Tracking Systems for Displacement of Solar Concentrator Implanted in Tunisia

Safa Skouri and Salwa Bouadila

CONTENTS

2.1 INTRODUCTION: BACKGROUND AND DRIVING FORCES

Sun-tracking systems play an important role in the development of high solar concentration applications that convert solar energy into electrical or thermal energy. Different methods of sun following have been surveyed and evaluated to keep the solar panels, solar concentrators, telescopes, or other solar systems orthogonal to the sun's beams (Gao et al., 2018).

Diffuse radiation is rarely used in high concentration systems; only direct radiation is concentrated. Therefore, it is necessary to specify the direct radiation by the orientation mechanisms. We will therefore have to change the position of the solar concentrator. The objective of this chapter is the selection of a sun-tracking system for a solar concentrator system to improve its performance.

Solar tracker systems can be classified in two ways: in first step depending on the number of tracking axes, the single-axis solar tracker, the two-axis solar tracker, and the three-axis solar tracker are very seldomly used (Assaf, 2014), which differs related to the solar concentrated technology. A one-axis solar follower has single free degree of movement that allows the rotation of the mechanism around one axis. This type may have a horizontal axis or a vertical axis. The horizontal type is used in tropical areas where the sun is very high at noon and the days are short. The vertical type is used in high latitude regions where the sun is not very high but summer days can be very long. For this system, the angle of inclination is manually adjustable, and the sun tracking is automatically followed from east to west; at night, the followers take the horizontal position. The two-axis tracker has two free degrees of movement that allow the rotation around two axes. This type of horizontal and vertical two-axis tracker is used to control astronomical telescopes, and so there is a lot of software that can automatically predict and track the movement of the sun across the sky. The majority of concentrating technologies, especially solar parabolic concentrators, require two-axis solar-tracking systems. The imperative role of sun trackers in photovoltaic (PV) technology is given by Singh et al. (2017). They cited a review of various existing solar-tracking systems in terms of the controller used to design the system, and their economic assessment has been studied. Rahimi et al. (2015) compared between dual-axis sun-tracking and hybrid sun–wind-tracking PV panels. The deductions based on the research tests confirm that the overall daily output energy gain was increased by 49.83% compared with that of a fixed system. Moreover, an overall increase of about 7.4% in the output power was found for the hybrid sun–wind-tracking over the two-axis sun-tracking system. Tharamuttam and Ng (2017) propose an automatic microcontroller-based solar tracker with a hybrid algorithm for locating the sun's position. Hoffmann et al. (2018) presented a monthly profile analysis based on a two-axis solar-tracker proposal for PV panels. The tracker uses light-dependent resistors (LDRs) to identify sun movement direction and engines adjust the panel position, according to the control performed by an electronic device. Optical model and calibration of a sun tracker Volkov et al. (2016). Volkov et al. (2016) presented a method for optimization of the optical altazimuth sun-tracker model with output radiation direction aligned with the axis of a stationary spectrometer. Modeling and analyses of energy performances of PV greenhouses with sun-tracking functionality is given by Gao et al. (2019) In this study, simulation models of typical greenhouses with high-density and low-density PV layouts are discussed. Four special sun-tracking positions are found in the model of equivalent global irradiance. Simulation models are also built in terms of PV modules and interior irradiance. Simulations are conducted using the climate data of Delft, the Netherlands. Results show that high-density PVs under no-shading sun tracking generate 6.91% more energy than that under conventional (quasi-perpendicular) sun tracking.

2.2 SOLAR IRRADIATIONS IN TUNISIA

Tunisia is located in North Africa. The country has two climate regions: the well-watered north and the semi-arid south. An abrupt southern turn of its shoreline gives Tunisia two faces on the Mediterranean Sea with a 1298 km coastline.

Figure 2.1a shows the monthly period variation of solar altitude angle in Tunisia made along the year according to the calculations using the MATLAB® program.

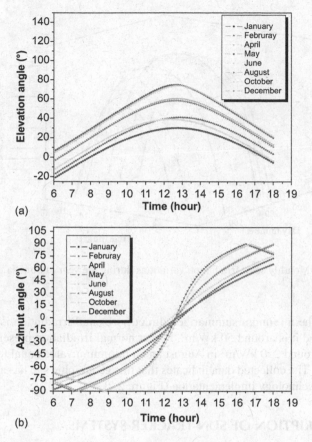

(a)

(b)

FIGURE 2.1 Monthly solar angle variations during the year in Tunisia: (a) monthly elevation-angle variation during the year in Tunisia; and (b) monthly azimuth-angle variation during the year in Tunisia.

The maximum value of the local solar altitude angle, where in Tunisia, reached 75° in summer season (June 21) and a minimum value to 30° in winter season (December 22). Figure 2.1b shows the period variation of solar azimuth angle in Tunisia for every month made along the year according to the calculations using the MATLAB software program. The local largest solar azimuth angle occurs as 90° (June 21) and the smallest solar azimuth angle occurs as 62° (December 22).

The global horizontal irradiance (GHI) is calculated from direct normal irradiation (DNI), direct horizontal irradiation (DHI), and solar elevation, which are determined by the next correlation:

$$GHI = DHI + DNI \cdot \sin(h) \qquad (2.1)$$

The maximum amount of irradiation is reached in July, and it presents a summarized direct horizontal irradiation around 175 kW/m², which represent 60% of the global

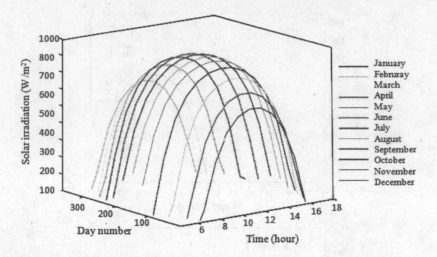

FIGURE 2.2 Monthly solar irradiation variations during the year in Tunisia.

irradiation. The minimum summarized direct horizontal irradiation is recorded in December, and it is around 50 kW/m². Direct normal irradiation presented an optimum value around 250 kW/m² in August and a minimum value equal to 125 kW/m² in December. The collected data indicates that this geographic site has a solar potential for CSP technology implementation (Figure 2.2).

2.3 DESCRIPTION OF SUN-TRACKER SYSTEMS

In this chapter, we have studied three sun-tracking systems where we cited the advantages and controversies of every system.

2.3.1 LDR SUN-TRACKING SYSTEM

The first sun-tracking system is an azimuth-altitude dual-axis tracker, which displaced the solar concentrator reflector facing the sun by two screw–nut actuators. It is considered an active tracking system. This tracker uses LDRs mounted on the solar concentrator edges: an azimuth sensor over the solar system and elevation sensor on the side of the reflector. The LDR sensor measures the light intensity due to its photo-diode made of selenide cadmium material. The LDR sensor is able to identify the relative position of the light source. This tracker uses a programmable interface controller (PIC) microcontroller 16F876. Then the compiled program has been simulated with ISIS software and loaded by IC_PROG software into the PIC. The ULN 2003 electrical circuit allows the control of the power switches. An electrical control card allows the visualization of the PIC interaction with different components. The LDR sensor sends an electrical signal to the PIC microcontroller. It reads the signal that can be inferior to 2.5 V during the day and over than 2.5 V at night. The motor shuts down when it reaches the end position during the day or the initial position overnight.

2.3.2 FIRST DATE- AND TIME-BASED SUN-TRACKING SYSTEM

The date- and time-based sun-tracking system is a programmed dual-axis tracker. It is a passive sun-tracking system. The solar collector is oriented toward the sun with two direct-current (DC) electric motors powered by an electrical control unit based on the signal delivered by a control module. A control program was processed on a computer using HP-VEE software. The program uses the data inputs (solar angles values) obtained from the previous program and delivers the outputs (motor angular steps) to an Agilent-3970 A channel through a HP 34903A control module. The HP control module is connected to an electronic control circuit (ECC). The ECC is a commercial electrical unit used for the parabolic satellite antenna displacement. The ECC has a control system in a closed loop for DC motors, and the rotation frequency of the motor is adjustable by a voltage signal delivered from a control knob. This system has a chopper and a speed measuring system that delivers a voltage proportional to the engine rotation frequency.

2.3.3 SECOND DATE- AND TIME-BASED SUN-TRACKING SYSTEM

In order to ameliorate the tracker accuracy, the previous mechanical structure was modified, and the screw–nut actuators mechanical system was replaced by a new reduction gear unit (WG075-40-E-F) transmission mechanism. The new mechanism will permit two rotation movements with advanced mobility degree.

The choice of the gear box is related to the design. It is a worm gear speed reducer type that allows a significant rotational speed reduction (up to 100) and increases its mechanical torque; meanwhile, the non-reversible aspect of the gear provides a position locking for the SPC. All the mentioned characteristics of the transmission mechanism will enhance the position stability and accuracy movement of the SPC.

A low rotational speed combined with an important mechanical torque provides a better precision and a good stability of the sun-tracking system.

Each gear unit is driven by a DC motor that put the parabola mechanism in rotation around the two axes. To ensure the displacement of the parabola in the exact position, two DC motors are controlled according to a program at every predefined time-step.

The monitoring and control unit (MCU) is divided into two subsystems: an ECC and a programmable logic controller (PLC) connected via an adjustment circuit. The MCU calculates the azimuth and the elevation angles from astronomical and geometrical parameters. Two angular position sensors on both rotation axes were integrated. The sensors will transmit the instant position of the SPC during the day. When the controller PLC defines the set point, it allows the rotation of the motor until the generated pulsation reaches the set-point value. Two encoders will calculate the number of revolutions, which is proportional to the rotation degree of the SPC structure. In order to drive the motors in both directions, we used the ECC, which is already designed for such a motor. It is reliable, simple, and available in the local market at a cheap price.

The PLC Siemens S71200-type CPU 1214C was chosen due to the system needs. This controller is equipped with 14 digital input channels, two analog input channels, and an ethernet port to insure the computer connection. Due to the outdoor use of the tracking system, the PLC should operate under severe climatic conditions (heat, humidity, dust, wind, etc.). The chosen PLC doesn't need an extra power supply unit; it is directly powered by the local electrical grid (220 V/50 Hz). In fact, the synchronization between the delivered impulse by the ECC and the calculated impulse via the PLC will reduce the position errors and will allow an accurate angular position.

2.4 COMPARATIVE STUDY

A comparative study was conducted to identify the adequate solution for the tracking system. An experimental study has been done to determine the accuracy of each sun-tracking system, and the price estimation of those three systems has been carried.

The present work is based on having an advanced accuracy and stability of a solar-tracking system able to adjust a solar concentrator. Comparisons between previous and current research are analyzed. These results are compared and summarized as follows.

The LDR sun-tracking system described in this chapter was characterized by a cheap price, the spare parts are available in the local market, but this follower becomes less accurate in cloudy days. It is more adequate for PV panels than the present application. The first date- and time-based sun-tracking system is independent of the climatic conditions, and it presents an advanced accuracy but less movement stability due to the screw–nut actuators. For this, the second date- and time-based sun-tracking system is the adequate follower for this application. The proposed two-axis sun-tracking system was characterized by an advanced accuracy, a movement stability, and a fairly simple and low-cost electromechanical setup with low maintenance requirements and ease of installation and operation.

2.5 CONCLUSION

The tracking position of a mobile object has been the objective of several research studies in order to provide the adequate pursuit systems for specific applications. Concentrated solar technology efficiency relies partially on solar-tracking ability, and the development of such systems becomes essential.

However, this work presents a contribution toward that goal, with a comparative experimental study of three different sun-tracking systems, which is used to ensure the movement of a solar parabolic concentrator around two axes. The main findings of the present study are:

- Solar direct irradiation in Tunisia is characterized by an important amount (60%) of the global irradiation. Like any sunny region, this parameter promotes the implementation of CSP technologies.
- For an efficient use of the concentration technologies, it is very important to dispose a sun-tracking system, and for this reason three pilot sun-tracking systems have been designed, realized, and compared. A comparison among the three trackers led to the choice of the adequate one for such an application.

- The second date- and time-based sun-tracking system provided an advanced accuracy (2.5 mrad) and a better stability with relatively cheap price (1300 €). It is the adequate follower. The proposed two-axis sun-tracking system was characterized by movement stability, fairly simple low maintenance requirements, and ease of installation and operation.

REFERENCES

Assaf, E.M., 2014. Design and implementation of a two axis solar tracking system using PLC techniques by an inexpensive method. *Int. J. Acad. Sci. Res.* 2, 54–65.

Gao, Y., Dong, J., Isabella, O., Santbergen, R., Tan, H., Zeman, M., 2018. A photovoltaic window with sun-tracking shading elements towards maximum power generation and non-glare daylighting. *Appl. Energy* 228, 1454–1472. https://doi.org/10.1016/j.apenergy.2018.07.015

Gao, Y., Dong, J., Isabella, O., Santbergen, R., Tan, H., Zeman, M., Zhang, G., 2019. Modeling and analyses of energy performances of photovoltaic greenhouses with sun-tracking functionality. *Appl. Energy* 233–234, 424–442. https://doi.org/10.1016/j.apenergy.2018.10.019

Hoffmann, M., Molz, R.F., Kothe, V., Oscar, E., Nara, B., 2018. Monthly profile analysis based on a two-axis solar tracker proposal for photovoltaic panels. *Renew. Energy J.* 115, 750–759. https://doi.org/10.1016/j.renene.2017.08.079

Rahimi, M., Banybayat, M., Tagheie, Y., Valeh-e-sheyda, P., 2015. An insight on advantage of hybrid sun—Wind-tracking over sun-tracking PV system. *Energy Convers. Manag.* 105, 294–302. https://doi.org/10.1016/j.enconman.2015.07.086

Singh, R., Kumar, S., Gehlot, A., Pachauri, R., 2017. An imperative role of sun trackers in photovoltaic technology: A review. *Renew. Sustain. Energy Rev.* https://doi.org/10.1016/j.rser.2017.10.018

Tharamuttam, J.K., Ng, A.K., 2017. Design and development of an automatic solar tracker Jerin Kuriakose Tharamuttam. *Energy Procedia* 143, 629–634. https://doi.org/10.1016/j.egypro.2017.12.738

Volkov, S.N., Samokhvalov, I.V., Du, H., Kim, D., 2016. Journal of quantitative spectroscopy & radiative transfer optical model and calibration of a sun tracker. *J. Quant. Spectrosc. Radiat. Transf.* 180, 101–108. https://doi.org/10.1016/j.jqsrt.2016.04.020

3 Solar Drying Technology
Sustainable and Low-Carbon Energy Technology

Ramkishore Singh

CONTENTS

3.1 INTRODUCTION

The fruits, vegetables, medicinal plants, and some types of seeds are major source of nutrients for the human diet because they are rich in vitamins and minerals. These products contain over 80% of water at the time of harvesting, which makes them perishable. Because these products are used regularly and demanded around the year, they need to be made available in the market. It is not possible to keep these products fresh and in their original condition in the natural environment due to significant fluctuation in the weather conditions and hardly remain in favor of such perishable products [1]. Different preservation methods including canning, refrigeration, and drying are used to curtail spoilage of the products. Refrigeration and drying are two of most commonly used methods that have their own advantages and disadvantages. Refrigeration is used to preserve products for meeting the demands of fresh products in their original conditions. However, the refrigeration process needs to be continuous from its initial storage until consumption of products. On the other hand, drying is required only once and significantly low energy consuming process. The dried product can be stored at ambient environmental conditions for a sufficiently long duration.

Moreover, it is much easier to pack and is cost effective to transport dried products compared to refrigerated products. Despite the fact that drying is required once, still the drying process is a highly energy-intensive process and accounts for approximately 10%–20% of total industrial energy consumed in developed countries. Thermal energy is required to remove water and other solvents in perishable items by reducing the moisture content to its saturation level. The energy required to remove the water content is equivalent to the latent heat of evaporation. Preservation of perishable products through drying assists reducing post-harvest losses in a cost-effective way and provides access to food to undernourished people worldwide [2]. The drying process does not change the fiber, carbohydrates (provide energy), or protein content. In the drying process, hot air is utilized to create a moisture concentration gradient within the product that is to be dried. The created moisture gradient causes moisture to move from the inner part of the product to its surface, and eventually hot air carries the moisture away from the surface of the product. The temperature of the supplied hot air plays a vital role in the effective drying of the product and varies with the type of product. A higher temperature of the drying air than the required value leads to physical and chemical changes, as well as damages capillaries at the surface of the product to be dried. The damages in the capillaries eventually prevent moisture movement from inner layers to the surface. Sometimes, the surface becomes hard, leaving higher moisture contents in the inner part of the product and deteriorates the quality of the product. The supply of drying air at a controlled temperature in the desirable range helps in improving the shelf life of the products. Drying of perishable products not only helps in avoiding bacterial and fungal growth but also minimizes the cost of packaging, storage, and transportation.

Drying is also an energy-consuming process and uses thermal energy that can be provided by different conventional and non-conventional fuels. However, fossil-fuel-based heat is prevalently used in the current drying industry because this can be used in any environmental conditions and is easily controllable and manageable. Heat sources used in a few drying industries for drying different products are listed in Table 3.1 [3].

However, using fossil fuels for drying in remote rural areas may not be economically viable. Moreover, practicing fossil-fuel-based drying pollutes environments locally as well as globally [4]. In developing countries like India, over 80% of the

TABLE 3.1
Heat Source for Different Drying Applications

S. No.	Commodities	Heat Source for Drying	Application
1	Milk powder, casein	Hot air in spray driers, fluidized beds	Dairy industries
2	Coffee beans, tea leaves, cocoa, nuts, rice, spices, corn etc.	Firewood and propane or oil	Agri-crops
3	Seasoning of wood and timber	Steam using petroleum fuels	Saw industries
4	Fruits and vegetables	Hot air from electricity	Food processing industries
5	Fabric conditioning	Steam	Textile industries

perishable agri-products are produced by small and marginal farmers, and they are not strong enough financially to afford the high operational cost of drying [1,5]. As a result, they are reluctantly selling their products immediately after harvesting in the market at lower prices. The drying system is easy to operate, maintain, and runs on freely available renewable energy sources; it is most suitable for these remotely located rural farmers to improve their returns in horticulture farming.

Solar energy is available freely and abundantly in most of the areas and is practiced for generations to preserve agri-products, medicinal herbs, and many other products. However, traditionally open sun drying is most prevalent in most of the areas, which has issues of dust accumulation on the products as well as a change in color. A solar dryer consists of an absorber and drying chamber. The energy collected by the absorber is transferred to drying chamber either by natural convection mode or by force convection. The system that works in natural convection is called a passive dryer, and the force convection mode dryer is called an active system [6]. Recently, a few different types of solar dryers were developed around the globe for direct and indirect drying of perishable products. This chapter aims to provide a brief discussion about the market demand of the dried products, benefits of drying of perishable products, principles of drying, and advancements in the solar drying technologies.

3.2 MARKET POTENTIAL

India is an agriculture-based economy and the second largest producer of fruits and vegetables in the world after China. India's horticulture annual production crossed 300 million MT. As per spice, tea, coffee, and the national fisheries development boards, the annual production of spices, tea, coffee and fish production is over 6 million MT, 1.25 million MT, over 0.3 million MT, and approximately 10 million MT, respectively. The country also exports a surplus production to different parts of the world, mainly to the United States, the Netherlands, United Kingdom, Germany, and the United Arab Emirates. The country's total export of fruits and vegetables, during 2018–2019, was of Rs. 10236.93 crores, out of which 47% (Rs. 4817.35 crores/USD$ 692 million) came from fruit and approximately 53% (Rs. 5419.48 crores/US$ 777 million) from vegetables. In addition, the country exported cereals worth Rs. 56,841.08 crore and floriculture of worth Rs. 571.38 crores/USD$ 81.94 million in the same year [7]. India is also a prominent exporter of vegetables and fruits after processing. During the years 2018–19, the country exported 2,48,121.88 MT (worth Rs. 2,473.99 crores/USD$354.65 million) of processed vegetables. Some of the products, which are supplied in dehydrated form, are truffles, asparagus, garlic powder, garlic flakes, potatoes, Gram, Gram daal, and onions.

The demand of dehydrated vegetables in the global market is significantly high. In the years 2016–17, the global revenue from the dehydrated business was approximately over US$ 50000 million, which is expected to cross approximately US$ 90000 million by the end of 2028 [8]. As per NOVONOUS, the dried and preserved vegetables market of India is expected to grow at a CAGR of 16% by the year 2020. The supportive agro-climatic conditions, potential domestic market, cost competitiveness, and government support are some of the key factors that will drive the growth of this industry.

3.3 WHY DO PERISHABLE PRODUCTS NEED DRYING?

Post-harvest losses are global issues that need urgent attention. Particularly in India, the post-harvest losses reach up to 40% due to inadequate infrastructure and unfavorable climatic conditions. The losses are significantly high in perishable products, and therefore such products need to be preserved until consumed. The preservation is a process of prevention of decay/spoilage of perishable products, which allows products to be stored in safe and usable conditions for future consumption. In the preservation process, quality, edibility, and nutritive values should remain intact. As shown in Figure 3.1, different preservation methods are applied to ensure long-term and short-term supply of perishable products for meeting the demand of dry and fresh products. The products are harvested in wet and warm conditions and start deteriorating after 3–4 days if kept in ambient conditions. The products are consumed either in fresh form or in dry form and hence preserved in warm or cold conditions, respectively. The products need to be refrigerated for later consumption in the fresh condition. The refrigeration is an energy-consuming process which needs to be continuously in operation to maintain the desirable temperature range for storage of products. On the other hand, products are dehydrated and stored. The dehydration is also an energy-consuming process but needs to be done once, and no further energy is consumed. However, best process for long-term storage is storing in a cold environment after drying and specifically used for seeds and germ plasm storage.

Further, Figure 3.2 explains favorable environmental conditions for bacteria, fungi, microbes, and insect growth in maize stored, having different moisture contents. The product does not deteriorate and remains safe if dehydrated to a moisture content less than 8%. Products having moisture contents in the range of 4%–8% can be stored at an ambient temperature safely for sufficient longer period. As the moisture content grows, growths of different harmful elements initiated. The equilibrium

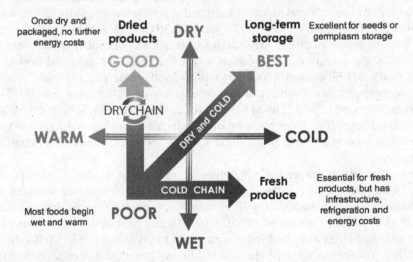

FIGURE 3.1 Representation of post-harvest and preservation conditions for perishable products. (From Bradford, K.J. et al., *Trends Food Sci. Technol.*, 71, 84–93, 2018.)

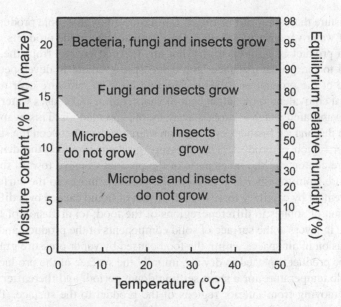

FIGURE 3.2 Favorable and unfavorable environmental conditions for the growth of bacteria, fungi, insects, and microbes in perishable maize. (From Bradford, K.J. et al., *Trends Food Sci. Technol.*, 71, 84–93, 2018.)

relative humidity also plays a significant role in the safe storage of dried products. There is a significantly high possibility of gaining moisture again by the dried product if stored in an environment of higher relative humidity. Therefore, it is recommended to pack the dried product in a leak-proof manner to avoid moisture gain. Figure 3.2 specifically demonstrates the behavior of maize; however, other perishable products behave more or less similarly.

3.4 PRINCIPLE OF AIR DRYING

In the drying process, energy is mainly consumed for taking the moisture content out from the product to be dried and transforming that liquid water into vapor. In general, 2258 kJ thermal energy is required to evaporate 1 kg of water at atmospheric pressure (101.3 kPa) and the drying process. The energy used for drying is an independent type of fuel used for heating the drying air. All products exhibit certain equilibrium moisture contents while getting exposed to air at a particular temperature and humidity. The products tend to lose or gain moisture and attain the equilibrium. The moisture content in the product may contain free moisture and bound moisture and is related directly to the drying rate. The free moisture can be removed simply by applying a pressure gradient. However, bound moisture is held by capillary forces in microporous regions (i.e., dead-end pores) of products and held strongly by negatively charged particles surfaces and hydration water associated with a mineral charge balancing unit. Products that contain free loose moisture (regarded as unbound moisture) are called non-hygroscopic. On the other hand, hygroscopic products contain bound moisture that is trapped in closed capillaries. In hygroscopic products, partial pressure of water vapour varies with the moisture content in the products.

When a product is exposed to dry hot air, thermal energy from the hot air is transferred to the product surface, and a pressure gradient is developed between the surface of the product and the surrounding air. The moisture near the surface absorbs heat equivalent to the latent heat of vaporization and causes water to evaporate. The evaporated water is carried away by the supplied air. This evaporation of water from the surface further develops a pressure gradient between the surface and closer inner region of product. This pressure gradient provides the driving force for moisture contents in the inner parts of the product to move toward the surface. The moisture from the interior regions of the products moves to the surface by: (a) liquid movement by capillary forces; (b) diffusion of liquid caused by a difference in concentration of solutes in different regions of the food; (c) diffusion of liquid that is absorbed in layers at the surface of solid components of the product; and (d) water vapor diffusion in air spaces within the food caused by vapor pressure gradients.

Once the product is placed in dry air, initially the surface of the product heats up the wet bulb temperature for a short settling down period, and thereafter the moisture starts moving from interior regions of the product to the surface. The rate at which moisture moves from the inner part to the surface remains equal to the rate of evaporation at the surface, and hence the surface remains wet. This period of drying is known as a constant drying rate period and continues until the critical moisture condition is not achieved.

The process of drying in different phases can be understood from Figure 3.3. Phase I and Phase II are the initial constant rate drying and falling rate periods, respectively. The phases shown in Figure 3.3 give a general view. AB, BC, and CE denote the time spent to achieve the drying temperature by initially heating up the product, the constant rate drying period, and the falling rate drying period (when moisture flow from the inner parts to the surface decreases continuously), respectively. Point C is a critical point and indicates that the surface is not saturated

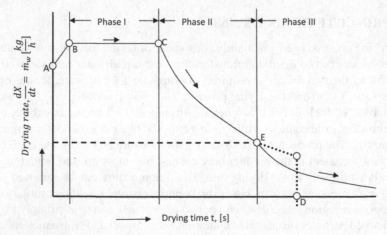

FIGURE 3.3 Drying rate curves for Phase I, II, and III. (From Belessiotis, V. and Delyannis, E., *Sol. Energy*, 85, 1665–1691, 2011.)

anymore. The products that have very low initial moisture contents near to the critical point do not show the constant drying rate period (Phase I) and directly reach Point C. From this point, the falling rate starts (Phase II). Only the internal liquid movement controls moisture diffusion while water at the surface depletes continuously. Thereafter, a second falling period, that is, Phase III, can be observed, when the moisture content continues to decrease until equilibrium is achieved and drying stops. For the majority of the products, drying stops before Phase III is reached. From Point E, very little moisture remains inside the product that slowly moves toward the surface by diffusion. Point D indicates the state of desirable moisture content in the product and no further drying is needed.

The drying time of each period depends on the nature of the product and the drying conditions. The surfaces of different agriproducts are different in terms of roughness and porosity. Hence, products dry out at different drying rates. Thus, critical moisture point varies for products and depends on the amount and preprocessing as well as the drying rate. Three parameters, that is, (a) adequate dry bulb temperature, (b) low relative humidity of the air, and (c) suitable air velocity are essentially critical for successful drying of the product. The drying period in these phases for hygroscopic products depends on the initial moisture content and the desirable moisture content for safe storage.

The drying process can be easily understood from the psychometric chart and demonstrated in Figure 3.4. Appropriately, hot and dry air from the heating section is directly sent to the drying chamber where the product to be dried comes in contact with the supplied hot air and heated for enabling drying. The chart shows different conditions of drying air, for example, cool/moderate/warm ambient air at different temperatures heated by solar or any other type of heating means to an appropriate temperature level and supplied for the drying chamber. The hot dry air works more effectively than hot humid air [6].

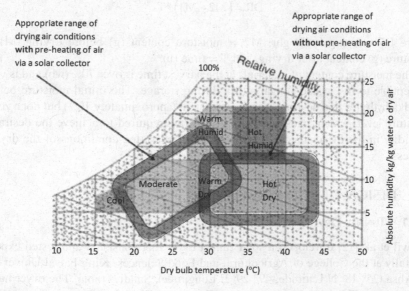

FIGURE 3.4 Indicative psychometry of conditions for the inclusion in a solar dryer of an air heating solar collector. (From Tomar, V. et al., *Sol. Energy*, 154, 2–13, 2017.)

3.5 DRYING KINETICS

The efficiency of the complete drying process is explained by the drying kinetics. The drying air flow rate and air temperature in the dryer are the main parameters that influence drying kinetics. As discussed earlier, the drying rate is crucial for hygroscopic products, for example, agri-products for long-term preservation. The moisture present in the product is either in the form of free moisture or bound moisture and directly influences the drying rate. The hygroscopic materials contain bound moisture in the closed capillaries that are expressed either in a dry or wet basis. In general, the mathematical calculations for moisture contents are performed on a dry basis most conveniently, but the wet basis is used commonly for agricultural products.

The moisture content in the wet (W) basis is the weight of moisture per unit of wet material and expressed by:

$$W = \frac{m_w}{m_w + m_d} \text{ kg per kg of mixture}$$

and on dry basis (X) is expressed as the ratio of water content to the weight of dry material as follows:

$$X = \frac{m_w}{m_d} \text{ kg of water per kg of dry material}$$

where m_w is the mass of water and m_d is the mass of dry solid.

The moisture removal rate or drying rate is calculated by the equation,

$$DR = (M2 - M1)/T$$

where DR = drying rate in ghr, M2 = moisture content (g) before drying, M1 = moisture (g) content after drying, and T = time (h).

The moisture content of materials at the harvest time is over 70% (wb) and is very susceptible to insect and fungal attack during storage. The initial moisture before dried dehydrates to a desirable moisture content (approximately 10%) but not a zero-moisture level, which leads to burning. The time required to achieve the desirable humidity in the dried material depends on the operating conditions of the drying process.

3.6 DESIGN OF SOLAR DRYERS

3.6.1 TUNNEL DRYER

Eltawil et al. [12] developed a solar tunnel dryer that consisted and tested experimentally at the College of Agricultural and Food Sciences, King Faisal University, Al Ahsa (25° 18′ N Latitude, 49° 29′ E Longitude), Saudi Arabia. The dryer had a photovoltaic (PV) module, charge controller, battery, DC fan, DC speed controller, and a flat-plate solar air collector (1.20 m length, 1.0 m width, and 2 mm thick) built

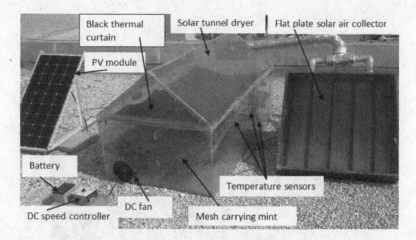

FIGURE 3.5 Experimental setup of solar tunnel dryer enhanced by a PV system and flat-plate solar air collector for drying mint under shade. (From Eltawil, M.A. et al., *J. Clean. Prod.*, 181, 352–364, 2018.)

by the galvanized corrugated absorber plate and a drying chamber made of transparent Plexiglass sheets of 2 mm thick (Figure 3.5). The drying chamber (2 m length × 1 m width) also had a perforated stainless-steel mesh mounted at 25 cm from the bottom for the material to be dried. The design was made portable and simple for easy assembling and disassembling. The DC fan was connected in the front side of the even span-type greenhouse solar collector for circulating the air. The absorber sheet was painted black from the inside, and the back side was insulated using 10-cm thick fiber glass [12]. The solar collector was oriented toward the south and inclined at 30° from the horizontal plane. An auxiliary heating system inbuilt tunnel was used to connect the solar collector and drying chamber. The tunnel was used to supply heat air from the collector to the drying chamber. The studied dryer works in a mixed drying mode, that is, the direct heating (the drying chamber is transparent and allows direct solar radiation) and indirect heating modes (air is heated in solar collector and supplied to the drying chamber.

Mewa et al. [13] developed and installed a Hohenheim-type low-tunnel solar tunnel dryer at the EwasoNg'iro North Development Authority (ENNDA) premises, Isiolo county, Kenya. The dryer was 18 m long, 1.96 m wide, and covered by the transparent cover of UV-stabilized sheet inclined at 15° from the horizontal. One end of the plastic sheet was fixed with the metal tube for allowing loading and unloading by rolling of the plastic sheet up and down. The dryer was placed at an elevated platform and ensured that it was not shaded by a building or trees. Also, the orientation of the system was kept suitable in order to capture the incident solar radiation more efficiently (Figure 3.6) [13]. The base of the collector and drying chamber were made in several small sections from plain metal sheets and wooden frames; thereafter, all the sections were joined together in series. The bottom and sides of the drying chamber and solar collector were adequately insulated using glass wool insulation. A solar PV module was integrated to power the installed fan to provide the required air flow

FIGURE 3.6 Pictorial view of the solar tunnel dryer from the collector end. (From Mewa, E.A. et al., *Renew. Energy*, 139, 235–241, 2019.)

within the tunnel dryer. The absorber surface in the collector was painted black to absorb the maximum incident radiation.

Half of the dryer area was used as a solar collector to heat the air. The other half area was used as a drying chamber, where the product was placed to dry. The tunnel dryer could reduce the moisture content of the drying product, that is, beef, to a level between 2.32% and 9.56% (dwb) in 11 h while traditional sun drying only reduced the moisture level up to 24.76% (dwb). The effective moisture diffusivity values for solar-tunnel-dried beef samples were estimated between 2.282 to 2.536×10^{-10} m^2/s, which was higher than the values for open sundried samples [13]. Figure 3.7 shows

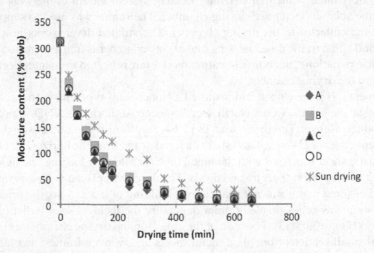

FIGURE 3.7 Moisture content vs. drying time graph during beef drying at different sections of the solar tunnel dryer (A–D) compared to open sun drying. (From Mewa, E.A. et al., *Renew. Energy*, 139, 235–241, 2019.)

the drying behavior of the beef samples in different sections of the dryer and compared with the open sun-drying samples. Faster and uniform drying in all the sections rather than open sun-drying can be observed clearly.

3.6.2 INDIRECT-TYPE SOLAR DRYER

Direct-type dryers have a drawback of direct exposure of the product to sunlight, which leads to a significant change in the natural product under drying. As a result, work has been initiated to explore the alternative indirect-type dryers. Further, researchers made significant efforts in recent years in order to enhance the performance of indirect-type forced-convection solar dryers and develop different designs [14].

The indirect solar dryer offers several advantages over the traditional drying method [15,16]. Figure 3.8 represents a schematic layout of a simple indirect solar dryer. The air is heated by a solar collector connected to a drying chamber that contains the agricultural product. The product remains under shade and isolated from the ambient air. The drying process occurs by the exchange of water between the product and the flowing hot air. The use of indirect solar dryers preserves the dried product's active principles, color, and protects it from dust and insects.

Recently, an indirect-type forced-convection dryer has been developed and tested in Iraq for drying fruits and vegetables. The dryer was made of two identical double-pass solar air collectors covered by a single glass and drying cabinet. An air blower was used to blow air into the solar drying cabinet. Further, Kadam and Samuel [18] developed a solar dryer using a V-grooved solar collector and tested for drying cauliflower. Karim and Hawlader [19] assessed the performance of the V-grooved fins and flat-plate solar collectors for crop drying. They observed 7%–12% higher efficiency for the V-grooved collector than the flat-plate collectors. Banout et al. [20] studied a double-pass solar dryer (DPSD), and performance of the

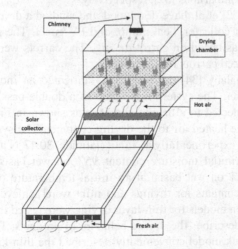

FIGURE 3.8 Schematic layout of an indirect solar dryer. (From Zoukit, A. et al., *Renew. Energy*, 133, 849–860, 2019.)

system was tested for drying red chillies. The performance of the system was compared with a typical cabinet dryer and a traditional open-air sun dryer. Mohanraj and Chandrasekar [21] fabricated and tested an indirect-mode forced-convection solar dryer for drying copra. The system was able to reduce the moisture content of the copra from 51.8% to 7.8% and 9.7% in 82 h placed in the bottom and top trays in the cabinet, respectively. Singh and Kumar [22] had developed the direct, indirect, and mixed-mode solar dryers and tested for no-load steady-state conditions in natural and forced circulation modes of the drying. The tests on the developed systems were performed in indoor environmental conditions for absorbed thermal energy, and the air flow rate varied between 300 and 800 W/m² and 1–3 m/s, respectively. Also, correlations established the convective heat transfer coefficient from the absorber plate to the flowing air in terms of dimensionless numbers for each drying mode (i.e., natural and forced convections). El-Sebaii et al. [23] developed an indirect-type natural convective solar dryer. They also studied the performance of the developed system for drying grapes, figs, apples, green peas, tomatoes, and onions. The dryer takes 60 and 72 h to dry seedless grapes with and without thermal storage material, respectively.

Mathematical modeling for solar dryers is important to design and develop effective solar dryers. Performance of dryers, particularly solar dryers, depends on the design of the system, which is influenced by geographical coordinates, solar radiation, relative humidity, and air flow rate at the location and drying kinetics of the product to be dried. Therefore, mathematical modeling has been given significant attention in recent years, and a few researchers have focused their research on the mathematical modeling and drying kinetics for various agricultural products and the thin layer drying of mint [24,25] figs [26,27], grapes [28], and pistachios [29]. The drying characteristics for chillies pepper using open sun and solar drying were investigated and reported in Ref. [30]. The drying kinetics in a forced convection for thin layer silk cocoon and orange peels were also studied and reported by [31] and by Ben Slama and Combarnous [32], respectively.

Recently, Sevik [33] et al. have developed and studied a dryer that consists of a double-pass solar air collector, heat pump, and PV panel. The performance of the developed system was tested on carrot drying. The carrots were sliced and dried. The system dried sliced carrots in 220 min.

El-Sebaii and Shalaby [34] designed and fabricated an indirect-mode forced-convection solar dryer. The system was made of a double-pass v-corrugated plate solar air heater connected to a drying chamber as shown in Figure 3.9. A blower was used to force the heated air to the drying chamber. The system's thermal performance was tested experimentally at Tanta (latitude, 30.47°N and longitude, 31°E) for drying thymus (initial moisture content 95% on wet basis) and mint (initial moisture content 85% on wet basis) at an initial temperature of 29°C. The desirable final moisture contents for thymus and mint were achieved in 34 h and 5 h, respectively. Fourteen models for thin-layer drying were tested in order to identify a suitable model that describes the drying behavior of products. They concluded that the Midilli and Kucuk model conveniently described the thin-layer solar drying of mint. However, the Page and modified Page models were found to be the best among others for describing the drying curves of thymus.

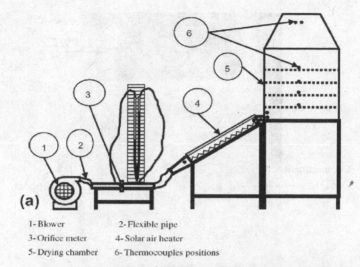

1- Blower   2- Flexible pipe
3- Orifice meter   4- Solar air heater
5- Drying chamber   6- Thermocouples positions

FIGURE 3.9 (a) A schematic diagram of the experimental setup and (b) photograph of the experimental setup. (From El-Sebaii, A.A. and Shalaby, S.M., *Energy Convers. Manag.*, 74, 109–116, 2013.)

3.6.3 INDIRECT FORCED CONVECTION SOLAR DRYER
WITH AUXILIARY HEATING DEVICE

Recently, Wang et al. [35] developed the indirect type of solar dryer that was made by integrating evacuated tube solar air collector, three phase inverter fan, auxiliary heating device, drying chamber and automatic control system. The indirect forced convection solar dryer (IFCSD) with auxiliary heating device was tested at the Solar Energy Research Institute, Kunming (latitude 25.02°N; longitude 102.43°E), China. The auxiliary heating device was integrated to stabilize the hot air temperature throughout the drying period, particularly at the time of insufficient sunshine and cloudy weather conditions. The system was tested for drying mango in sliced form and to study the effects of different drying air temperatures on the dryer performance and the drying kinetics of the product. The results indicated that the temperatures at four different locations in the drying chamber were varied within a narrow

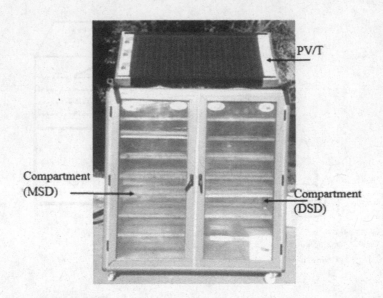

FIGURE 3.10   Solar dryer design. (From Fterich, M. et al., *Sol. Energy*, 171, 751–760, 2018.)

band. The average thermal efficiency of the system was estimated in the range of 30.9%–33.8%. The specific moisture extraction rate was estimated at 1.67 kg water/kWh at 52°C drying temperature. The effective moisture diffusivity was also estimated using Fick's diffusion equation for mango slices and found in the range of 6.41×10^{-11} to 1.18×10^{-10} m²/s, over a temperature range of 40°C–52°C.

Fterich et al. [36] examined the performance of a forced-convection mixed-solar dryer, for drying tomatoes. The studied system has a PV-thermal (PV/T) air collector and drying chamber (Figure 3.10). The air enters in the aluminum tubular canals located under the PV panel and spreads simultaneously into an upper gap, which enables the system to exchange heat in both the faces of the PV panel and also cool the PV cells by extracting heat. Because the dryer completely runs on solar energy without any grid connection, it enables farmers to improve their economic conditions by storing and selling dried tomatoes in the dryer. The dyer had two drying trays, and hot air was flown in to the trays. The product moisture content dropped from 91.94% to 22.32% for Tray 1 and to 28.9% for Tray 2, whereas in the open sun, the moisture drop was estimated only to 30.15%. It was also observed that integration of PV/T improves the efficiency of the drying process. The drying was observed faster in the tested system over open sun drying.

3.6.4   AIR-CONDITIONER-ASSISTED SOLAR DRYER

Chandrasekar et al. [3] developed and tested an integrated solar dryer that uses hot air from the condenser of a split air conditioner (see Figure 3.11). The performance of the system was tested for the drying of sultana grapes at Tiruchirappalli (78.6°E & 10.8°N), Tamil Nadu, India. On integration of the air conditioner, the drying time

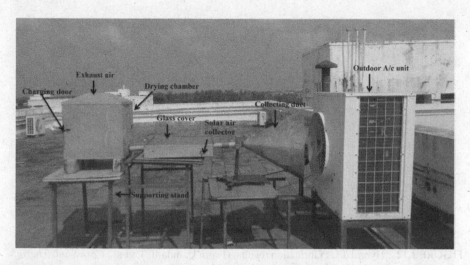

FIGURE 3.11   AC-assisted solar dryer. (From Chandrasekar, M. et al., *Renew. Energy*, 122, 375–381, 2018.)

was observed 16.7% shorter than open sun drying and a 13% higher efficient than conventional indirect solar dryer. The drying characteristic of seedless grapes in the developed dryer was described suitably by an exponential model. The model predicted the moisture ratio in good agreement with experimental values.

3.6.5 HYBRID SOLAR-BIOMASS DRYER

The intermittent nature of solar energy is one of the drawbacks for continuous and effective drying. Therefore, researchers started looking at options for developing continuous hybrid-type dryers. Recently, a 5-MT solar-biomass hybrid dryer was developed at Kwame Nkrumah University of Science and Technology in Kumasi, Ghana for drying maize [37]. The hybrid dryer uses agri-biomass along with solar energy for drying. The system was observed to maintain high temperature and low humidity as well as dry faster compared to only solar dryer. The hybrid dryer was developed for targeting nucleus farmer aggregators, farmer-based organizations, poultry farmers, post-harvest service providers, seed companies, and other stakeholders in the maize value chain in Ghana. Also, mortalities of adults and immature *Sitophilus zeamais* (Motschulsky), *Tribolium castaneum* (Herbst), and *Cryptolestes ferrugineus* (Stephens) were investigated.

Experimental results demonstrated that the dryer was able to generate temperatures above 50°C and dry maize more effectively compared to open sun drying. The economic viability of the hybrid solar-biomass dryer could be possible only if onsite available agri-biomass, for example, maize cobs and husks are used instead of wood in biomass combustor.

Further, Hamdani et al. [38] developed another solar-biomass hybrid dryer at Samudra University, Langsa City, Aceh, Indonesia. The developed hybrid dryer consisted of biomass-fueled air heaters, solar air heater cum drying chamber of length

FIGURE 3.12 Hybrid drier fabrication results. (From Hamdani, T. et al., *Case Stud. Therm. Eng.*, 12, 489–496, 2018.)

260 cm and width of 80 cm, a fan, and an exhaust chimney (Figure 3.12). The solar air heater cum dryer was covered with glass covers. The fan blew air first through pipes and then was heated in the biomass combustor and dried the product.

The system performance was tested for drying fish. The drying was performed using solar energy between 09:00 and 16:00 h and thereafter continued with biomass combustion until 06:00. The system maintained the drying temperature between 40°C and 50°C. The fabrication cost of the hybrid system for 100 kg fish drying capacity was estimated at $1,870. Also, an internal rate of return (IRR), net present value (NVP), and breakeven point were calculated at 18.61%, $21.091, and 2.6 years, respectively, for a dry fish production capacity of 12,000 kg/yr and selling price of $3.3/kg.

3.6.6 Energy-Storage-Integrated Dryer

Thermal storage integration is another option that was assessed by researchers recently in order to resolve the issues of intermittent operation and variable temperature in the solar dryer. Excess thermal energy during the sunshine hours is stored as a thermal storage component and used for drying later in the night/non-sunny hours. Recently, El Khadraoui et al. [39] developed a novel indirect-type solar dryer by integrating a thermal storage component as shown in Figure 3.13. The thermal storage component used paraffin wax, a phase-change material, for storing thermal energy in the form of latent heat. The drying chamber with the dimensions $0.8 \times 0.8 \times 1.2$ m³ was made of galvanized iron and insulated at all sides using 0.03-m-thick layer of polyurethane. Six easily removable drying trays made of wooden frames and wire mesh were positioned in the drying chamber at equal vertical spacing of 0.15 m. The drying tests on the system were performed with and without a thermal storage component. Test results indicated that the system with a thermal storage component raised the ambient air temperature by 4°C–16°C higher during night drying. Also, 17%–34.5% lower values of relative humidity were observed in the drying chamber. Daily efficiency and exergy with the thermal storage component were estimated at 33.9% and 8.5%, respectively.

FIGURE 3.13   Photograph of the indirect solar dryer: (1) solar energy accumulator, (2) solar air panel, and (3) drying chamber. (From El Khadraoui, A. et al., *J. Clean. Prod.*, 148, 37–48, 2017.)

Enibe [40] studied indirect-type solar drying having a phase change material (PCM) component at Nsukka, Nigeria. The solar dryer consisted of a single-glazed flat-plate solar collector integrated with paraffin wax (a phase-change thermal storage material) and a drying chamber. The thermal storage component was developed in modules, which were equip-placed across the solar absorber. The test results indicated that the system can be used successfully for crop drying.

Abubakar et al. [41] tested mixed-mode solar crop dryers with 12 different thermal storage materials under the climate of Zaria in Nigeria. The collector length, collector area, height of the drying chamber, chimney height, length of the drying chamber, and width of the drying chamber of the developed system were 0.65 m, 0.30 m^2, 0.9 m, 0.7 m, 1.64 m, and 0.43 m, respectively. The average drying rates, collector, and drying efficiencies of the solar crop dryers with and without thermal storage for June and August 2016 were estimated 2.71 × 10^{-5} kg/s and 2.35 × 10^{-5} kg/s, 67.25% and 40.10%, 28.75% and 24.20%, respectively. The system demonstrated a 13% higher efficiency with the storage materials.

Further, Devahastin and Pitaksuriyarat [42] investigated the feasibility of PCM-integrated solar drying for conserving and utilizing excess solar heat when solar is not adequate/not available for drying. The extractable energy from the thermal storage component was estimated using experimental results is 1920 and 1386 kJ min/kg for inlet air velocity of 1 and 2 m/s, respectively, which eventually led to 40% and 34% energy savings, respectively.

Further, an indirect-forced convection-type solar dryer with a packed-bed-type PCM thermal storage was developed by Esakkimuthu [43]. Test results revealed that the system at low mass flow rate was able to utilize the maximum stored energy from the thermal storage component and extended the hours of operation. Reyes [44] tested a hybrid solar dryer with the PCMs for mushroom drying. The efficiency of the accumulator solar panel was estimated to be varied in the range of 10%–21%. Results indicated a significant reduction in electrical energy use after integrating the storage unit. Shalaby and Bek [45] experimentally studied the solar dryer with and without PCM. Test results from their study indicated that after 2:00 p.m., a 3.5°C to 6.5°C higher drying air temperature with thermal storage integration was compared to without thermal storage. Earlier, Aiswarya, and Divya [46] studied an indirect solar dryer with the latent heat (i.e., paraffin wax) and sensible energy storage (pebbles) mediums. They integrated a PCM storage unit at the inner bottom of the drying chamber.

Jain and Tewari [47] developed an indirect and natural draft solar drying system that consisted mainly of a flat-plate collector, a packed bed for the thermal storage, a drying chamber, and a natural draft system. They observed that with inclusion of thermal storage, the system raised the ambient air temperature by 6°C in the drying chamber after the sunshine hours till midnight. The payback of the system was calculated 1.5 years.

Most recently, Bhardwaj et al. [48] experimentally investigated an IFCSD with sensible and latent heat storage materials (Figure 3.14) in the Himalayas (latitude 30.91°N) climate. They used iron scrap mixed with gravel and copper tubes using engine oil as the sensible thermal storage material placed in the solar collector. Tests were also performed with paraffin RT-42 as a latent heat storage material included in the drying chamber. The drying experiments were performed for the drying

FIGURE 3.14 Photographic view of experimental test setup. (From Bhardwaj, A.K. et al., *Sol. Energy*, 177, 395–407, 2019.)

of *Valeriana jatamansi* (a medicinal herb) in order to reduce its moisture content from the initial value of 89% to 9%. Tests were also performed using both types of energy storage materials simultaneously, and the overall drying rate was observed at 0.051 kg/hr, which was almost double that of 0.028 kg/hr and 0.018 kg/hr without any energy storage medium and traditional shade drying, respectively. The system reduced the moisture content to the saturation level of 9% in 120 h with the thermal storage component. The drying times to achieve the same level of moisture content without thermal storage and in shade were estimated at 216 and 336 h, respectively. The dried material was observed in good quality in terms of essential oil and bio-medical compounds. The results also indicated rehydration capacity, and total Valepotriates were 7.11% and 3.47% as compared to 6.18% and 3.31%, respectively, achieved in traditional shade drying. The energy and exergy efficiencies for the system with sensible heat storage were estimated at 26.10% and 0.81%, respectively.

Recently, Iranmanesh et al. [49] investigated the performance of a solar cabinet drying system equipped with a heat-pipe-evacuated tube solar collector and latent-heat-type thermal storage system (Figure 3.15), which was designed and constructed at the Institute of Science and High Technology and Environmental Sciences, Kerman. They have also used the CFD model to study the performance of the solar collector and drying efficiency of the system. Experimental results were obtained for the solar cabinet dryer, a laboratory scale system equipped with a heat-pipe-evacuated tube solar collector, and storage tank with PCM.

The drying process and thermal performance of the system were assessed for apple slices of 5 mm thickness and at three different air flow rates. With thermal storage material, the drying time observed was about 9.37%, 9.67%, and 10.02%

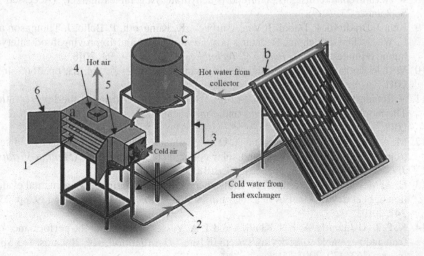

FIGURE 3.15 The schematic of solar cabinet dryer including: (a) dryer: 1—sample trays and cabinet, 2—fan, 3—chassis, 4—exhaust channel, 5—heat exchanger, 6—doors; (b) evacuated tube solar collector; and (c) storage tank and PCM container inside. (From Iranmanesh, M. et al., *Renew. Energy*, 145, 1192–1213, 2020.)

lower at the air flow rates of 0.025, 0.05, and 0.09 kg/s, respectively, than the system without PCM. However, inclusion of the PCM increased the input thermal energy by about 1.72% and 5.12% at the air flow rates of 0.025 and 0.05 kg/s, respectively. Lower input energy was observed on further increase in the air flow rate. The maximum overall drying efficiency estimated was 39.9% for the system with PCM at the air flow rate of 0.025 kg/s [49].

REFERENCES

1. E. C. López-Vidaña, L. L. Méndez-Lagunas, and J. Rodríguez-Ramírez, "Efficiency of a hybrid solar-gas dryer," *Sol. Energy*, vol. 93, pp. 23–31, 2013.
2. V. Tomar, G. N. Tiwari, and B. Norton, "Solar dryers for tropical food preservation: Thermophysics of crops, systems and components," *Sol. Energy*, vol. 154, pp. 2–13, 2017.
3. M. Chandrasekar, T. Senthilkumar, B. Kumaragurubaran, and J. P. Fernandes, "Experimental investigation on a solar dryer integrated with condenser unit of split air conditioner (A/C) for enhancing drying rate," *Renew. Energy*, vol. 122, pp. 375–381, 2018.
4. A. Lingayat, V. P. Chandramohan, V. R. K Raju, "Design, development and performance of Indirect type solar dryer for Banana drying," *Energy Procedia*, vol. 109, pp. 409–416, 2017.
5. C. K. K. Sekyere, F. K. Forson, and F. W. Adam, "Experimental investigation of the drying characteristics of a mixed mode natural convection solar crop dryer with back up heater," *Renew. Energy*, vol. 92, pp. 532–542, 2016.
6. O. V. Ekechukwu and B. Norton, "99/02111 Review of solar-energy drying systems II: An overview of solar drying technology," *Fuel Energy Abstr.*, vol. 40, no. 3, p. 216, 2003.
7. http://apeda.gov.in/. (Accessed on September 23, 2019).
8. www.futuremarketinsights.com/reports/dehydrated-vegetables-market. (Accessed on September 23, 2019).
9. Kent J. Bradford, P. Dahal, J. Van Asbrouck, K. Kunusoth, P. Bello, J. Thompson and F. Wue, "The dry chain: Reducing postharvest losses and improving food safety in humid climates," *Trends Food Sci. Technol.*, vol. 71, pp. 84–93, 2018.
10. V. Belessiotis and E. Delyannis, "Solar drying," *Sol. Energy*, vol. 85, pp. 1665–1691, 2011.
11. V. Tomar, G. N. Tiwari, and B. Norton, "Solar dryers for tropical food preservation: Thermophysics of crops, systems and components," *Sol. Energy*, vol. 154, pp. 2–13, 2017.
12. M. A. Eltawil, M. M. Azam, and A. O. Alghannam, "Energy analysis of hybrid solar tunnel dryer with PV system and solar collector for drying mint (*Mentha Viridis*)," *J. Clean. Prod.*, vol. 181, pp. 352–364, 2018.
13. E. A. Mewa, M. W. Okoth, C. N. Kunyanga, and M. N. Rugiri, "Experimental evaluation of beef drying kinetics in a solar tunnel dryer," *Renew. Energy*, vol. 139, pp. 235–241, 2019.
14. K. E. J. Al-Juamily, A. J. N. Khalifa, and T. A. Yassen, "Testing of the performance of a fruit and vegetable solar drying system in Iraq," *Desalination*, vol. 209, nos. 1–3 Spec. Iss., pp. 163–170, 2007.
15. O. Prakash, A. Kumar, and V. Laguri, "Performance of modified greenhouse dryer with thermal energy storage," *Energy Rep.*, vol. 2, pp. 155–162, 2016.
16. A. Kumar, R. Singh, O. Prakash, and Ashutosh, "Review on global solar drying status," *Agric. Eng. Int. CIGR J.*, vol. 16, no. 4, pp. 161–177, 2014.

17. A. Zoukit, H. El Ferouali, I. Salhi, S. Doubabi, and N. Abdenouri, "Takagi Sugeno fuzzy modeling applied to an indirect solar dryer operated in both natural and forced convection," *Renew. Energy*, vol. 133, pp. 849–860, 2019.

18. D. M. Kadam and D. V. K. Samuel, "Convective flat-plate solar heat collector for cauliflower drying," *Biosyst. Eng.*, vol. 93, no. 2, pp. 189–198, 2006.

19. M. A. Karim and M. N. A. Hawlader, "Development of solar air collectors for drying applications," *Energy Convers. Manag.*, vol. 45, no. 3, pp. 329–344, 2004.

20. J. Banout, P. Ehl, J. Havlik, B. Lojka, Z. Polesny, and V. Verner, "Design and performance evaluation of a double-pass solar drier for drying of red chilli (*Capsicum annum* L.)," *Sol. Energy*, vol. 85, no. 3, pp. 506–515, 2011.

21. M. Mohanraj and P. Chandrasekar, "Drying of copra in a forced convection solar drier," *Biosyst. Eng.*, vol. 99, no. 4, pp. 604–607, 2008.

22. S. Singh and S. Kumar, "Development of convective heat transfer correlations for common designs of solar dryer," *Energy Convers. Manag.*, vol. 64, pp. 403–414, 2012.

23. A. A. El-Sebaii, S. Aboul-Enein, M. R. I. Ramadan, and H. G. El-Gohary, "Experimental investigation of an indirect type natural convection solar dryer," *Energy Convers. Manag.*, vol. 43, no. 16, pp. 2251–2266, 2002.

24. E. K. Akpinar, "Drying of mint leaves in a solar dryer and under open sun: Modelling, performance analyses," *Energy Convers. Manag.*, vol. 51, no. 12, pp. 2407–2418, 2010.

25. I. Doymaz, "Thin-layer drying behaviour of mint leaves," *J. Food Eng.*, vol. 74, no. 3, pp. 370–375, 2006.

26. G. Xanthopoulos, S. Yanniotis, and G. Lambrinos, "Study of the drying behaviour in peeled and unpeeled whole figs," *J. Food Eng.*, vol. 97, no. 3, pp. 419–424, 2010.

27. S. J. Babalis and V. G. Belessiotis, "Influence of the drying conditions on the drying constants and moisture diffusivity during the thin-layer drying of figs," *J. Food Eng.*, vol. 65, no. 3, pp. 449–458, 2004.

28. D. R. Pangavhane, R. L. Sawhney, and P. N. Sarsavadia, "Design, development and performance testing of a new natural convection solar dryer," *Energy*, vol. 27, no. 6, pp. 579–590, 2002.

29. A. Kouchakzadeh, "The effect of acoustic and solar energy on drying process of pistachios," *Energy Convers. Manag.*, vol. 67, pp. 351–356, 2013.

30. T. Y. Tunde-Akintunde, "Mathematical modeling of sun and solar drying of chilli pepper," *Renew. Energy*, vol. 36, no. 8, pp. 2139–2145, 2011.

31. P. L. Singh, "Silk cocoon drying in forced convection type solar dryer," *Appl. Energy*, vol. 88, no. 5, pp. 1720–1726, 2011.

32. R. Ben Slama and M. Combarnous, "Study of orange peels dryings kinetics and development of a solar dryer by forced convection," *Sol. Energy*, vol. 85, no. 3, pp. 570–578, 2011.

33. S. Şevik, "Design, experimental investigation and analysis of a solar drying system," *Energy Convers. Manag.*, vol. 68, pp. 227–234, 2013.

34. A. A. El-Sebaii and S. M. Shalaby, "Experimental investigation of an indirect-mode forced convection solar dryer for drying thymus and mint," *Energy Convers. Manag.*, vol. 74, pp. 109–116, 2013.

35. W. Wang, M. Li, R. H. E. Hassanien, Y. Wang, and L. Yang, "Thermal performance of indirect forced convection solar dryer and kinetics analysis of mango," *Appl. Therm. Eng.*, vol. 134, pp. 310–321, 2018.

36. M. Fterich, H. Chouikhi, H. Bentaher, and A. Maalej, "Experimental parametric study of a mixed-mode forced convection solar dryer equipped with a PV/T air collector," *Sol. Energy*, vol. 171, no. June, pp. 751–760, 2018.

37. A. Bosomtwe et al., "Effectiveness of the solar biomass hybrid dryer for drying and disinfestation of maize," *J. Stored Prod. Res.*, vol. 83, pp. 66–72, 2019.

38. T. Hamdani, A. Rizal, and Z. Muhammad, "Fabrication and testing of hybrid solar-biomass dryer for drying fish," *Case Stud. Therm. Eng.*, vol. 12, pp. 489–496, 2018.

39. A. El Khadraoui, S. Bouadila, S. Kooli, A. Farhat, and A. Guizani, "Thermal behavior of indirect solar dryer: Nocturnal usage of solar air collector with PCM," *J. Clean. Prod.*, vol. 148, pp. 37–48, 2017.

40. S. O. Enibe, "Performance of a natural circulation solar air heating system with phase change material energy storage," *Renew. Energy*, vol. 2, pp. 69–86, 2002.

41. S. Abubakar, S. Umaru, M. U. Kaisan, U. A. Umar, B. Ashok, and K. Nanthagopal, "Development and performance comparison of mixed-mode solar crop dryers with and without thermal storage," *Renew. Energy*, vol. 128, pp. 285–298, 2018.

42. S. Devahastin and S. Pitaksuriyarat, "Use of latent heat storage to conserve energy during drying and its effect on drying kinetics of a food product," *Appl. Therm. Eng.*, vol. 26, pp. 1705–1713, 2006.

43. Esakkimuthu, S., Hassabou, A. H., Palaniappan, C., Spinnler, M., Blumenberg, J., and Velraj, R., "Experimental investigation on phase change material based thermal storage system for solar air heating applications," *Sol. Energy*, vol. 88, pp. 144–153, 2013.

44. A. Reyes, D. Negrete, A. Mahn, and F. Sepúlveda, "Design and evaluation of a heat exchanger that uses paraffin wax and recycled materials as solar energy accumulator," *Energy Convers. Manag.*, vol. 88, pp. 391–398, 2014.

45. S. M. Shalaby and M. A. Bek, "Experimental investigation of a novel indirect solar dryer implementing PCM as energy storage medium," *Energy Convers. Manag.*, vol. 83, pp. 1–8, 2014.

46. M. S. Aiswarya and C. R. Divya, "Economic analysis of solar dryer with PCM for drying agricultural products." *Int. Res. J. Eng. Technol.*, vol. 4, pp. 1948–1953, 2015.

47. D. Jain and P. Tewari, "Performance of indirect through pass natural convective solar crop dryer with phase change thermal energy storage," *Renew. Energy*, vol. 80, pp. 244–250, 2015.

48. A. K. Bhardwaj, R. Kumar, and R. Chauhan, "Experimental investigation of the performance of a novel solar dryer for drying medicinal plants in Western Himalayan region," *Sol. Energy*, vol. 177, November 2018, pp. 395–407, 2019.

49. M. Iranmanesh, H. Samimi Akhijahani, and M. S. Barghi Jahromi, "CFD modeling and evaluation the performance of a solar cabinet dryer equipped with evacuated tube solar collector and thermal storage system," *Renew. Energy*, vol. 145, pp. 1192–1213, 2020.

4 Experimental and Economic Performance of Two Solar Dryer Systems in Tunisia

Aymen El Khadraoui and Salwa Bouadila

CONTENTS

4.1 INTRODUCTION

Tunisia is located in the Mediterranean region. Tunisia has between 2860 and 3200 h of sunshine per year and receives a daily average solar energy of 4.8 kWh m^{-2} day^{-1} [1]. Today the fossil energy situation of Tunisia indicated several weakness aspects due to its limited character. Thus, the need for energy alternatives becomes inevitable. The exploitation of different renewable energy resources seems to be one of the potential solutions. Among them, solar energy is considered a suitable option because of its value-added benefits, such as eco-friendly, economic, easy installation, and so on [2]. Therefore, solar energy is used in different domains in Tunisia, such as sanitary water heating [3], air conditioning of buildings [4,5], and in drying [6]:

The use of the solar energy in the drying process of agricultural products has become one of the most popular and economic investments. Open sun drying is the traditional method of preserving the agricultural products employed in most of the developing countries. However, this technique has problems of contamination with the dust, rain, soil, and insects [7,8]. Various solar dryers have been designed and developed to overcome the problems of open sun drying. These solar drying systems are, quite simply, low-cost technology and allow the dried products to be dried under hygienic conditions [9]. Depending upon the heat transfer mechanisms to the crop, the solar dryers can be classified in three types: the direct solar dryers, the indirect solar dryers, and the mixed mode solar dryers. It is for that reason that in the last years the interest in using the solar dryers has increased.

A number of studies that have been reported on the solar drying system have been conducted in The Thermal Process Laboratory, and The Research and Technology Center of Energy, Tunisia. Farhat et al. [1] validated the Passamia and Saravia model [10,11] on the red pepper under tunnel greenhouse and in open-air. Fadhel et al. have studied solar drying grapes in three different processes: open sun drying, natural convection solar dryer, and under tunnel greenhouse. These tests show that the solar tunnel greenhouse drying is satisfactory and competitive to a natural convection solar drying process [12]. Experiments inside a wind tunnel were conducted to study the drying of red pepper in open sun and greenhouse conditions [13] where solar radiation was simulated by a 1000 W lamp, for different external parameters (incident radiation, ambient temperature, and air velocity). A simple drying model of red pepper related to the water evaporation process was developed and verified. Further studies were conducted by Fadhel et al. [14] to dry hot red peppers in three different processes—open sun drying, natural convection solar dryer, and under tunnel greenhouse—and it was concluded that the solar tunnel greenhouse dryer must be improved to become competitive with the solar air dryer.

In this chapter, we introduce two models of the solar dryer: solar greenhouse dryer and indirect solar dryer. The two solar drying systems were tested under the climatic conditions of Borj Cedria (North of Tunisia). As a reference crop for drying, a red pepper was chosen due to its high prevalence in the Tunisian cuisine. The economic aspects of drying have been investigated in this study.

4.2 MATERIALS AND METHODS

4.2.1 DESCRIPTION OF THE SITE

The experimental study was performed during September in the Research and Technology Center of Energy (CRTEn) in Borj-Cedria. The site is a sunny region situated on the Mediterranean coast of North Africa, near the city of Tunis (Tunisia), with the latitude 36°43′N and the longitude 10°25′E. The ambient temperature can exceed 40°C in July and August and rarely exceeding above 20°C in winters

FIGURE 4.1 Photograph of the indirect solar dryer.

days [15]. In fact, Tunis, the capital of Tunisia, receives a global solar radiation intensity varied between 1800 and 2600 kW h/m²/year (Figure 4.1). It presents 350 sunny days per year with a daily insolation range from 5 h in December to 13 h in July.

4.2.2 DESCRIPTION OF THE INDIRECT SOLAR DRYER

An indirect solar dryer has been designed, constructed and installed at the CRTEn in Borj Cedria, North of Tunisia. The solar dryer is divided into two major parts: (1) solar flat-plate solar collector and (2) a drying chamber. A pictorial view of the dryer developed in this study is shown in Figure 4.1.

4.2.2.1 Flat-Plate Solar Collector

The solar collector (Figure 4.2) was fabricated, using the locally available materials in the CRTEn in Borj Cedria. It consists of an insulator, absorber, and cover glass. The length, width, and total volume of the collector are 2 m, 1 m, and 0.28 m³, respectively. The 0.004-m-thick transparent glass cover was placed 0.05 m apart from the absorber. A copper corrugated plate of 0.001-m thick used as the absorber plate was placed 0.04 m apart from the insulator. The solar air collector back and sides were insulated with a 0.05-m layer of polyurethane, with heat conductivity 0.028 W/m K, to decrease thermal losses.

There are two air gaps between the glass cover and the absorber and between the absorber and the insulator through which the ambient air is sucked by a centrifugal

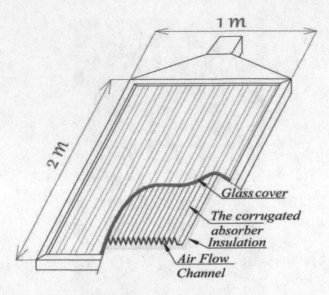

FIGURE 4.2   Schematic diagram of the solar air collator.

fan from lower side of the collector to the greenhouse. The solar collector was oriented full south and inclined 37° to the horizontal plane.

4.2.2.2   The Drying Chamber

The drying chamber was fabricated from a galvanized iron with dimensions $0.8 \times 0.8 \times 1.2$ m³ (length × width × height). The drying chamber was insulated at all sides using a layer of polyurethane as an insulating material with thickness 0.03 m. The drying chamber had six drying trays positioned at equal vertical spacing of 0.15 m. The drying trays were made of a wooden frame on all four sides and a wire mesh on the bottom to uphold the samples. The trays can be easily removed to load or unload the drying product from the door, which represents one side of the drying chamber.

4.2.3   Description of the Solar Greenhouse Dryer

The solar-greenhouse-forced convection drying system has been installed at the CRTEn in Borj Cedria (North of Tunisia). The system essentially consists of two parts: (1) a flat-plate solar air collector (described in the Section 4.2.2.1), and (2) an experimental east–west oriented chapel-shaped greenhouse (Figure 4.3).

The experimental greenhouse (Figure 4.4) occupies a floor area equal to 14.8 m², 3.7 m wide, 4 m long, and 3 m high at the center. The greenhouse walls and roof are covered by Plexiglass with 0.003 m of thickness. To exhaust the moist air from the greenhouse, it was equipped with two centrifugal fans (0.0833 m³s⁻¹, 0.3 kW, 220 V, 50 Hz, 1380 min⁻¹). This drying system has a loading capacity of about 80 kg of peppers. It could accommodate four trays in stacks with a total drying area of 40 m², and the dimension of one tray is 2.5 m × 4 m.

FIGURE 4.3   Pictorial view of the solar greenhouse dryer.

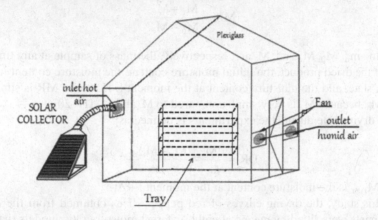

FIGURE 4.4   A schematic diagram of the solar greenhouse dryer.

4.2.4   EXPERIMENTAL PROCEDURE

In the drying experiments, the red pepper was used as the test sample. During the drying days, the weather was generally sunny and no rain appeared. The meteorological parameters as well as the mass and the product temperature are measured during the drying operation.

Several K-type thermocouples were fixed at different locations of the two solar dryers and in the open sun to measure the temperature. The incident solar radiation was measured with a Kipp and Zonen pyranometer in a range of 0–1000 W/m^2. The air velocity was measured by an anemometer in the measurement range of 0–20 m/s. The relative humidity was periodically measured by a HMP155A sensor. All climatic and measured parameters are sample recorded every 10 min using a CR5000 data logger (Campbell Scientific Inc).

For each drying test, the red peppers ("Baklouti" variety) were washed with water to remove skin dirt and then cut longitudinally in two slices. After the removal of stems and seeds, the products were spread on a perforated metal grid, the skin against the grid. The product samples were weighed periodically using a digital balance (RADWAG, 200 g capacity). At the end of each drying experiment, the sample was placed in an oven drying at 120°C for 12 h and then weighed to determine its dry matter.

4.2.5 MATHEMATICAL MODELING OF DRYING CURVES

The moisture content during the drying experiments was calculated using the following equation:

$$M = \frac{m_t - m_d}{m_d} \tag{4.1}$$

The moisture content of the drying sample at time t can be transformed to be the moisture ratio (MR). This MR is expressed as [16]:

$$MR = \frac{M_t - M_e}{M_0 - M_e} \tag{4.2}$$

where m_t, m_d, M_0, M_e and, M_t are, respectively, the mass of sample at any time, the mass of the dried product, the initial moisture content, the moisture content in equilibrium state, and the moisture content at the moment t. However, MR is simplified to M_t/M_0, because M_e is very small compared to M_t and M_0 [17–20].

The drying rate during the experiments is defined as:

$$DR = \frac{dM}{dt} = \frac{M_t - M_{t+\Delta t}}{\Delta t} \tag{4.3}$$

where $M_{t+\Delta t}$ is the moisture content at the moment $t + \Delta t$.

In this study, the drying curves of red pepper slices obtained from the drying experiments were fitted by means of eight different moisture ratio models that were widely used in most food and biological materials, which are presented in Table 4.1.

TABLE 4.1
Mathematical Models Applied to the Drying Curves

Model No.	Model Name	Model	References
1	Page	$MR = \exp(-k\, t^n)$	[21]
2	Midilli and Kucuk	$MR = a \exp(-k\, t^n) + b\, t$	[22]
3	Two-term exponential	$MR = a \exp(-k\, t) + (1-a)\exp(-k\, a\, t)$	[23]
4	Diffusion approach	$MR = a \exp(-k\, t) + (1-a)\exp(-k\, b\, t)$	[24]
5	Modified Henderson and Pabis	$MR = a \exp(-k\, t) + b \exp(-g\, t) + c \exp(-ht)$	[25]
6	Wang and Singh	$MR = 1 + at + bt^2$	[26]

* a, b, c, h, g, k, k_0, k_1 are the empirical constants in drying models; t is the drying time (h).

The regression analysis was performed using the MATLAB program. The correlation coefficient (R^2), reduced squared-chi (χ^2), and the root mean square error (RMSE) were used as criteria for adequacy of the fit. The higher R values and the lower χ^2 and RMSE values indicate the better goodness of fit.

These statistical analysis values can be calculated as follows:

$$R^2 = \frac{\sum_{i=1}^{n}(MR_i - MR_{pre,i}) \cdot \sum_{i=1}^{n}(MR_i - MR_{exp,i})}{\sqrt{\left[\sum_{i=1}^{n}(MR_i - MR_{pre,i})^2\right] \cdot \left[\sum_{i=1}^{n}(MR_i - MR_{exp,i})^2\right]}} \tag{4.4}$$

$$\chi^2 = \frac{\sum_{i=1}^{n}(MR_{exp,i} - MR_{pre,i})^2}{N - n} \tag{4.5}$$

$$RMSE = \sqrt{\frac{\sum_{i=1}^{N}(MR_{exp,i} - MR_{pre,i})^2}{N}} \tag{4.6}$$

where $MR_{exp,i}$ is the moisture ratio value obtained from experiment results, $MR_{pre,i}$ is the predicted moisture ratio value by using the model, N is the coefficient number used in the model, and n is the number of experimental data.

4.3 RESULTS AND DISCUSSIONS

4.3.1 DRYING OF RED PEPPERS IN THE SOLAR GREENHOUSE DRYER

The solar drying of red pepper slices in the solar greenhouse dryer and in the open sun was carried out under the weather conditions of Borj Cedria, Tunis. The weather conditions during the typical experimental run are shown in Figure 4.5. The solar radiation on the horizontal plane outside the greenhouse varied from 66.98 W/m² to 808 W/m² during the drying period. The Plexiglass cover transmitted about 85% of the incident solar radiation. In the two days of this experimental run, the solar radiation increased sharply from the morning till the noon but was considerably decreased in the afternoon. In Figure 4.5, the ambient air temperature ranged from 21.25°C to 32.8°C, the temperature of the drying air inside the greenhouse from 29.21°C to 49.88°C, and the temperature of the drying air at the outlet of the solar collector from 27.87°C to 54.68°C. The airflow temperature inside the greenhouse is an important parameter in the characterization of the dryer performance. Observing Figure 4.5, we can find significant differences between the temperatures of the air inside and outside the greenhouse for most periods of the day. The temperature difference varied in the range of 6°C–19.96°C. The high temperature is more favorable for drying due to the increase of the evaporating capacity of the drying air. From these results, it can be deduced that the drying rate of pepper in the solar greenhouse dryer was higher than that of the open sun drying.

4.3.1.1 Drying Curves

The temperature and relative humidity of the drying air obtained inside the greenhouse (drying chamber) during the experimentation are presented in Figure 4.6.

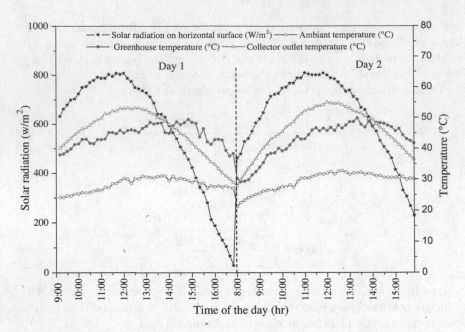

FIGURE 4.5 Variations of solar radiation, ambient temperature, greenhouse temperature, and outlet collector temperature during the red pepper drying.

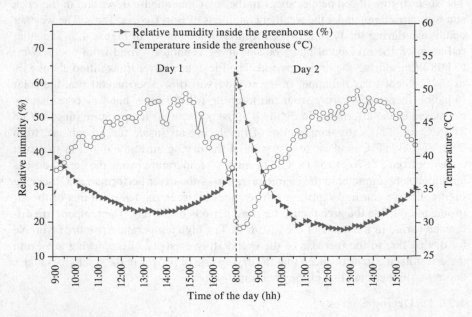

FIGURE 4.6 Variations of relative humidity and temperature inside the greenhouse during the red pepper drying.

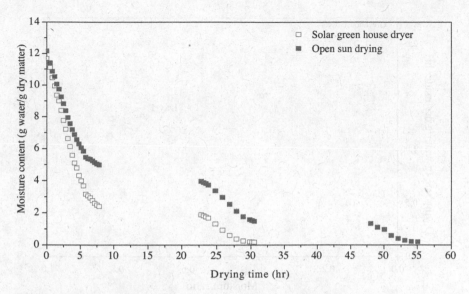

FIGURE 4.7 Variation of red pepper moisture ratio with drying time.

The variations of the temperature and relative humidity were caused by the variations in the daily solar radiation (Figure 4.5). The relative drying air humidity inside the greenhouse decreased with increasing solar radiation, whereas the drying air temperature increased. The drying air temperature was maximum at noon while the relative humidity was the minimum. The relative humidity of the drying air under the greenhouse ranged from 17.6% to 62.6% (Figure 4.6).

Figure 4.7 presents a comparison of the moisture ratio of red pepper dried in the solar greenhouse dryer and in the open sun. The interruptions of the lines in this figure represent the night periods of the drying operation. As it would be expected, during the first day of drying there was a rapid moisture removal from the product, which later decreased with an increase in the drying time. From this figure, it can also be seen that the moisture ratio decreases continually with drying time. It is observed that the moisture removal from the inside greenhouse crop was significantly higher during all the crop drying hours. The inside greenhouse crop took only 17 h to reach at 17 db moisture content level, whereas the outside crop took 24 h to attain 19 db moisture content level. The present system (the solar greenhouse dryer) practically shortens the drying time of red peppers by one day.

The drying rate of red peppers during drying was estimated based on Eq. (4.2), and its changes with MR are shown in Figure 4.8. As expected, the drying rate fluctuated very much and decreased with decreasing the moisture ratio. It can be observed that all the curves present no constant drying rate period, which shows that the important part of the drying process occurred in the falling rate period [27]. In the falling rate period, the material surface is no longer saturated with water, and the drying rate is controlled by the diffusion of the moisture from the interior of solid to the surface [28]. It was also clear from Figure 4.9 that the drying in the open sun takes more time as compared to the solar greenhouse dryer because of the lesser rate of water evaporated during the drying experiments.

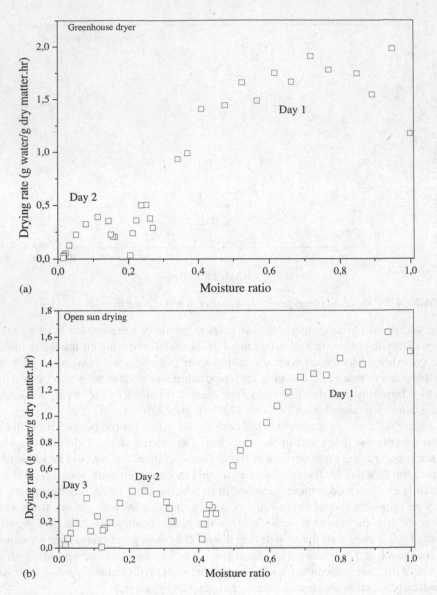

FIGURE 4.8 Variation of drying rate with moisture ratio of red pepper (a) inside solar greenhouse dryer and (b) open sun drying.

4.3.1.2 Fitting of Drying Curves

The moisture content data at the different experimental mode were converted to the most useful moisture ratio expression. The resulted moisture ratio curves [MR = f(t)] were regressed against various thin-layer drying models listed in Table 4.1, using the MATLAB program. The results of statistical analysis undertaken on these models for the forced solar drying and the open sun drying of red peppers are given in

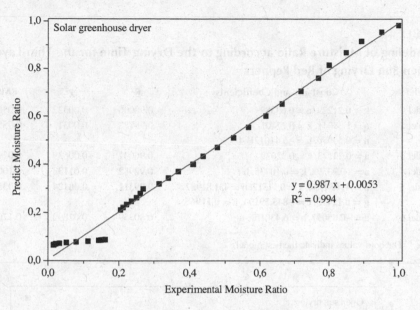

FIGURE 4.9 Comparison of experimental and predicted dimensionless moisture ratio for solar greenhouse drying.

Tables 4.2 and 4.3, respectively. Based on the range and average values of the statistical parameters for each model, it can be concluded that the Modified Henderson and Pabis model was the best descriptive model for the thin-layer-forced solar drying and the open sun drying of red peppers, as shown in Tables 4.2 and 4.3. For the solar greenhouse drying of red peppers the Modified Henderson and Pabis model gave $R^2 = 0.99443$, $\chi^2 = 0.00115$, and RMSE = 0.03396. For the open sun drying of red peppers, the Modified Henderson and Pabis model gave $R^2 = 0.99311$, $\chi^2 = 0.00124$, and RMSE = 0.03521.

TABLE 4.2
Modeling of Moisture Ratio according to the Drying Time for the Thin-Layer-Forced Solar Drying (Solar Greenhouse Dryer) of Red Peppers

Model	Constants and Coefficients	R^2	χ^2	RMSE
Model 1	k = 0.13966, n = 1.22088	0.98072	0.0035	0.05916
Model 2	a = 1.00573, k = 0.13942	0.99109	0.00173	0.04156
	n = 1.25547, b = 0.00274			
Model 3	a = 1.79062, k = 0.28565	0.98102	0.00345	0.05869
Model 4	a = 1, k = 0.19133, b = 1	0.9786	0.004	0.06321
Model 5	**a = 0.21565, k = 0.03915, b = 30.82204**	**0.99443**	**0.00115**	**0.03396**
	g = 0.55141, c = −30.05483, h = 0.56556			
Model 6	a = −0.12189, b = 0.00319	0.89725	0.01786	0.13364

Note: The bold values indicate the best model.

TABLE 4.3

Modeling of Moisture Ratio according to the Drying Time for the Thin-Layer Open Sun Drying of Red Peppers

Model	Constants and Coefficients	R^2	χ^2	RMSE
Model 1	k = 0.21229, n = 0.63382	0.98006	0.00322	0.05677
Model 2	a = 1.08041, k = 0.27401	0.98172	0.0031	0.05572
	n = 0.54979, b = −6.41012E-4			
Model 3	a = 0.22173, k = 0.32039	0.96007	0.00639	0.07993
Model 4	a = −0.93392, k = 0.10435, b = 1	0.92982	0.01133	0.10642
Model 5	**a = 414.81934, k = 0.11512, b = 431.81982**	**0.99311**	**0.00124**	**0.03521**
	g = 0.12415, c = −845.59194, h = 0.11965			
Model 6	a = −0.05097, b = 6.43015E-4	0.80359	0.02892	0.17005

Note: The bold values indicate the best model.

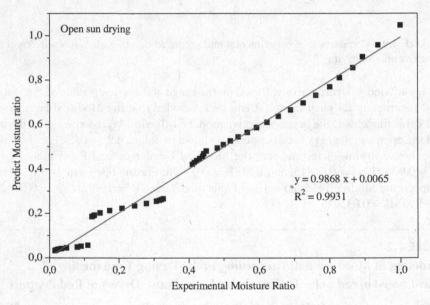

FIGURE 4.10 Comparison of experimental and predicted dimensionless moisture ratio for open sun drying.

The fitting performance of the suitable model was also illustrated by curves of the predicted values versus the experimental values presented in Figures 4.9 and 4.10. A good agreement was observed between the experimental and the predicted moisture ratio values.

4.3.1.3 Economic Performance of the Solar Greenhouse Dryer

Based on the climatic conditions in Tunisia, the greenhouse can be exploited in the warmer months of the year for drying agricultural products. For the economic evaluation, the cost of greenhouse functioning as the dry cabinet is not included in the total cost of the solar greenhouse dryer. It is assumed that each year the solar greenhouse dryer is used to dry peppers in June–September. Approximately, 444 kg of dry red peppers will be produced annually.

The total capital cost for the solar dryer (C_T) is estimated as the sum of the material cost C_m of the dryer and the labor cost for the construction C_l as shown in Eq. (4.7).

$$C_T = C_m + C_l \tag{4.7}$$

The annual costs are calculated using Eq. (4.8) proposed by Audsley and Wheeler [29]:

$$C_{annual} = \left[C_T + \sum_{i=1}^{N} \left(C_{main,i} + C_{op,i} \right) \omega^i \right] \left[\frac{\omega - 1}{\left(\omega^N - 1 \right)} \right] \tag{4.8}$$

where C_{annual} is the annual cost of the system and $C_{main,i}$ and $C_{op,i}$ are the maintenance cost and the operating cost at the year i, respectively. The value w is expressed as:

$$\omega = \frac{100 + i_{in}}{100 + i_f} \tag{4.9}$$

where i_{in} and i_f are the interest rate and the inflation rate in percent, respectively.

The annual cost per unit of the dried product is calculated using Eq. (4.10).

$$Z = \frac{C_{annual}}{M_{dry}} \tag{4.10}$$

where M_{dry} is the dried product obtained from this solar greenhouse dryer per year.

The payback period (n_p) is calculated from:

$$n_p = \frac{C_T}{M_{dry}P_d - M_f P_f - M_{dry}Z} \tag{4.11}$$

where M_{dry} is the annual production of the dry product (kg), M_f is the amount of the fresh product per year (kg), P_d is the price of the dry product (DT/kg), and P_f is the price of the fresh product (DT/kg).

Table 4.4 shows the economic evaluation of the solar greenhouse dryer for drying red peppers. It is noted that the payback period estimated using Eq. (4.11) is about 1.02 years.

4.3.2 DRYING OF RED PEPPERS IN THE INDIRECT SOLAR DRYER

The indirect solar dryer with phase change material [PCM] was tested outdoors during October. During the drying experiments, the weather was generally sunny and no rain appeared. The hourly variations of the solar radiation, ambient temperature, and drying chamber temperature are presented in Figure 4.11.

TABLE 4.4

Data on Costs and Economic Parameters

Cost of Dryer	**2000 DT**
Annual electricity cost for fans	15 DT/year
Capacity of dryer	80 kg
Price of fresh peppers	0.5 DT/kg
Price of dried peppers	10 DT/kg
Expected life of dryer	20 years
Interest rate	8%
Inflation rate	5%

Note: 1 US Dollar = 2.7 DT.

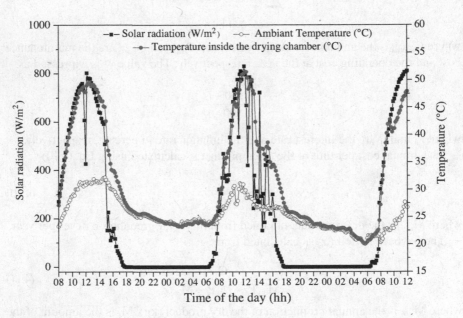

FIGURE 4.11 Variations of solar radiation, ambient temperature, and the drying chamber temperature during the red pepper drying.

The solar radiation for all three days reached 802, 830, and 810 W/m², respectively. The ambient temperature varies from 20°C to 32°C. The solar radiation and the ambient temperature strongly affected the air temperature inside the drying chamber. This is because the hot drying air inside the drying chamber was drawn from the ambient air and heated by the solar radiation. When the solar radiation falls down, the ambient and the temperature inside the drying chamber will be decreased. During the daytime, the temperature inside the drying cabinet was found to be much higher than the ambient temperature. The maximum air-drying temperature was 51.5°C at 12:00 h on the second day.

4.3.2.1 Drying Curves

The hourly variation of the moisture content is shown in Figure 4.12. The moisture content of the peppers was reduced from 9.64 g water/g dry matter to 0.13 g water/g dry matter after 52 h. It is clearly seen in Figure 4.12 that the moisture removal is higher in initial period (Day 1) due to evaporation of the moisture from the surfaces, and the rate of the moisture removal decreases gradually.

The drying rate of the pepper versus the drying time and the moisture content are illustrated in Figure 4.13. It is apparent that the drying rate decreased continuously with the drying time. The constant drying rate period was not observed.

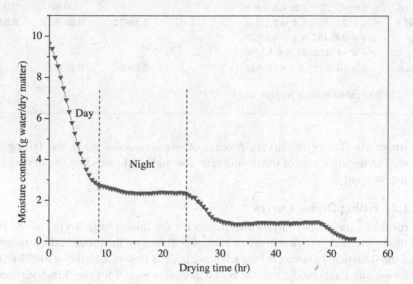

FIGURE 4.12   Variation of red pepper moisture content with drying time.

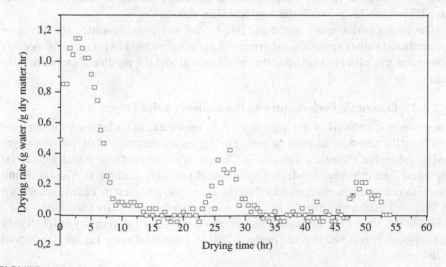

FIGURE 4.13   Variation of red pepper drying rate with drying time.

TABLE 4.5

Modeling of Moisture Ratio according to the Drying Time for the Thin-Layer-Forced Solar Drying (Indirect Solar Dryer) of Red Peppers

Model	Constants and Coefficients	R^2	χ^2	RMSE
Model 1	k = 0.19667, n = 0.74435	0.97865	0.00215	0.04632
Model 2	a = 1.10949, k = 0.25129	0.98092	0.00196	0.04424
	n = 0.68742, b = 1.37888			
Model 3	a = 0.28124, k = 0.28164	0.97649	0.00236	0.04858
Model 4	a = −272239.28897, k = 0.10184, b = 1	0.95969	0.00405	0.06364
Model 5	**a = 226.98586, k = 0.12541,**	**0.98972**	**0.00108**	**0.03287**
	b = 244.19226, g = 0.13623,			
	c = −470.06726, h = 0.13087			
Model 6	a = −0.05723, b = 6.95684E-4	0.83096	0.01572	0.12537

Note: The bold values indicate the best model.

The important part of the drying process of peppers occurred in the falling-rate period. The negative sign of the drying rate overnight indicates that the drying process has stopped.

4.3.2.2 Fitting Drying Curves

The results of statistical analysis undertaken for the forced solar drying of red peppers by the indirect dryer are given in Table 4.5. Based on the range and average values of the statistical parameters for each model, it can be concluded that the Modified Henderson and Pabis model was the best descriptive model for the thin-layer-forced solar drying and the open sun drying of red peppers, as shown in Table 4.5. For the solar greenhouse drying of red peppers, the Modified Henderson and Pabis model gave $R^2 = 0.98972$, $\chi^2 = 0.00108$, and RMSE = 0.03287.

The fitting performance of the suitable model was also illustrated by curves of the predicted values versus the experimental values presented in Figure 4.14. A good agreement was observed between the experimental and the predicted moisture ratio values.

4.3.2.3 Economic Performance of the Indirect Solar Dryer

The economic analysis of the solar devices is important to understand the application from the commercial point of view. For the economic analysis of the solar dryer and based on the climatic conditions in Tunisia, it is assumed that this dryer can be exploited 7 months/year for drying agricultural products (210 days). The costs and the main economic parameters based on the economic situation in Tunisia are shown in Table 4.5. Approximately 67 kg of dry red peppers are annually produced.

Table 4.6 shows the economic evaluation of the solar indirect dryer for drying red peppers. It is noted that the payback period estimated using Eq. (4.11) is about 8.264 years.

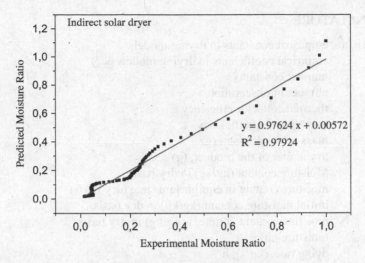

FIGURE 4.14   Comparison of experimental and predicted dimensionless moisture ratio for indirect solar drying.

TABLE 4.6
Data on Costs and Economic Parameters

Cost of Dryer	2200 DT
Annual electricity cost for fans	27 DT/year
Capacity of Dryer	7 kg
Price of fresh pepper	0.5 DT/kg
Price of dried pepper	10 DT/kg
Expected life of dryer	20 years
Interest rate	8%
Inflation rate	5%

Note: 1US Dollar = 2.7 DT.

4.4   CONCLUSION

The technology of solar drying is very economical and efficient. This paper aims to present two new solar dryers: a solar greenhouse dryer and an indirect solar dryer. The two solar drying systems are designed, constructed, and tested for drying of red peppers ("Baklouti" variety). An economic evaluation was calculated using the criterion of payback period. The payback period of the solar greenhouse dryer was found to be 1.6 years. Also, it was found at 8.2 years for the indirect solar dryer. The solar greenhouse drying is advantageous over to the indirect solar dryer. Indeed, the solar greenhouse drying presents a big drying capacity and doesn't require a large initial investment.

NOMENCLATURE

a, b, c, g, h, n	empirical constants in drying models
k, k_0, k_1	empirical coefficients in drying models (s^{-1})
n	number constants
N	number of observations
η_c	thermal collector efficiency
η_d	overall drying efficiency
m	mass of the product, (g)
m_d	dry matter of the product, (g)
M	Moisture content (kg/kg) in dry basis
M_e	moisture content in equilibrium state (dry basis)
M_0	initial moisture content (kg/kg) in dry basis
M_t	moisture content at time, t (kg/kg) in dry basis
MR	moisture ratio
DR	dying rate, (kg/kg h^{-1})
MR_{exp}	experimental moisture ratio
MR_{pre}	predicted moisture ratio
R^2	correlation coefficient
RMSE	root mean square error
w_R	total uncertainty in measurement of result
w_1, w_2,..., w_n	uncertainties in independent variables
w_{tm}	total uncertainty in the measurement of time of mass loss values
w_{ml}	total uncertainty in the measurement of mass loss values
w_{mq}	total uncertainty in the measurement of the moisture quantity
w_{DR}	total uncertainty in the calculated of drying rate
w_{MR}	total uncertainty in the calculated of moisture rate
t	time (s, h, min)
C_l	labor cost for construction of dryer (DT)
C_{annual}	annual cost of the system (DT)
C_m	material cost of dryer (DT)
$C_{maint, i}$	maintenance cost (DT)
$C_{op, i}$	operating cost (DT)
C_T	total capital cost of dryer (DT)
i_{in}	interest rate (%)
i_f	inflation rate (%)
n_p	life span of the dryer (year)
P_d	price of the dry product (DT/kg)
P_f	price of the fresh product (DT/kg)
Z	drying cost (DT/kg)
W	the weight of the water evaporated from the product (Kg)
ω	parameter defined in Eq. (4.12)
χ^2	chi-square

REFERENCES

1. Farhat, A., Kooli, S., Kerkeni, C., Maalej, M., Fadhel, A., & Belghith, A. (2004). Validation of a pepper drying model in a polyethylene tunnel greenhouse. *International Journal of Thermal Sciences*, 43, 53–58.
2. Kalaiarasi, G., Velraj, R., & Swami, M. V. (2016). Experimental energy and exergy analysis of a flat plate solar air heater with a new design of integrated sensible heat storage. *Energy*, 111, 609–619. doi:10.1016/j.energy.2016.05.110.
3. Hazami, M., Naili, N., Attar, I., & Farhat, A. (2013). Solar water heating systems feasibility for domestic requests in Tunisia: Thermal potential and economic analysis. *Energy Conversion and Management*, 76, 599–608. doi:10.1016/j.enconman.2013.07.079.
4. Mehdaoui, F., Hazami, M., Naili, N., & Farhat, A. (2014). Energetic performances of an optimized passive solar heating prototype used for Tunisian buildings air-heating application. *Energy Conversion and Management*, 87, 285–296. doi:10.1016/j.enconman.2014.07.024.
5. Abbassi, F., Dimassi, N., & Dehmani, L. (2014). Energetic study of a Trombe wall system under different Tunisian building configurations. *Energy and Buildings*, 80, 302–308. doi:10.1016/j.enbuild.2014.05.036.
6. El Khadraoui, A., Kooli, S., Hamdi, I., & Farhat, A. (2015). Experimental investigation and economic evaluation of a new mixed-mode solar greenhouse dryer for drying of red pepper and grape. *Renewable Energy*, 77, 1–8.
7. Labed, A., Moummi, N., Aoues, K., & Benchabane, A. (2016). Solar drying of henna (*Lawsonia inermis*) using different models of solar flat plate collectors: An experimental investigation in the region of Biskra (*Algeria*). *Journal of Cleaner Production*, 112, 2545–2552.
8. Perea-Moreno, A. J., Juaidi, A., & Manzano-Agugliaro, F. (2016). Solar greenhouse dryer system for wood chips improvement as biofuel. *Journal of Cleaner Production*, 135, 1233–1241.
9. Saleh, A., & Badran, I. (2009). Modeling and experimental studies on a domestic solar dryer. *Renewable Energy*, 34, 2239–2245.
10. Passamia, V., & Saravia, L. (1997). Relationship between a solar drying model of red pepper and Kinetics of pure water evaporation (I). *Drying Technology*, 15 (5), 1419–1432.
11. Passamia, V., & Saravia, L. (1997). Relationship between a solar drying model of red pepper and pure water evaporation (II). *Drying Technology*, 15 (5), 1433–1445.
12. Fadhel, A., Kooli, S., Farhat, A., & Belghith, A. (2005). Study of the solar drying of grapes by three different processes. *Desalination*, 185, 535–541.
13. Kooli, S., Fadhel, A., Farhat, A., & Belghith, A. (2007). Drying of red pepper in open sun and greenhouse conditions: Mathematical modeling and experimental validation. *Journal of Food Engineering*, 97 (53), 1094–1103.
14. Fadhel, A., Kooli, S., Farhat, A., & Belghith, A. (2014). Experimental study of the drying of hot red pepper in the open air, under greenhouse and in a solar drier. *International Journal of Renewable Energy & Biofuels*, 2014, 1–14.
15. Bouadila, S., Lazaar, M., Skouri, S., Kooli, S., & Farhat, A. (2014). Energy and exergy analysis of a new solar air heater with latent storage energy. *International Journal of Hydrogen Energy*, 39, 15266–15274.
16. Şahin, U., & Öztürk, H. K. (2016). Effects of pulsed vacuum osmotic dehydration (PVOD) on drying kinetics of figs (*Ficus carica* L). *Innovative Food Science and Emerging Technologies*, 36, 104–111.
17. Yaldiz, O., Ertekin, C., & Uzan, H. E. (2001). Mathematical modeling of thin layer solar drying of sultana grapes. *Energy Conversion and Management*, 26, 457–465.

18. Midilli, A., & Kucuk, H. (2003). Mathematical modelling of thin layer drying of pistachio by using solar energy. *Energy Conversion and Management*, 44, 1111–1122.
19. Jain, D., & Pathare, P. B. (2007). Study the drying kinetics of open sun drying of fish. *Journal of Food Engineering*, 78, 1315–1319.
20. Lakshmi, D. V. N., Muthukumar, P., Layek, A., & Nayak, P. K. (2018). Drying kinetics and quality analysis of black turmeric (*Curcuma caesia*) drying in a mixed mode forced convection solar dryer integrated with thermal energy storage. *Renewable Energy*, 120, 23–34.
21. Page, G. (1949). Factors influencing the maximum rates of air-drying shelled corn in thin layers. M.S. Dissertation, Lafayette, IN Purdue University.
22. Midilli, A., Kucuk, H., & Yapar, Z. (2002). A new model for single layer drying. *Drying Technology*, 20, 1503–1513.
23. Doymaz, I. (2004). Effect of dipping treatment on air drying of plums. *Journal of Food Engineering*, 64, 465–470.
24. Kassem, A. S. (1998). Comparative studies on thin layer drying models for wheat. In: 13th international congress on agricultural engineering, vol. 6, 2–6 February, Morocco.
25. Karathanos, V. T. (1999). Determination of water content of dried fruits by drying kinetics. *Journal of Food Engineering*, 39, 337–344.
26. Akpinar, E., Midilli, A., & Bicer, Y. (2003). Single layer drying behaviour of potato slices in a convective cyclone dryer and mathematical modeling. *Energy Conversion and Management*, 44, 1689–1705.
27. Mennouche, D., Bouchekima, B., Boubekri, A., Boughali S., Bouguettaia, H., & Bechki, D. (2014). Valorization of rehydrated Deglet-Nour dates by an experimental investigation of solar drying processing method. *Energy Conversion and Management*, 84, 481–487.
28. Akpinar, E. K. (2010). Drying of mint leaves in a solar dryer and under open sun: Modelling, performance analyses. *Energy Conversion and Management*, 51, 2407–2418.
29. Audsley, E., & Wheeler, J. (1978). The annual cost of machinery calculated actual cash flows. *Journal of Agricultural Engineering Research*, 23, 189–201.

5 Performance Enhancement of Solar PV System by Using Nano Coolants

Kashif Irshad

CONTENTS

5.1 INTRODUCTION

It is notable that a global atmospheric deviation and environmental change was caused by ozone-depleting substance outflows whereby a high level of emanations is because of consuming nonrenewable energy sources (Irshad et al. 2019). To lessen the natural effects of these gases, photovoltaic (PV) innovation can be viewed as a perfect arrangement. Nonetheless, one of the principle issues that farthest point the broad utilization of PV frameworks is the ascending in temperature of PV panels. It was found that 5%–20% of the solar irradiance coming to the PV surface is changed over into electrical energy, while the remaining radiations are either reflected back or consumed by the PV cell as heat (Zhou et al. 2015). Retained heat can build its temperature up to 70°C. The warmness of a PV module diminishes output efficiency by 0.4%–0.5% per 1°C (Eldin et al. 2015) over its evaluated temperature (which by and large is 25°C) because of the constriction of the band gap and an expanded number of carriers. The electrical power yield decreases due to a decrease in open-circuit voltage and an increase in the saturation current caused by an expanded number of carriers (Li et al. 2018). This is the reason the idea of "cooling of PV" has turned out to be so significant (Teo et al. 2012). An effective method for improving efficiency and diminishing the rate of thermal degradation of a PV module is by lessening the working temperature of its surface. This can be accomplished by either by utilizing the customary strategy for cooling by water (Nižetić et al. 2016) and air (Amelia et al. 2016), with characteristic or cooling by phase change material (Preet 2018), or by what is called nanofluid cooling, which was presented by Choi and Estman (1995) in 1995. As compared to normal suspension, the thermal conductivity of nanofluids is higher. The micrometer-sized suspensions normally have a few times lesser thermal conductivity as compared to nanofluids over base fluid. In order to carry these fluids (water or nanofluid), a photovoltaic/thermal (PV/T) system was coupled with a PV module. This heat-absorbing unit captures heat when fluid is circulated inside it and thus improves PV system efficiency (Nižetić et al. 2016). Lelea et al. (2014) evaluated the thermal conductivity of the fluid (i.e., water and ethylene) containing nanoparticles as Al_2O_3 and CuO on the performance of PV/T system. The results clearly demonstrate that PV module temperature decreases significantly by using nanofluid as compared to water.

Nanofluids are basically blended of strong particles, such as metallic oxides, metals, or carbon nanotubes of under 100 nm estimate at any rate in one measurement (nanoparticles) scattered in the fluid liquids, such as water and polyethylene glycol as the base fluid. These fluids can be utilized in the PV/T system as an optical filter or as a coolant (Yazdanifard et al. 2017). The PV/T framework utilizing nanofluid as a coolant can deliver obviously better outcomes than the water-cooled framework. Al-Waeli et al. (2017b) performed an exploratory investigation, and they found that the cooling of the PV module by means of SiC expanded the electrical productivity by 24.1% overall and thermal efficiency by 88.9% and 100.19% when contrasted with the water-cooled PV/T framework. Xu and Kleinstreuer (2014) proposed nanofluid-based silicon PV/T frameworks as a helpful choice for household applications because its general efficiency came to up to 70% (11% electrical proficiency and 59% thermal effectiveness).

This chapter reviews the effectiveness of PV frameworks being cooled by different nanofluids. The regular methods for cooling PV framework by means of nanofluids are expressed in detail alongside the parameters affecting the efficiency of the PV/T frameworks, for example, irradiance, absorption and stream rate of nanofluid, size of nanoparticles, and geometry of smaller-scale channels. The effect of different factors, for example, the sort of nanoparticles and base liquid on the framework effectiveness are examined. At last, challenges and scope of future work is discussed in detail.

5.2   PARAMETERS AFFECTING NANOFLUID-ASSISTED PHOTOVOLTAIC/THERMAL SYSTEM EFFICIENCY

There are many factors on which the performance of nanofluid-assisted PV/T system efficiency depends.

5.2.1 Type and Size of Nanoparticle

Nanoparticle type is also one of the most important parameters to be considered before selecting any nanofluids for cooling of the PV system. For example, for met-alloid nanoparticles, the nanofluid viscosity increases, which decreases the specific heat capacity, thus resulting in nonfavorable conditions. Hasan et al. (2017) saw that cooling the PV/T by impinging SiC, TiO_2, and SiO_2 nanofluids and unadulterated water improved the most extreme power yield by 62.5%, 57%, and 55% and a half when contrasted with the conventional PV module. Al-Shamani et al. (2016) tried SiO_2, TiO_2, and SiC-based nanofluids for the cooling reason to break down the pro-ficiency advancement. Following the test results, the SiC/water nanofluid beat the rest of the nanofluids. At 1000 W/m² irradiance and 0.170 kg/s mass flow rate, the SiC/water nanofluid-based PV/T framework demonstrated 13.529% electrical pro-ductivity, although TiO_2/water and SiO_2/water nanofluid-based PV/T frameworks portrayed 10.978% and 10.302% electrical effectiveness individually. The PV/T framework using water exclusively for cooling moved toward 9.608% electrical pro-ductivity. Various experimentations have been led to inspect the impact of a mag-netic field on the behavior of nanofluids (Khairul et al. 2016; Sheikholeslami 2017). It was found that an alternating magnetic field was generated around the channel by using ferro-nanoparticles, which improve the effectiveness of the framework. Test results additionally portrayed that the exchanging magnetic field improved the framework execution while the consistent field did not create any critical efficiency up-gradation when contrasted and the no-field condition. The framework effective-ness was observed to be 71.91% when there was no field connected, while, the pro-ductivity went up to 73.58% within the sight of the exchanging magnetic field (50 Hz) in the event of 1 wt% and 1100 W/m² (Ghadiri et al. 2015). The efficiency of the nanofluid is impacted by the magnetic field type and nanoparticle shape.

Further, the size of the nanoparticles also plays an important role because due to a smaller size, a higher rate of heat transfer occurs due to a larger area of the surface of each particle. The heat capacity of the nanoparticle is low, apart for that hav-ing higher thermal conductivity. At higher temperatures, nanoparticles are steady in the base liquid, and the chances of agglomeration in the base fluid is also reduced (Elsheikh et al. 2018). In the turbulent area of fluid flow, by increasing the size of the nanoparticle exergy and energy, the performance of the system was enhanced by this effect and is opposite in the laminar area of fluid flow. Due to the difference in results of Yazdanifard et al. (2017), which portray that nanoparticle size has no effect on efficiency improvement, Al-Shamani et al. (2014) saw that the heat transfer of the nanofluid diminished with a reduction in size of the nanoparticle. There is as yet a requirement for further experimentation to decisively portray the impacts of nanoparticles measured on the productivity of the PV modules.

5.2.2 Nanoparticle Concentration

Work done to investigate the effect of concentration of nanoparticles on PV/T system performance shows a variation in results. Khanjari et al. (2016) saw that expanding volumetric concentration of the nanoparticle (from 1% to 5%) expanded the coefficient of heat transfer and accordingly the overall effectiveness (from 1.33% to 11.54% for silver and 0.72% to 4.26% for alumina). Manikandan and Rajan (2016) outfitted sand for the chilling of the PV/T framework so as to improve the proficiency. They tried 0.5, 1, and 2 vol% concentration, and the accumulation efficiency proportion for these fixations was observed to be 3.6%, 11.2%, and 26.9% while the sunlight gathering efficiency improved by 9% and 16.5% for 0.5% and 2% separately. Radwan et al. (2016) found that as the concentration increases, efficiency increases. Sardarabadi and Passandideh-Fard (2016) likewise analyzed that by expanding the mass portion of nanoparticles from 0.05 to 10 wt%, the thermal efficiency of the framework expanded multiple times. Ghadiri et al. (2015) found the maximum overall efficiency of the framework was observed to be 75.93% and 80.58% when the ferrofluid concentration was expanded from 1 to 3 wt% individually. However, a few scientists saw conflicting outcomes. Karami and Rahimi (2014) inspected that expanding nanoparticle concentration decreases the effectiveness on account of agglomeration or bunching of the suspended particles. Radwan et al. (2016) uses SiC and Al_2O_3 nano particles for cooling of PV panel and found best outcomes at 0.1 wt% among (0.01, 0.1, 0.5 wt%). So as to acquire the best outcomes, there is dependably a need to decide the ideal nanoparticle concentration in base liquids as opposed to utilizing a high-volume division of nanofluid (George et al. 2019). In any case, rather than mixing a similar sort of nanoparticle concentration, mixing an alternate sort of nanoparticle can enhance the effectiveness of the PV panel in a progressively proficient manner (Chen et al. 2016).

5.2.3 Solar Irradiance

As the intensity of solar irradiance falling on the horizontal surface of the PV module increases, the temperature of the PV cell surface increases as more heat starts forming on the upper layer of the module. Khanjari et al. (2017) researched factors that influence the productivity of the PV framework chilled by nanofluids (Al_2O_3/water) through computational fluid dynamics (CFD) analysis. The result shows that as the solar irradiance increases from 200 W/m^2 k to 800 W/m^2 k, the electrical proficiency of framework diminished from 11.4% to 10.23% for alumina nanofluid and 11.41% to 10.12% for pure water. The thermal efficiency of a PV/T system increases from 76% to 91% by using alumina nanofluid while from 65% to 79% when pure water was used. Ghadiri et al. (2015) found that the productivity of the PV framework increases from 78.60% to 80.58% and 73.58% to 75.93% for 1 wt% and 3 wt%, separately, with diminishing solar irradiance from 1100 W/m^2 to 600 W/m^2.

5.2.4 Configuration and Method of Circulation

When cooling the PV module by means of a nanofluid, the flow technique is likewise of much importance. Active convection cooling ought to be utilized to get ideal outcomes, because if the flow is done by means of passive strategy, the increase in light intensity would result in the decrement of electrical efficiency and improvement in thermal effectiveness on the grounds that natural convection isn't that productive. Thus, forced circulation of the nanofluid can further improve the effectiveness of the PV module as associated with passive cooling (Al-Waeli et al. 2017a). Further, Hassan et al. (2004) suggested that wide channels cause instability in lateral heat transfer, whereas narrow channels increase the heat transfer coefficient. Also, by increasing the roughness of the pipe contact area, it further increases the heat transfer rate. So as to accomplish a higher efficiency of the framework, uniformity of the temperature distribution inside the channel and maintaining the least pressure drop inside the channel is recommended.

5.3 CHALLENGES

The potential difficulties for the utilization of nanofluids may incorporate the expense of nanoparticle creation, agglomeration, nonstable nature, and power to pump up the fluid and drop in pressure. The purity of high level and low imperfection are prerequisites fundamental to nanotechnology applications. For precise control of nanoparticle properties, the assembling procedure must be reliable, and thus the expense of nanoparticle creation is high. The energy required for nanofluids flowing in a total framework including expansion, straight tubing, and elbows were experimentally calculated by Routbort et al. (2011). The result shows that the energy required for pumping increases as the nanoparticle was added in the base fluid. Further, Lee and Mudawar (2007) found that viscosity and pressure drop increases as the nanoparticle is added in the fluid, which results in high costing and an extra pumping power requirement. Razi et al. (2011) led a test study utilizing CuO/oil nanofluid, and their results demonstrated that as compared with base fluid, the pressure drop increases. More consideration ought to be given to the choice of nanoparticles in high-temperature applications.

In the long run, the presence of nanoparticles in a nanofluid may lead to erosion and corrosion of equipments. The impacts of nanofluid stream on erosion and corrosion of metal surfaces was studied by Celata et al. (2014). They led their trials for TiO_2, Al_2O_3, SiC, and ZrO_2 nanoparticles with water as the base liquid, where the nanofluids stream in funnels with three distinct materials: stainless, aluminum, and copper. They suggested that the nanofluids have no impact on the erosion of the stainless pipe, while the aluminum pipe has the most noteworthy erosion. They additionally discovered that ZrO_2 and TiO_2 nanoparticles lead to most noteworthy erosion while SiC nanoparticles result in the least erosion.

5.4 SCOPE OF FUTURE WORK

The preceding review demonstrates that the utilization of nanofluids in solar applications is still in its early stages. Nanofluids can be utilized in numerous fields of PV/T systems. Here, a few recommendations are introduced for future work.

The effect of using various types of nanofluids having a different fraction of volume and particle size on the rate of cooling and performance enhancement of the PV/T system should be experimentally evaluated. Application of the nanofluid for enhancing the efficiency of a parabolic trough collector has not been investigated experimentally; thus, it is an area that requires proper attention. Further application of nanofluids in solar pond heat transfer enhancement can be used for creating a salient gradient. Lastly, a nanofluid can also be used in the solar thermoelectric system for creating a higher temperature gradient across the hot and cold side of the thermoelectric module, which will result in generating higher electrical power.

5.5 CONCLUSIONS

Nanofluids are propelled liquids containing nano-sized particles that have risen during the most recent two decades. Nanofluids are utilized to improve framework execution in numerous solar thermal systems. This study has given the handy ramifications of electrical yield and the thermal efficiency of a solar PV cell with different cooling methods utilized and their results on efficiency. A proposed arrangement has been discussed in this paper to handle the varieties in efficiency without much expense and is effectively structured by utilizing a custom blend of water, nanofluids, and bricks of clay as coolants on the rear of the solar PV module. The cooling performance of nanofluids escalates as the concentration of particles increases up to certain optimum level after that performance starts decreasing due to agglomeration and clustering of nanoparticles. Further entropy generation impedes by adding a nanofluid inside the system. This chapter further uncovers that the use of nanofluids in the solar thermal system is yet in its early stages. Consequently, a few recommendations are exhibited to build up the utilization of nanofluids in various solar thermal systems. Lastly, the most significant difficulties on the utilization of nanofluids in solar thermal system, including generation cost, clustering and agglomeration issues, instable behavior, high power for pumping, and corrosion, are referenced. These difficulties might be decreased with the advancement of nanotechnology later on.

REFERENCES

Al-Shamani, A.N., Sopian, K., Mat, S., Hasan, H.A., Abed, A.M., and Ruslan, M.H., 2016. Experimental studies of rectangular tube absorber photovoltaic thermal collector with various types of nanofluids under the tropical climate conditions. *Energy Conversion and Management*, 124, 528–542.

Al-Shamani, A.N., Yazdi, M.H., Alghoul, M.A., Abed, A.M., Ruslan, M.H., Mat, S., and Sopian, K., 2014. Nanofluids for improved efficiency in cooling solar collectors— A review. *Renewable and Sustainable Energy Reviews*, 38, 348–367.

Al-Waeli, A.H.A., Chaichan, M.T., Kazem, H.A., and Sopian, K., 2017a. Comparative study to use nano-(Al_2O_3, CuO, and SiC) with water to enhance photovoltaic thermal PV/T collectors. *Energy Conversion and Management*, 148, 963–973.

Al-Waeli, A.H.A., Sopian, K., Chaichan, M.T., Kazem, H.A., Hasan, H.A., and Al-Shamani, A.N., 2017b. An experimental investigation of SiC nanofluid as a base-fluid for a photovoltaic thermal PV/T system. *Energy Conversion and Management*, 142, 547–558.

Amelia, A.R., Irwan, Y.M., Irwanto, M., Leow, W.Z., Gomesh, N., Safwati, I., and Anuar, M.A.M., 2016. Cooling on photovoltaic panel using forced air convection induced by DC fan. *International Journal of Electrical and Computer Engineering*, 6 (2), 526–534.

Celata, G.P., D'Annibale, F., Mariani, A., Sau, S., Serra, E., Bubbico, R., Menale, C., and Poth, H., 2014. Experimental results of nanofluids flow effects on metal surfaces. *Chemical Engineering Research and Design*, 92 (9), 1616–1628.

Chen, M., He, Y., Huang, J., and Zhu, J., 2016. Synthesis and solar photo-thermal conversion of Au, Ag, and Au-Ag blended plasmonic nanoparticles. *Energy Conversion and Management*, 127, 293–300.

Choi, S.S. and Eastman, A.A., 1995. Enhancing thermal conductivity of fluids with nanoparticles. In: *International mechanical engineering congress and exhibition*, San Francisco, CA, 12–17 November 1995, ASME Publications FED, pp. 99–105.

Eldin, A.H., Refaey, M., and Farghly, A., 2015. A review on photovoltaic solar energy technology and its efficiency. In: *17th International Middle-East Power System Conference (MEPCON'15)*, Mansoura University, Egypt, pp. 1–9.

Elsheikh, A.H., Sharshir, S.W., Mostafa, M.E., Essa, F.A., and Ahmed Ali, M.K., 2018. Applications of nanofluids in solar energy: A review of recent advances. *Renewable and Sustainable Energy Reviews*, 82, 3483–3502.

George, M., Pandey, A.K., Abd Rahim, N., Tyagi, V.V., Shahabuddin, S., and Saidur, R., 2019. Concentrated photovoltaic thermal systems: A component-by-component view on the developments in the design, heat transfer medium and applications. *Energy Conversion and Management*, 186, 15–41.

Ghadiri, M., Sardarabadi, M., Pasandideh-Fard, M., and Moghadam, A.J., 2015. Experimental investigation of a PVT system performance using nano ferrofluids. *Energy Conversion and Management*, 103, 468–476.

Hasan, H.A., Sopian, K., Jaaz, A.H., and Al-Shamani, A.N., 2017. Experimental investigation of jet array nanofluids impingement in photovoltaic/thermal collector. *Solar Energy*, 144, 321–334.

Hassan, I., Phutthavong, P., and Abdelgawad, M., 2004. Microchannel heat sinks: An overview of the state-of-the-art. *Microscale Thermophysical Engineering*, 8, 183–205.

Irshad, K., Habib, K., Saidur, R., Kareem, M.W., and Saha, B.B., 2019. Study of thermoelectric and photovoltaic facade system for energy efficient building development: A review. *Journal of Cleaner Production*, 209, 1376–1395.

Karami, N. and Rahimi, M., 2014. Heat transfer enhancement in a hybrid microchannel-photovoltaic cell using Boehmite nanofluid. *International Communications in Heat and Mass Transfer*, 55, 45–52.

Khairul, M.A., Doroodchi, E., Azizian, R., and Moghtaderi, B., 2016. Experimental study on fundamental mechanisms of ferro-fluidics for an electromagnetic energy harvester. *Industrial and Engineering Chemistry Research*, 55 (48), 12491–12501.

Khanjari, Y., Kasaeian, A.B., and Pourfayaz, F., 2017. Evaluating the environmental parameters affecting the performance of photovoltaic thermal system using nanofluid. *Applied Thermal Engineering*, 115, 178–187.

Khanjari, Y., Pourfayaz, F., and Kasaeian, A.B., 2016. Numerical investigation on using of nanofluid in a water-cooled photovoltaic thermal system. *Energy Conversion and Management*, 122, 263–278.

Lee, J. and Mudawar, I., 2007. Assessment of the effectiveness of nanofluids for single-phase and two-phase heat transfer in micro-channels. *International Journal of Heat and Mass Transfer*, 50 (3–4), 452–463.

Lelea, D., Calinoiu, D.G., Trif-Tordai, G., Cioabla, A.E., Laza, I., and Popescu, F., 2014. The hybrid nanofluid/microchannel cooling solution for concentrated photovoltaic cells. In: *AIP Conference Proceedings*, Thessaloniki, Greece, pp. 122–128.

Li, G., Shittu, S., Diallo, T.M.O., Yu, M., Zhao, X., and Ji, J., 2018. A review of solar photovoltaic-thermoelectric hybrid system for electricity generation. *Energy*, 158, 41–58.

Manikandan, S. and Rajan, K.S., 2016. Sand-propylene glycol-water nanofluids for improved solar energy collection. *Energy*, 113, 917–929.

Nižetić, S., Čoko, D., Yadav, A., and Grubišić-Čabo, F., 2016. Water spray cooling technique applied on a photovoltaic panel: The performance response. *Energy Conversion and Management*, 108, 287–296.

Preet, S., 2018. Water and phase change material based photovoltaic thermal management systems: A review. *Renewable and Sustainable Energy Reviews*, 82, 791–807.

Radwan, A., Ahmed, M., and Ookawara, S., 2016. Performance enhancement of concentrated photovoltaic systems using a microchannel heat sink with nanofluids. *Energy Conversion and Management*, 119, 289–303.

Razi, P., Akhavan-Behabadi, M.A., and Saeedinia, M., 2011. Pressure drop and thermal characteristics of CuO-base oil nanofluid laminar flow in flattened tubes under constant heat flux. *International Communications in Heat and Mass Transfer*, 38 (7), 964–971.

Routbort, J.L., Singh, D., Timofeeva, E.V., Yu, W., and France, D.M., 2011. Pumping power of nanofluids in a flowing system. *Journal of Nanoparticle Research*, 13 (3), 931–937.

Sardarabadi, M. and Passandideh-Fard, M., 2016. Experimental and numerical study of metal-oxides/water nanofluids as coolant in photovoltaic thermal systems (PVT). *Solar Energy Materials and Solar Cells*, 157, 533–542.

Sheikholeslami, M., 2017. Numerical simulation of magnetic nanofluid natural convection in porous media. *Physics Letters, Section A: General, Atomic and Solid State Physics*, 381, 494–503.

Teo, H.G., Lee, P.S., and Hawlader, M.N.A., 2012. An active cooling system for photovoltaic modules. *Applied Energy*, 90 (1), 309–315.

Xu, Z. and Kleinstreuer, C., 2014. Concentration photovoltaic-thermal energy co-generation system using nanofluids for cooling and heating. *Energy Conversion and Management*, 87, 504–512.

Yazdanifard, F., Ameri, M., and Ebrahimnia-Bajestan, E., 2017. Performance of nanofluid-based photovoltaic/thermal systems: A review. *Renewable and Sustainable Energy Reviews*, 76, 323–352.

Zhou, J., Yi, Q., Wang, Y., and Ye, Z., 2015. Temperature distribution of photovoltaic module based on finite element simulation. *Solar Energy*, 111, 97–103.

6 Global Trends of Biofuel Production and Its Utilization

Renu Singh

CONTENTS

6.1 INTRODUCTION

The high consumption of fossil fuels across the world is imposing various environmental issues. There is a need to shift the dependency on gasoline to renewable sources of energy. Biofuels are liquids, solids, and gaseous fuels that are derived from organic matter and have the potential to reduce CO_2 emission in the transportation sector with an objective of energy security. Biofuels are the leading solutions for transport fuel, which is chiefly in road transport, while marine and aviation have equal scope. The railway is another sector where biofuels are taking an important stake. Utilization of biofuels in restaurants and farm operations are now common in some countries. Biofuel production and utilization not only answer the problem of the energy crisis but also give solutions to environmental issues, such as global warming, waste utilization, and recycling. Production and utilization of biofuels in the remote areas are leading to energy self-sufficiency and employment opportunities. The aspects related to biofuels can be categorized into social (employment, food security, land holding), environmental (greenhouse gas emissions, air quality, water quality), and economic (energy security, finance, and fuel trade). Predominantly bioethanol and biodiesel are the main types of biofuels that are the substitutes to patrol and diesel, respectively. Bioethanol can be produced from sugar-based crops, cellulosic biomass, and hence categorized into generation according to the type of biomass used for the production through fermentation. Lignocellulosic-based biomass from agriculture and forest sources need pretreatment before fermentation, which is energy- and time-intensive. Biodiesel is produced by vegetable oils and animal fats by transesterification process. The transesterification process varies slightly in accordance with the substrate and process conditions, such as temperature, pressure, and time. Nowadays, a fully recycled closed system of transesterification is developed where chemicals and water utilized during the process can be recycled. Another category of biofuels is advanced biofuels, which excludes bioethanol and biodiesel. This is cellulosic ethanol, renewable diesel, and vegetable oil after hydrotreatment.

6.2 PRODUCTION OF BIOFUELS

Globally, the production of biofuels has reached up to 126 billion liters in 2014 (Figure 6.1) and the highest producers were the USA and Brazil. As per estimation, America produced 95.1 billion liters of liquid biofuels (Figure 6.2). Since 2000, the global biofuel annual growth rate was about 15%, whereas the average biomass supply growth rate is much lower (2.3%). America produced 95.1 billion liters of biofuels from corn chiefly. Twenty-eight European Union countries, Argentina, China, and Indonesia are the top producer of biofuels.

Biofuel production is dependent on the land availability for the raw material from crops, such as maize, wheat, sugar beet, sugarcane, and vegetable oils. About 2.9% of the world land area is engaged in the chief biofuel crop cultivation, which produced 4.2 billion tons of biofuel crop out of which approximately 122 million tons were used for biofuel production. Distillers dried grain with soluble (DDGS) obtained during biofuel production is an important component that can be utilized for animal feed and oil cake. In the year 2014, 75.3 million tons of DDGS were produced, which supported the economics of biofuels. Table 6.1 shows the bioethanol and biodiesel production from different crops with DDGS generation (WBAS 2017).

The bioethanol production system can be categorized on the basis of raw material, namely simple sugars, starch, and lignocellulosic material. Bioethanol from simple sugars, especially from sugarcane and sorghum juices, are the

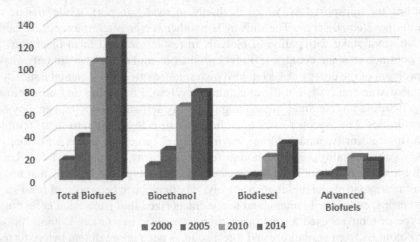

FIGURE 6.1 Global liquid biofuel production trend. (All values in billion liters. Source: IEA Key World Energy Statistics and REN21 GSR 2017.)

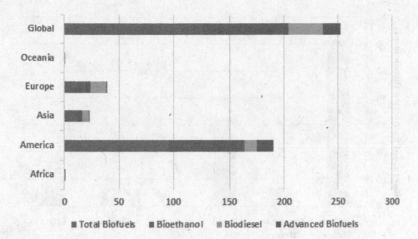

FIGURE 6.2 Liquid biofuel production in continents in 2014. (All values in billion liters. Source: IEA Key World Energy Statistics and REN21 GSR 2017.)

TABLE 6.1

Bioethanol and Biodiesel Production from Different Feedstocks

		Production (Mt)	Biofuels (Mt)	% of Biofuels Use	
Bioethanol	Wheat	720	2.62	0.4	2.64
	Maize	1014	53.2	5.2	48.8
	Other grains	299	4.16	1.4	0.00
	Sugar beet	257	1.11	0.4	0.00
	Sugarcane	1812	25.3	1.4	0.00
Biodiesel	Palm oil	61	23.2	38	23.9
	Vegetable oils	113	14.1	12	0.00
Total		4276	124		75.3

easiest route of production through fermentation. Starch from grains needs an enzymatic step known as saccharification before fermentation. Conversion of lignocellulosic is the toughest and most energy-intensive. All these processes have clear potential in greenhouse gas reduction. Recently, a lot of research has been done for raw material processing, efficient enzyme development, and genetically modified organisms. Utilization of microalgae and continuous systems with engineered immobilized cells are the recent developments in the technology.

Ethanol production may increase with the growth rate of 14% from 120 billion liters in 2017 to 131 billion liters in 2027. The projected production is shown in Figure 6.3. To fulfill the domestic needs, 50% increased quantity is expected

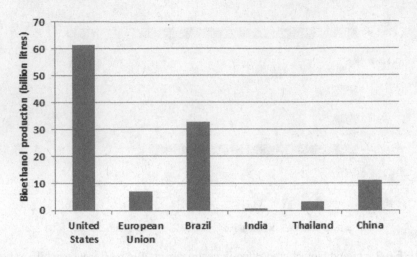

FIGURE 6.3   Bioethanol production by 2027 (billion liters).

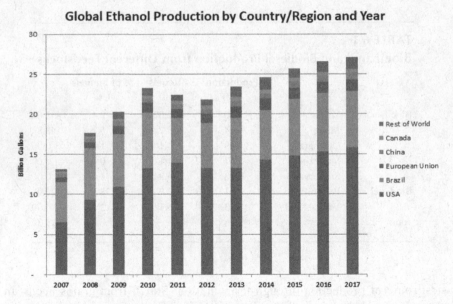

FIGURE 6.4   Global ethanol production in different countries from 2007 to 2017. (Source: https://afdc.energy.gov/data/categories/biofuels-production.)

to come from Brazil. Other contributors are China, India, Thailand, and the Philippines with 10%, 9%, 12%, and 5% in global increase. The United States is expected to be the largest producer followed by Brazil, China, and the European Union (Figure 6.4). Production of ethanol is toward stagnation or decrease in developed countries with the increase in developing countries. United States ethanol production may decrease after a few years with lower international and domestic gasoline demand in the country. Brazil has a favorable differential taxation system and gasohol blending policies that may lead to increased bioethanol production. China mainly produces from maize and cassava for the domestic use, whereas the European Union is dependent on sugar beet and coarse grain. Thailand and India's ethanol production is based on cassava, sugarcane, and molasses. In the future, sugarcane and coarse grains will remain chief ethanol feedstock. By 2027, ethanol production will use 18% and 15% of world's sugarcane and maize, respectively. Approximately 0.3% of global ethanol production will come from biomass (OECD-FAO 2018).

The biodiesel production system is categorized into three sections: upstream, mainstream, and downstream. Feedstocks type and quality are governed in the upstream strategy, whereas the mainstream addresses the chemical, mechanical, and process parameters. The innovative technology of separation of biodiesel and glycerin and its purification comes under downstream processes. *Jatropha curcas, calophyllum inophyllum, Nicotiana tabacum, Ceiba Pentandra, Hevea brasiliensis*; microalgae, cyanobacteria, unicellular microorganisms such as bacteria, filamentous fungi, and yeasts; waste frying oil, soap stocks, and spent bleaching earth oil are the feedstocks for the biodiesel production. Modern mainstream strategies for biodiesel refinery are the use of cosolvents under chemical strategies and improved conventional impeller agitation systems and non-impeller novel agitation systems under mechanical strategies (Tabatabaei et al. 2015). Modern trends for downstream processes in biodiesel production are

1. Biodiesel-glycerin separation (decantation)
   NaCl-assisted gravitational settling
   Electrocoagulation
   Stand-alone membrane modules
2. Biodiesel purification
   Membrane separation technology
   Extraction by ionic liquids
3. Glycerin purification
   Glycerin prepurification procedures followed by the distillation method
4. Biodiesel wastewater treatment
   Commercial electrospun polystyrene membrane
   Commercial chitosan flakes treatment through absorption
5. Alcohol recovery
6. Biodiesel additives

Biodiesel production is more affected by the policies rather than the market. The European Union is the major contributor to the production. The global production may reach 39.3 billion liters by 2027 with a 9% increase from the year 2017. Vegetable oil is the major feedstock that will remain the same in the future. Moreover, waste oil and tallow will be the important raw material for biodiesel in the United States and European Union. The United States produced 6.9 billion liters and 7.2 billion liters in the years 2017 and 2019, respectively, which is expected to increase in the future (OECD-FAO 2018). The major players in biodiesel production are Brazil, Indonesia, and Thailand. Indonesia, Philippines, and Malaysia are the growing producers of biodiesel.

6.3　UTILIZATION OF BIOFUELS

As per estimation, biofuels may provide 27% of the energy need in the transport sector by decreasing 2.1 Gt CO_2 emissions (IEA 2019a). The production of biofuels may require 100 Mha by 2050 with 65 EJ energy output. The major challenges related to the cost of biofuels can be addressed by increased scale and efficiency.

6.4　ELECTRICITY GENERATION FROM BIOFUELS

Germany was the first country that produced electricity (15 GWh) from liquid biofuel in 2001. Therefore, electricity generation from liquid biofuels is considered the new technology, but gradually there is a substantial increase in this field in a few countries. About 6298 GWh is produced by 14 countries in 2018. Among these countries, Italy reportedly produced the highest 4299 GWh electricity. According to an estimate, electricity generation from solid biofuels grew from 94.3 to 184.2 TWh from the year 1990 to 2018 with a 2.4% average annual growth rate. In 2018, solid biofuels produced 6.4% of the electricity, which was the fourth largest source after hydro, wind, and solar power. The United States and the United Kingdom produced 45.7 and 24.9 TWh electricity from solid biofuels, respectively. Figure 6.5 shows the electricity generation percent share from solid biofuels in different countries (IEA 2019b).

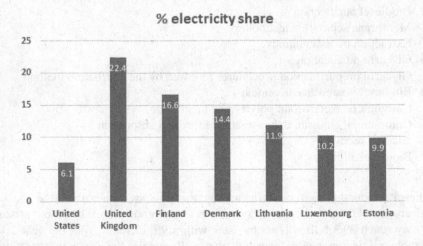

FIGURE 6.5　Electricity generation percent share from solid biofuels in different countries.

6.5 BIOFUELS FOR THE TRANSPORT SECTOR

Biofuels are used in aviation and shipping transportation. Approximately 150,000 flights used biofuels in 2018, but its availability is low in the shipping industry. Biofuels provided 3% energy to road transport in 2017. In 2018, transport biofuel production increased up to 7% with a 3% growth rate in the next five years. To reduce the life cycle carbon intensity, biofuels have an important role to play. The production of ethanol in the United States is expected to stabilize because of relatively low investment and limited feedstock. Brazil has the target to reach 18% stake of biofuels by 2030, which requires speedy production rate. The European Union has scaled up the cap on edible crop-based biofuels to 3.8% to achieve a 2030 energy target (iea.org/etp/tracking2017/transportbiofuels/). Asian countries, namely, China, India, and Thailand, have advanced biofuel projects. In addition, the aviation industry long-term decarbonization targets can be achieved by using biofuels.

6.6 CONCLUSIONS

The long-term stable policy across the world for the expansion of biofuel producing industry and reduction of cost of production is necessary. Framework for the utilization of biofuels in the road, aviation, and marine sectors are very much required. In fact, international collaboration for the expansion of social, environmental, and economic sustainability through biofuel utilization should be anticipated.

REFERENCES

IEA (International Energy Agency), 2019a, Biofuels for transport roadmap. https://webstore. iea.org/technology-roadmap-biofuels-for-transport-foldout

IEA (International Energy Agency), 2019b, Renewables Information 2019, IEA, Paris, https:// www.iea.org/reports/renewables-information-2019.

OECD-FAO, 2018, Organisation for Economic Co-operation and Development (OECD) and the Food and Agriculture Organization (FAO) Agricultural Outlook 2018–2027, 192–206. http://www.agri-outlook.org/

Tabatabaei, M., Karimi, K., Sárvári Horváth, I., Kumar, R., 2015, Recent trends in biodiesel production. *Biofuel Research Journal* 7:258–267.

WBAS (World Bioenergy Association Statistics), 2017, www.worldbioenergy.org

7 Biofuel
An Alternative Fuel for Fossil Fuel

Guruwendra Singh and Prashant Baredar

CONTENTS

7.1   INTRODUCTION

Due to the increasing demand for fuel oils day to day, it is essential to look for alternative fuel sources. Fossil fuel reserves are rapidly declining, and their harmful effect on the environment is a reason for serious concern. Development of an optional fuel that is reliable, sustainable, and gainful is the need of the day [1]. Biofuels were professed as one of the most sustainable options for petroleum-derived fuels for transportation just 10 to 15 years ago, but with a number of socioenvironmental questions, the positive image of biofuel has significantly changed [2]. During the twentieth century, there has been major industrial development to extract reasonably obtainable fossil fuel feedstock, such as petroleum, coal, and asphalt. These feedstocks are used in a variety of industries to produce different products in the chemical and fertilizer industries, manufacturing industry (for synthetic fiber, plastics, etc.), medicinal industry, etc., to fit the rising demand of the population [3]. Nowadays, the petroleum resources are not observed as sustainable and arguable in the view of the environment and ecology. The combustion of petroleum products is a large contributor to raising the level of carbon dioxide in the atmosphere, which is directly linked with the global warming noticed in recent years [4].

Considering the current scenario of the world, fossil fuel is being a chief source of energy with its role of approximately 80% in meeting energy demands. It is important to understand that these fossil fuel resources are limited on the planet. Moreover, there are several disadvantages to being dependent on fossil reserves. First, the sources of these fossil fuels and oil reserves are reducing very rapidly. Second, they are responsible for the emission of harmful gases, which can pose several negative consequences, such as, loss of biodiversity, receding of glaciers, rise in sea level, and climate change. Third, due to the increasing demand for petroleum products, it is also affecting the world's economic performance as their unparallel rise in the prices of crude oil. It is, therefore, crucial to analyze the scope of other alternative energy sources to meet the energy requirement [5,6].

Biofuels obtained from vegetable oils are extremely viscous and the source of a lot of maintenance problems. Past several studies shows that blending of bio-fuels with the Petrol or diesel without change the existing petrol and diesel engines. As a result, a great deal of CO_2 and other greenhouse gas (GHG) emissions were seized in a great amount [1]. Fossil fuels have supplied most of the energy required for the developed nations, yet the GHG emissions that result threaten to gravely affect natural systems through human-made climate change. Biofuels produced from forming commodities may decrease our dependence on fossil fuels and mitigate human-based carbon emissions [3]. Increasing the use of biofuels would support a number of policy motives, such as energy security, environmental benefits, economic benefits, and fuel value [7]. Biomass can also be referred to as a sustainable and economical form of storage device for energy and that could be used at any instance. Biofuels that are derived primarily from biomass are known as solid, liquid, and gaseous fuels. Biomass has been accounted for as fourth largest accessible energy resources of the world [1,6]. Due to this, a number of countries have made policies to sustain biofuel research and describe production targets by financial support and governmental incentives [8].

However, a biofuel derived from agricultural products may decrease our dependence on conventional fuels and diminish the human-made carbon emissions. Yet strengthening and development are the main causes of habitat modification, climate change, and deduction in biodiversity [3]. The increasing amount of GHGs in the atmospheric system will have a deep effect in terms of climate change. Flue gas emissions from industries contain HCs, NOx, PM, SO_2, CO_2, and CO; nearly all of these gaseous emissions are GHGs [9,10]. In the drive toward initial sustainable resources of energy, biomass-based energy keeps a distinct benefit. It can create liquid fuels as straight alternatives for petrol and diesel. Moreover, biomass accessibility is comparatively stable throughout the year [11]. In addition, biomass can be converted to heat, power, and other low-volume, high-worth goods and chemicals [10,12]. The biomass-based energy can also likely be carbon neutral and can be derived locally, thereby assembling the key sustainability criterion [13]. The deduction of carbon dioxide from the environment happened mainly because of the photosynthesis process, in which flora naturally consumes carbon dioxide in the presence of sunlight and liberates oxygen [14,15]. Though due to the fast development of industries, plant life alone is incapable of reducing the amount of carbon dioxide from the environment naturally [16]. A definite sum of GHGs present in the atmospheric system helps to soak up thermal radiation from the earth's surface and then reemits the radiation back to the earth as shown in Figure 7.1. The presence of CO_2 in the atmosphere is significant because it maintains the greenhouse effect. GHGs trap the energy from the sun and hold the temperature of the earth warm and appropriate for sustaining life on earth. Absent of this, we cannot envisage life on the earth because the mean temperature of the earth goes down, which is incapable of sustaining life [17]. Sustainability means the fulfillment of the requirement of the present generations without compromising the ability of future generations to meet their own needs [16,18].

Biofuels, which are derived from biomass, are referred to as gas, liquid, and solid fuels. The biofuel crop is divided into three types as first, second, and third generations, depending on the complex and chemical nature of the biomass [19]. The fuel produced from the crop plants is generally known as the first-generation biofuel. The fuels derived from energy plants and agricultural by-products are called second-generation biofuels, which consist of biodiesel and bioethanol. The feedstock of second generations of biofuels requires fertile lands for the growth of its crops. Third-generation biofuel includes seaweeds and Cyanobacteria, which can be used for the production of biogas, bioethanol and biobutanol. The third-generation biofuel crop creates a large biomass in a specific time period, and it doesn't require land for growth. A number of scientific papers and articles are available, which have concern against the universal drive toward the biofuel economy, usually focusing on the potential impacts on food security [20]. A number of species of algae are used in the production of biofuel, which has the highest capacity to open a new way for an alternate sources of fossil fuel. Presently biodiesel, ethanol, alcohols, lipids, fatty acids, carbohydrates, triglycerides, cellulose, and the biomass of organisms are measured as the main biofuel resources. International attempts have been ended for the identification of desired strains of algae. A number species of algae available that have the ability to increase the amount of biomass production (in terms of lipids, proteins, carbohydrates) can be utilized as an optional source of bioenergy [21,22].

Utilization of biofuels, such as alcohols, biodiesel, seed, or vegetable oils, and biogas are achieving significant use in compression ignition (CI) engines because they are environmentally and locally suitable. It is also available in abundance, consistently extends to the whole world, technically possible, and can satisfy the global energy demand. Biogas can be derived locally from animal wastes and food crops that don't have any geographic limits [11,23]. We define biofuels as liquid fuel produced from biomass, which can be biodiesel (produced from plant oil) or bioethanol (produced from the fermentation of plant sugar), and the center of attention on its uses as a transportation fuel.

7.2   TYPES OF BIOFUELS

Biofuels, similar to fossil fuels, are found in numerous types and collect a number of diverse energy requirements. The group of biofuels is further divided into four generations; each one of four generations contains a number of different types of fuels that will be shown in this chapter. We have already discussed biofuels at length: what they are, what they're made of, their advantages, and disadvantages [24]. Now, we tackle the various types of biofuels and what they're good for. There are three main categories of biofuels, classed according to the complexity of the process through which they are obtained. The categories in question are called "generations," in the case of biofuels. First-generation biofuels are derived directly from the feedstock crops; the further down the line you go, the more complex the production process becomes [25].

7.2.1   FIRST-GENERATION BIOFUEL

There has been an uncontrollable rise in the oil prices observed in the last few decades that have allowed liquid biofuels to become to some extent cost-competitive. Due to this, there has been a surge around the world in research and production of liquid biofuels. Those biofuels that are normally derived from sugar seeds or grains are called first-generation biofuels. Two most important types of first-generations biofuels produced commercially are known as biodiesel and bioethanol. The production of biodiesel is done through the transesterification process of vegetable oils, residue oils, and fats. Biodiesel can be used as an alternative fuel for a diesel-fueled engine with some modifications. Ethanol can be used as a substitute fuel for gasoline (petrol) engine. Ethanol is produced from starch or sugars by fermentation [26].

But the commercial production of biofuels by agricultural products brings some disputes, such as first-generation biofuels have increased the cost of food crops like sugarcane, sugar beet, potatoes, wheat, and some other fruits. Therefore, a close examination has required for sustainable generation of first-generation biofuels [27]. First-generation biofuel plants are grown on tillable land. By using yeast fermentation or transesterification the starch, sugar, or vegetable oil got from the plant that is converted into ethanol or biodiesel [26].

7.2.2 Second-Generation Biofuel

The impact of second-generation biofuel on the environment in terms of carbon dioxide is negligible because of the production of second-generation biofuels from biomass in an extra sustainable approach (Figure 7.2). Lignocellulosic material refers mainly to "plant biomass" in the context of biofuel generations, which produces the majority of the sufficient and cheap nonliving materials available from crops [28,29]. Presently, there are numerous types of technical difficulties required to rise above before the feedstock's potential can be used for the production of second-generation biofuel, because the production of such biofuel is not economical generally. Crop feedstocks represent amongst the amplest and less significant biological sources on the globe, and they look like a capable source of feedstocks for fuels and natural resources. For the production of heat and electricity, plant biomass can easily be burnt; this is the most significant property of crop biomass. Plant biomass has huge potential to produce liquid biofuels, because in plant biomass, approximately 75% of their plant cell walls are composed of polysaccharides. These polysaccharides shows a important source of possible sugars. Polysaccharides are an important class of biological polymers which are used to store energy in organism. In the conventional vegetable plant such as cereals is contains so much sugar in the branches of cereals plant residue as there is having in the starch of the granules [28].

7.2.3 Third-Generation Biofuels

Third-generation biofuels are produced from algae. The existing process of algae can generate oil. This oil can be further processed and advanced into diesel and a few contents of petrol generally. By a direct carbon metabolic process of the genome, metabolic engineering can produce ethanol as a finished product [30]. The fabrication method of algal biomass can be attained in both photobioreactors and open channel ponds. The third-generation biofuel production method is less steady than the other biofuel production methods, which is a disadvantage of the third-generation biofuel production method [31].

7.2.4 FOURTH-GENERATION BIOFUELS

The fourth-generation biofuels are derived by using the non-arable ground in the same way as third-generation biofuels. However, fourth-generation biofuels do not require the destruction of the biofuel as in third-generation biofuels. The fourth-generation class of biofuels includes photobiological solar fuels and electro fuels, and some of these fuel types are carbon neutral [32].

7.3 BIOFUEL FEEDSTOCKS

Biomass feedstocks for energy generation can be produced from crops grown openly for energy or from crop parts, residues, giving out wastes, and materials from animal and human actions.

7.3.1 FIRST-GENERATIONS BIOFUEL FEEDSTOCKS

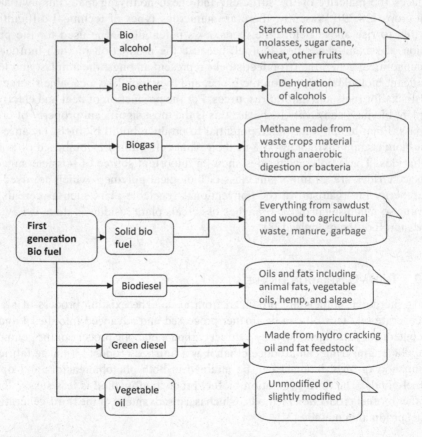

FIGURE 7.1 First-generation biofuel feedstocks. (From Types of biofuel. http://biofuel.org.uk/types of biofuels.)

7.3.2 Second-Generations Biofuel Feedstocks

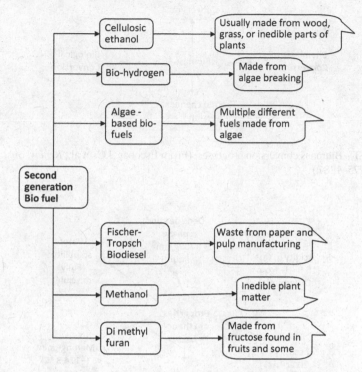

FIGURE 7.2 Second-generation biofuel feedstocks. (From Types of biofuel. http://biofuel. org.uk/types of biofuels.)

7.4 THE CONVERSION PROCESS FOR BIOFUELS

At the present time, important attention has been intense on useful approaches related to conventional biomass adaptation to fuels. Several types of alcohols, for example, methanol, ethanol, propanol, butanol, iso-pentanol, and glycerol, derived via the fermentation of biomass or by triglycerides transesterification (glycerol) can be incorporated into petrol and biodiesel, either directly or after chemical/catalytic change into extraefficient additives [33] (As seen in Figure 7.3).

7.4.1 Ethanol Conversion Process

Ethanol is a combustible, colorless chemical compound. It is also known as "ethyl alcohol" or "grade alcohol," one of the alcohols that is largely often found in alcoholic beverages. Normally it is frequently referred to as alcohol. The molecular formula of ethanol is C_2H_6O, diversely represented as EtOH, C_2H_5OH, or as ethanol empirical formula C_2H_6O [34,35]. The most important properties of C_2H_5OH are shown in Figure 7.4. (As seen in Figure 7.4).

Ethanol (C_2H_5OH) generation is frequently obtained by enzymatic hydrolysis of starch consisting of crops such as corn and wheat. Corn ethanol generation facilities

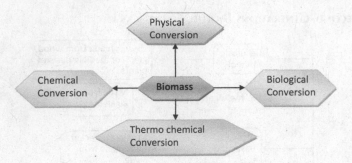

FIGURE 7.3 Biomass conversion processes. (From Escobar, J.C. et al., *Renew. Sust. Energ. Rev.*, 13, 1275–1287.)

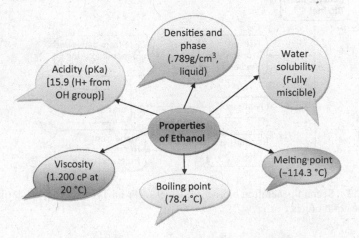

FIGURE 7.4 Properties of ethanol. (From Damiani, L. and Revetria, R., *Int. J. Renew: Energy Biofuels*, *2014*, 1–18, 2014.)

can be classified into two groups: wet and dry mill methods [36]. Dry mills are generally smaller in size on the potential basis and are constructed mostly to formulate ethanol only. Modern wet-milling processes are capable of producing 1 gallon of ethanol, expensing 35,150 Btu of thermal power and 2134 KWh of electricity. If the molecular strainer is used, the thermal energy input drops to 32,150 Btu/gal. The starch particle is set for ethanol fermentation by either wet milling or dry grinding. Wet grinding refines ethanol and creates different types of important co-products, for example, nutraceuticals, pharmaceuticals, organic acids, and solvents. The dry grinding refining process is mainly designed for the production of ethanol and animal feed [37].

The chemical formulae of starch consists of elongated chain polymers of glucose. The conventional fermentation machinery cannot ferment the macromolecular starch in ethanol. The macromolecular arrangements first break down into easier and smaller glucose arrangements. Starch feedstocks are bleached and mixed with water to create a smash that normally consists of 15%–20% starch. The smash is then cooked at more than its boiling point and treated consequently with two enzyme grounding [36,38].

Nowadays, there is nearly no commercial making of ethanol from cellulosic biomass, but there is much research happening. There are quite a lot of potentially significant benefits from developing a feasible and commercial cellulosic ethanol process [39]:

- Access to a much wider array of potential feedstock types (including waste cellulosic materials and devoted cellulosic crops plants such as grasses and trees) creates the path to more ethanol production levels.
- Greater avoidance of conflicts with land use for foodstuff and nourish production.
- A much greater dislodgment of fossil energy per liter of fuel, due to almost completely biomass-motorized systems.
- Much lower net well-to-wheels GHGs emissions than with grain-to-ethanol production powered principally by fossil energy [40].

7.4.2 OTHER FUEL CONVERSIONS

There are two very important types of processes available for producing liquid biofuels from biomass: thermochemical and biochemical processing. The conversion of biomass from a variety of products by thermal decay and chemical reformation with the presence of oxygen at different concentrations is called thermochemical conversion. The main advantages of thermochemical practicing are that it can significantly change all types of organic materials of the biomass in comparison to biochemical practicing, which mainly focuses on the conversion of the polysaccharides [41] (As seen in Figure 7.5). As Figure 7.5 shows, biorefineries are based on different types of feedstocks and their end products.

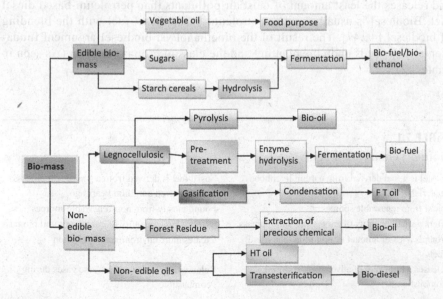

FIGURE 7.5 Generation of biofuel production from biomass. (From www.conserve-energy-future.com/advantages-and-disadvantages-of-biofuels.php.)

7.5 COMPARISON

7.5.1 BIOFUEL VERSUS FOSSIL FUEL

Biofuel as a renewable energy source and fossil fuel as a non-renewable energy source are the most noticeable fundamental differences between these two [42]. However, in further describing the difference between fossil fuel and biofuel, let us first describe the definition of both fuels. Biofuel has been gaining popularity for a few years, but fossil fuel has been used for a long time. Here is the main reason for interest in biofuel: the demand for fuel for energy requirements is increasing day to day. Due to this, it is very difficult to fulfill the energy demand of the world by conventional fuel. Hence, it is time to give greater attention to an optional fuel to complete the energy demand [43]. For satisfying our energy requirement, biofuel becomes a better alternative fuel for fossil fuel because biofuels are produced from feedstocks of natural biomass [44]. The difference between biofuel and fossil fuel is shown in Table 7.1 (As seen in Table 7.1).

7.5.2 BIOETHANOL VERSUS BIODIESEL

Together bioethanol and biodiesel are environmentally friendly fuels. They are manufactured from organic resources, such as jatropha, soybean, castor seed, algae, and palm oil. Potato, sugar cane, and sugar beet are also used as supply resources for bioethanol and its auxiliary substitute fuel products [42,45]. Biodiesel is a fuel prepared from crop oils, fats, or greases such as used restaurant grease. Biodiesel fuel can be used in diesel engines exclusive of changing the mechanics of the engine. Unadulterated biodiesel is nontoxic, eco-friendly, and releases the least amount of most air pollutants than petroleum-based diesel fuel. Biodiesel is usually sold as petroleum-based diesel fuel with the blending of biodiesel [46,47]. The result of the bioethanol vs. biodiesel argument fundamentally depends upon the planting and the class of crop production (As seen in Table 7.2).

TABLE 7.1
Biofuel Versus Fossil Fuel

Biofuel is a fuel derived from natural livelihood materials	Fossil fuel is derived from a geological process that occurred a million years ago
Found from renewable sources	Found mostly from non-renewable sources
Offers a small quantity of energy per unit biomass	Offers a large quantity of energy per unit biomass
Grounds a lesser amount of pollution than fossil fuels	Creates more environmental pollution
Releases a little sum of adverse gases during combustion	Releases a great sum of adverse gases during combustion

Source: Zaleta-Aguilar, A. et al., *Biomass Bioenerg.*, 116, 1–7, 2018.

TABLE 7.2
Comparison Between Bioethanol and Biodiesel

	Bioethanol	Biodiesel
Process	Dry-mill process: sugars, starch, and yeast are fermented. From starch, it is fermented into sugar; after that, it is fermented once more into alcohol.	Transesterification: glycerin and methyl esters, which are not used as fuel for engines, are kept away.
Environmental advantage	Both decrease GHG emissions, as biofuels are mainly produced from plants, which consume CO_2.	Both decrease GHG emissions, as biofuels are mainly produced from plants that consume CO_2.
Compatibility	Ethanol has to be blended with fossil fuels, such as petrol; therefore, it's only well-suited with certain petrol-powered automobiles.	Capable to drive in any diesel-powered engine.
Costs gallons per acre	Economical 420 gallons of ethanol can be produced per acre.	More costly, 60 gallons of biodiesel can be produced per acre soya beans. Price of soya bean oil would also be a boost if biodiesel production is enhanced.
Energy	Supplies 93% additional net energy per gallon.	Supplies only 25% additional net energy per gallon.
Greenhouse gas emissions	12% lower amount of GHG emissions than the making and ignition of normal diesel vehicle.	41% smaller amount of weight against the conventional gasoline engine.

Source: Zaleta-Aguilar, A. et al., *Biomass Bioenerg.*, 116, 1–7, 2018.

7.5.3 BIODIESEL VERSUS DIESEL

Biodiesel has the capacity to put back an important quantity of the petroleum used to power diesel engines. The emissions from biodiesel are dissimilar than petroleum-based diesel, and it is significant to appreciate how they are dissimilar with admiration to the levels released and the combustion ability of the particulates. One of the main pollutants released from engine exhaust is particulate matter (PM). Particulate matter emitted from tailpipes consists of a range of toxic impurities either fixed or adsorbed on its surface (As seen in Table 7.3) [48].

7.6 APPLICATION OF BIOFUEL

7.6.1 TRANSPORTATION

It is estimated that approximately 30% of total energy consumption in the US comes from the transportation sector, but other sectors, such as residential and commercial, use only 10% combined [49]. This shows that the population in developed countries use on an average three times more energy in transportation to heat their rooms and to cook their meals. Approximately 40% of all energy use in electricity generation does not include the above. Generally, transportation alone can consume

TABLE 7.3

Emission Comparison Between Biodiesel and Diesel

Emissions	Biodiesel (%)	Diesel (%)
CO_2	−78	100
CO	−41	100
NOx	+5	100
Particulate matter	−55	100
PAH	−80	100
NPAH	−90	100
Sulfates	−100	100

Source: Agarwal, A.K., *Prog. Energ. Combust. Sci.*, 33, 233–271, 2007.

approximately 64% of oil for 25% of energy demand by itself. A large part of this energy, nearly two-thirds, is burned to drive vehicles, and the remaining part is utilized in infrastructure, manufacturing, the maintenance service sector, and untreated fabric producing. By looking forward, we find that about of 70% of the energy consumed is for moving people from one place to other and that for the most part of this energy is utilized in personal vehicles, which is less efficient of the mode of transportation. Here only 12% of the energy consumption by a vehicle goes to moving it, and approximately 3% is actually used to transfer the people around. The remaining part of the energy is used up through the action of friction, heat, partial burning, and running bulky automobiles [50,51]. Studies are shows that we have to reach high oil demand at present and increases in future but because of limited oil reserve. It will remain for short time period, and when this is combined with various impacts on the environment, such as petroleum recovery, refining, processing, pollution through combustion, and GHG emission, the move for a substitute will be clear [14]. The main problems with another substitute—wind, solar, tidal—is that these all are available in a dilute form, which is very difficult to use practically. Producing a sufficient form of stored electricity from these renewable energy resources becomes very difficult. Most of the professionals think that useful advances in these technologies are best done before the next 10 to 15 years. Therefore, the main problems here are to discover a fuel that can replace the actual merits of fossil fuel as well as does not pollute the environment [40].

7.6.2 FOR THE GENERATION OF POWER

The major application of fuel on earth is the production of electricity. In 2008, the world formed approximately 20,261 TWh of energy in the form of electricity. Approximately 41% of that energy was extracted from coal, 4% from oil, 22% from natural gas, 17% from renewable energy, and 16% from other sources [52]. About 16,430 TWh were distributed to consumers from 20,261 TWh energy produced, and the remaining was utilized by plants themselves. It is understandable that a large amount of energy is used to generate electricity, because in this industrialized world, humans need electricity for things from running water to using the internet. A number of approximations suggest that approximately 40% of all GHG emissions appear from the generation of electricity with transportation [48]. The use of coal for the production of electricity is the main cause of the formation of SOx in the atmosphere. Hence, if humans are not going to use nuclear power, then a fresh, extra-renewable type of energy is required. That is the main reason for the world to move toward the production of biofuels, which is a clean and reliable form of fuels that release fewer pollutants as compared to fossil fuels [47].

7.6.3 THERMAL APPLICATION

Heating from natural gas is the most significant application, though it can also be used for energy production. Natural gas not only comes from fossilized plant material but also recently grown some specific plant materials [53]. For heating purposes, most of the biofuels are used in solid form. Solid biomass such as wood and timber are used in both artistic and realistic methods for energy production, such as heating homes and cooking by burning wood like natural gas. Because of economic and environmental issues in the extraction of fossil fuels, the world changed their attention to solid biofuels, partially in reaction to increasing the fuel prices as a result of increased research in the industry for enhanced competence, increased efficiency, and decreased emissions. Solid biomass grassfires can attain high efficiencies up to 91% [48].

7.7 ETHICAL ISSUES FOR BIOFUEL

The ethical examination will consider the comparative analysis of biofuels in explaining the difficulty of climate change, as well as distributive and procedural justice problems required by various policies on biofuels. A complete ethical investigation cannot be executed until there is an enhanced understanding of the environmental, economic, and social crashes of biofuels. We need to continue developing substitutes for the future, a substitute that is enhanced to doing what biofuels are fit to do. These five ethical philosophies and one ethical duty formed the core of the report's ethical framework [51]:

1. Biofuel growth should not be at the cost of people's necessary human rights, consisting of health, water, and food.
2. Biofuels should be shown more environmentally sustainable.
3. Biofuels should be given their contribution to reducing the net GHG emissions.
4. Biofuels should hold to justified trade principles.
5. The economics of biofuels should be circulated in a reasonable way.
6. If these five philosophies of ethical issues depend on certain considerations, such as absolute charge or whether there are even better substitutes, then there is a responsibility to produce such biofuels.

7.8 CHALLENGES FOR THE FUTURE

A committing international approach is rising in the course of sustainability initiatives that cover problems applicable to human rights as well as environmental sustainability (philosophy 2) and the reduction of GHG emissions (philosophy 3). Biofuel sustainability principles layout strategies for environmentally and socially suitable biofuel production and allocation. Certification is a rule mechanism to signal that such principles have been fulfilled by indicating compliance. The 433 Certification can be optional or compulsory; at present, the research education design (RED) calls for the optional sustainability system to be applied by EU member countries. Benchmarked 434 against the renewable transport fuel obligation (RTFO)-Meta standard, the Roundtable on Sustainable Biofuels principles emerge presently to have the most widespread set of sustainability criteria. The Roundtable on Sustainable Biofuels has the ultimate aims to operate as a certification system [52].

REFERENCES

1. Godbole, E. P., & Dabhadkar, K. C. (2016). Review of production of bio fuels. *IOSR Journal of Biotechnology and Biochemistry, 2*(6), 62–69.
2. Baudry, G., Delrue, F., Legrand, J., Pruvost, J., & Vallée, T. (2017). The challenge of measuring biofuel sustainability: A stakeholder-driven approach applied to the French case. *Renewable and Sustainable Energy Reviews, 69*, 933–947.
3. Danielsen, F., Beukema, H., Burgess, N. D., Parish, F., Brühl, C. A., Donald, P. F., ... & Fitzherbert, E. B. (2009). Biofuel plantations on forested lands: Double jeopardy for biodiversity and climate. *Conservation Biology, 23*(2), 348–358.
4. Naik, S. N., Goud, V. V., Rout, P. K., & Dalai, A. K. (2010). Production of first and second generation biofuels: A comprehensive review. *Renewable and Sustainable Energy Reviews, 14*(2), 578–597.
5. Chang, W. R., Hwang, J. J., & Wu, W. (2017). Environmental impact and sustainability study on biofuels for transportation applications. *Renewable and Sustainable Energy Reviews, 67*, 277–288.
6. Gaurav, N., Sivasankari, S., Kiran, G. S., Ninawe, A., & Selvin, J. (2017). Utilization of bioresources for sustainable biofuels: A review. *Renewable and Sustainable Energy Reviews, 73*, 205–214.
7. Su, Y., Song, K., Zhang, P., Su, Y., Cheng, J., & Chen, X. (2017). Progress of microalgae biofuel's commercialization. *Renewable and Sustainable Energy Reviews, 74*, 402–411.
8. Rodionova, M. V., Poudyal, R. S., Tiwari, I., Voloshin, R. A., Zharmukhamedov, S. K., Nam, H. G., & Allakhverdiev, S. I. (2017). Biofuel production: Challenges and opportunities. *International Journal of Hydrogen Energy, 42*(12), 8450–8461.
9. Meher, L. C., Sagar, D. V., & Naik, S. N. (2006). Technical aspects of biodiesel production by transesterification—A review. *Renewable and Sustainable Energy Reviews, 10*(3), 248–268.
10. Azadi, P., Malina, R., Barrett, S. R., & Kraft, M. (2017). The evolution of the biofuel science. *Renewable and Sustainable Energy Reviews, 76*, 1479–1484.
11. Babu, V., & Murthy, M. (2017). Butanol and pentanol: The promising biofuels for CI engines–A review. *Renewable and Sustainable Energy Reviews, 78*, 1068–1088.
12. Carneiro, M. L. N., Pradelle, F., Braga, S. L., Gomes, M. S. P., Martins, A. R. F., Turkovics, F., & Pradelle, R. N. (2017). Potential of biofuels from algae: Comparison with fossil fuels, ethanol and biodiesel in Europe and Brazil through life cycle assessment (LCA). *Renewable and Sustainable Energy Reviews, 73*, 632–653.

13. Huber, G. W., Iborra, S., & Corma, A. (2006). Synthesis of transportation fuels from biomass: chemistry, catalysts, and engineering. *Chemical Reviews, 106*(9), 4044–4098.

14. Oschatz, M., & Antonietti, M. (2018). A search for selectivity to enable CO_2 capture with porous adsorbents. *Energy & Environmental Science, 11*(1), 57–70.

15. Srivastav, A., Srivastav, & Nishida. (2019). *The science and impact of climate change.* Springer, Singapore.

16. Khan, S., Siddique, R., Sajjad, W., Nabi, G., Hayat, K. M., Duan, P., & Yao, L. (2017). Biodiesel production from algae to overcome the energy crisis. *HAYATI Journal of Biosciences, 24*(4), 163–167.

17. Wijffels, R. H., Barbosa, M. J., & Eppink, M. H. (2010). Microalgae for the production of bulk chemicals and biofuels. *Biofuels, Bioproducts and Biorefining: Innovation for a Sustainable Economy, 4*(3), 287–295.

18. Moore, T. R., Roulet, N. T., & Waddington, J. M. (1998). Uncertainty in predicting the effect of climatic change on the carbon cycling of Canadian peatlands. *Climatic Change, 40*(2), 229–245.

19. Peters, G. P., Marland, G., Le Quéré, C., Boden, T., Canadell, J. G., & Raupach, M. R. (2012). Rapid growth in CO_2 emissions after the 2008–2009 global financial crisis. *Nature Climate Change, 2*(1), 2.

20. Moore, T. R., Roulet, N. T., & Waddington, J. M. (1998). Uncertainty in predicting the effect of climatic change on the carbon cycling of Canadian peatlands. *Climatic Change, 40*(2), 229–245.

21. Drexhage, J., & Murphy, D. (2010). Sustainable development: From Brundtland to Rio 2012. *Background paper prepared for consideration by the High Level Panel on Global Sustainability at its first meeting* September 19, 2010. New York, NY.

22. Drexhage, J., & Murphy, D. (2010). Sustainable development: From Brundtland to Rio 2012. *Background paper prepared for consideration by the High Level Panel on Global Sustainability at its first meeting* September 19, 2010. New York, NY.

23. Kumar, N., & Sharma, P. B. (2005). Jatropha curcus-A sustainable source for production of biodiesel. *Journal of Scientific and Industrial Research, 64*(2005), 883–889.

24. Union, E. (2009). Directive 2009/28/EC of the European Parliament and of the Council of 23 April 2009 on the promotion of the use of energy from renewable sources and amending and subsequently repealing Directives 2001/77/EC and 2003/30/EC. *Official Journal of the European Union, 5,* 2009.

25. Kumar, N., & Chauhan, S. R. (2013). Performance and emission characteristics of biodiesel from different origins: A review. *Renewable and Sustainable Energy Reviews, 21,* 633–658.

26. Biofuel. Retrieved from https://en.wikipedia.org/wiki/Biofuel#cite_note-bio-7.

27. Zheng, C. (2017, August). Three generation production biotechnology of biomass into bio-fuel. In *AIP Conference Proceedings* (Vol. 1864, No. 1, p. 020107). New York, NY: AIP Publishing.

28. Zabaniotou, A., Ioannidou, O., & Skoulou, V. (2008). Rapeseed residues utilization for energy and 2nd generation biofuels. *Fuel, 87*(8–9), 1492–1502.

29. Gomez, L. D., Steele-King, C. G., & McQueen-Mason, S. J. (2008). Sustainable liquid biofuels from biomass: The writing's on the walls. *New Phytologist, 178*(3), 473–485.

30. Burton, T., Lyons, H., Lerat, Y., Stanley, M., & Rasmussen, M. B. (2009). *A review of the potential of marine algae as a source of biofuel in Ireland.* Dublin, Ireland: Sustainable Energy Ireland-SEI.

31. Dragone, G., Fernandes, B. D., Vicente, A. A., & Teixeira, J. A. (2010). Third generation biofuels from microalgae. In A. Méndez-Vilas (Ed.), Current research, technology and education topics in applied microbiology and microbial biotechnology (pp. 1355–1366). Badajoz, Spain: Formatex Research Center.

32. Aro, E. M. (2016). From first generation biofuels to advanced solar biofuels. *Ambio*, *45*(1), 24–31.
33. Types of biofuel. Retrieved from http://biofuel.org.uk/types of biofuels.
34. Escobar, J. C., Lora, E. S., Venturini, O. J., Yáñez, E. E., Castillo, E. F., & Almazan, O. (2009). Biofuels: Environment, technology and food security. *Renewable and Sustainable Energy Reviews*, *13*(6–7), 1275–1287.
35. Monir, M. U., Aziz, A. A., Kristanti, R. A., & Yousuf, A. (2018). Syngas production from co-gasification of forest residue and charcoal in a pilot scale downdraft reactor. *Waste and Biomass Valorization*, *11*(2), 1–17.
36. Damiani, L., & Revetria, R. (2014). Numerical exergetic analysis of different biomass and fossil fuels gasification. International Journal of Renewable Energy and Biofuels, *2014*, 1–18.
37. Zaleta-Aguilar, A., Rodríguez-Alejandro, D. A., & Rangel-Hernández, V. H. (2018). Application of an exergy-based thermo characterization approach to diagnose the operation of a biomass-fueled gasifier. *Biomass and Bioenergy*, *116*, 1–7.
38. Subbaiah, G. V., Gopal, K. R., & Hussain, S. A. (2010). The effect of biodiesel and bio-ethanol blended diesel fuel on the performance and emission characteristics of a direct injection diesel engine. *Iranica Journal of Energy & Environment*, *1*(3), 211–221.
39. Agarwal, A. K. (2007). Biofuels (alcohols and biodiesel) applications as fuels for internal combustion engines. *Progress in Energy and Combustion Science*, *33*(3), 233–271.
40. Applications of biofuels. Retrieved from http://biofuel.org.uk/applications of bio fuels.
41. Global Agricultural Information Network (GAIN) Report: BR17006 September 15, 2017.
42. Canakci, M., & Van Gerpen, J. (2001). Biodiesel production from oils and fats with high free fatty acids. *Transactions of the ASAE*, *44*(6), 1429.
43. Meher, L. C., Sagar, D. V., & Naik, S. N. (2006). Technical aspects of biodiesel production by transesterification—A review. *Renewable and Sustainable Energy Reviews*, *10*(3), 248–268.
44. Ghaly, A. E., Dave, D., Brooks, M. S., & Budge, S. (2010). Production of biodiesel by enzymatic transesterification. American *Journal of Biochemistry* and *Biotechnology*, *6*(2), 54–76.
45. Kumar, A., Kumar, N., Baredar, P., & Shukla, A. (2015). A review on biomass energy resources, potential, conversion and policy in India. *Renewable and Sustainable Energy Reviews*, *45*, 530–539.
46. Luque, R., Herrero-Davila, L., Campelo, J. M., Clark, J. H., Hidalgo, J. M., Luna, D.,... & Romero, A. A. (2008). Biofuels: A technological perspective. *Energy & Environmental Science*, *1*(5), 542–564.
47. Luque, R., Herrero-Davila, L., Campelo, J. M., Clark, J. H., Hidalgo, J. M., Luna, D.,... & Romero, A. A. (2008). Biofuels: A technological perspective. *Energy & Environmental Science*, *1*(5), 542–564.
48. Kulkarni, M. G., Gopinath, R., Meher, L. C., & Dalai, A. K. (2006). Solid acid catalyzed biodiesel production by simultaneous esterification and transesterification. *Green Chemistry*, *8*(12), 1056–1062.
49. Shahir, S. A., Masjuki, H. H., Kalam, M. A., Imran, A., & Ashraful, A. M. (2015). Performance and emission assessment of diesel–biodiesel–ethanol/bioethanol blend as a fuel in diesel engines: A review. *Renewable and Sustainable Energy Reviews*, *48*, 62–78.
50. Greyling, S., Marais, H., van Schoor, G., & Uren, K. R. (2019). Application of exergy-based fault detection in a gas-to-liquids process plant. *Entropy*, *21*(6), 565.
51. www.ncbi.nlm.nih.gov/books/NBK196458/
52. www.global-greenhouse-warming.com/ethics-of-biofuel.html
53. www.conserve-energy-future.com/advantages-and-disadvantages-of-biofuels.php

8 Alternative Fuel for Transportation
Hydrogen

Sumit Lonkar and Prashant Baredar

CONTENTS

8.1 INTRODUCTION

Transportation is a major energy-consuming sector. Fossil fuels are a major source of energy for all kinds of vehicles available in the market. Burning of carbon-based fuel increases the percentage of carbon dioxide in the atmosphere. Carbon dioxide is a major constituent of greenhouse gases, which leads to higher earth temperature and climate change. To overcome the problem of greenhouse gas emission, one needs to think about alternative fuels like biodiesel and hydrogen. Hydrogen combustion has a unique characteristic that produces water and heat as a by-product, which is non-polluting in nature.

$$H_2 + O_2 \rightarrow H_2O + Heat \qquad (8.1)$$

Hydrogen plays an important role as an alternative to conventional fuel, provided its technical problems of storage and production can be resolved satisfactorily and the cost could be brought down to acceptable limits.

8.2 HYDROGEN AS AN ALTERNATIVE FUEL

Hydrogen can become the most promising and sustainable fuel for the future. It can directly use in internal combustion engine with some modification in engine design. Hydrogen is the primary fuel for fuel-cell-based electric vehicles. Hydrogen can be compared with other fuel by considering its various properties as shown in Table 8.1.

Hydrogen has the least ignition energy compared to other fuels. Also, the autoignition temperature for hydrogen is highest compared to other fuels. The diffusion coefficient for hydrogen is the highest; hence, it disperses at a faster rate in the atmosphere. The flame velocity is higher, which leads to better combustion of fuel, and it produces the least exhaust gases. These properties make hydrogen more suitable for an internal combustion engine [1].

The standard heating value of hydrogen gas is 12.1 MJ/m³ compared with an average value of 38.3 MJ/ m³ for natural gas. The heating values of liquid hydrogen are 120 MJ/kg and 8400 MJ/m³, and the corresponding values for gasoline are 44 MJ/kg and 32,000 MJ/m³. Hence, for producing a specific amount of energy, liquid hydrogen is superior to gasoline on a weight basis but inferior on a volume basis [2–5].

TABLE 8.1
Comparison of Properties of Hydrogen with Methane and Gasoline

Properties	Units	Hydrogen	Methane	Gasoline
Flammability limits	Vol. %	4–75	5–15	1.0–7.6
Minimum ignition energy	mJ	0.02	0.29	0.24
Flame temperature	°C	2045	1875	2200
Auto ignition temperature	°C	585	540	230–500
Diffusion coefficient	10^{-3} m²/s	0.61	0.20	1.1–3.3

Source: Satyapal, S. et al., *Catal. Today*, 120, 246–256, 2007; Hill, P.G., *Combustion*, by Irvin Glassman, Academic Press, Orlando, FL; *Can. J. Chem. Eng.*, 66, 350–351, 1988; Heywood, J.B., *Internal Combustion Engine Fundamentals*, McGraw-Hill, New York, 1988; Morley, C., Gaseq: A chemical equilibrium program for Windows. *Ver. 0.79*, 2005; White, C.M. et al., *Int. J. Hydrogen Energ.*, 31, 1292–305, 2006.

8.3 GENERATION OF HYDROGEN FROM DIFFERENT ROUTES

Hydrogen is not readily available in the atmosphere. It is bonded with different elements, and the general chemical formula can be expressed as $A_X H_Y$, where A can be any chemical element. To use hydrogen for power production, one needs to isolate it. Hydrogen can be separated from hydrogen-rich fuels like hydrocarbons (i.e., coal, CH_4, CH_3OH, etc.) and from water (H_2O). Solar, wind biomass are the primary sources of energy and coal, petroleum, and methane these are secondary sources of energy. Hydrogen can be separated from primary as well as secondary sources by following a sequence of processes (Figure 8.1).

Steam reforming and partial oxidation are two technologies that are being used for separation of hydrogen from fossil fuel on a commercial level. Partial oxidation uses low-quality petroleum coke or residue from refineries. Gasification of low-grade coal is also attracting scientists and engineers for the generation of hydrogen.

Apart from this, there is research on the new technologies to get hydrogen from biomass and biological separation of hydrogen. However, commercialization in the near future is expected for biomass gasification. Water is abundantly available over the earth. Hydrogen separation from water with the help of electricity is known as electrolysis. Electrolysis of water is the most common process for the separation of hydrogen and oxygen.

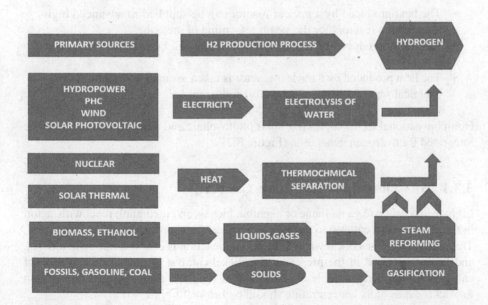

FIGURE 8.1 Different processes of separating hydrogen from primary and secondary sources. (From da Silva Veras T. et al., *Int. J. Hydrog. Energy*, 42, 2018–2033, 2017.)

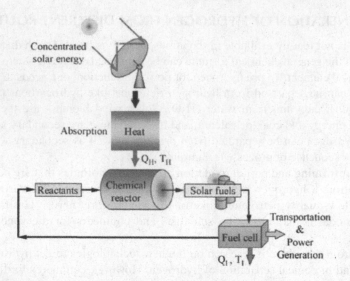

FIGURE 8.2 Solar fuel cell integrated cycle. (From Chen, H. et al., *Prog. Nat. Sci. Mater,* 19, 291–312, 2009.)

Nuclear energy is also helpful in the generation of hydrogen in the following ways:

- The heat produced by a nuclear reactor can be supplied to advanced high-temperature reactors for the steam reforming of methane.
- Electricity produced by a nuclear power plant can be used for electrolysis of water.
- The heat produced by a nuclear reactor is taken as input energy for thermo-chemical separation cycle (i.e., sulfur-iodine cycle).

Non-conventional technologies like solar photovoltaic and wind power plants can be integrated for hydrogen generation (Figure 8.2).

8.3.1 Steam Reforming of Natural Gas (CH₄)

Light hydrocarbon (e.g., methane or naphtha, biogas, and methanol) react with steam in the presence of a catalyst to produce a mixture of carbon monoxide and hydrogen. This mixture of gases is known as syngas. The reaction is endothermic in nature [8], and heat is supplied in the presence of a nickel catalyst. To avoid deactivation of catalyst clean, the vaporizable feed material is supplied. Sulfur is the main poison for nickel catalyst; its concentration should be below 0.1 ppm.

To increase the yield of hydrogen, excess steam is supplied. Hence, the molar ratio of steam to carbon should be 2.5–3 [9,10]. Generation of hydrogen is also affected by the temperature and pressure of the reaction. According to

Le Chatelier's principle for an endothermic reaction, temperature increases the concentration of the product of formation and pressure reduces the concentration of the product of formation. Steam reforming is generally conducted at temperatures of 800°C–900°C and pressures of approximately 20–40 bar (if the pressure is too low, additional energy is needed to compress outlet gases) and takes place as:

$$CH_4 + H_2 \leftrightarrow CO + 3H_2 \Delta h = 206 \, kJ/Mole \qquad (8.2)$$

After the reforming reaction, the gas is rapidly cooled down to about 350°C–450°C before it enters the water–gas shift reaction. This reaction is exothermic, and heat is released during the reaction. Here, shifting of an oxygen atom from water to CO takes place and forms CO_2, which can be separated and stored.

$$CO + H_2O \leftrightarrow CO_2 + H_2 \Delta h = -42 \, kJ/Mole \qquad (8.3)$$

In a steam reforming process, natural gas is first desulfurized. Then the clean natural gas is preheated and supplied to a reformer where it reacts with steam and produces a syngas (a mixture of hydrogen and CO). The syngas is then supplied to a water shift reactor where CO is converted to CO_2 and stored separately. Now hydrogen-rich syngas is purified in a pressure swing adsorption (PSA) plant.

First, the natural gas supplied into the system is desulfurized. Then the gas is mixed with steam and preheated before it is entered into the reformer. This consists of reactor pipes containing a nickel catalyst. Steam reacts with water in the presence of a nickel catalyst and forms a CO-rich syngas. The heat required for this process is produced by combusting some of the inlet gas (up to 25%), and the gas coming from the PSA plant. Excess steam used in the reformer avoids any carbon deposition over the catalyst. The CO-rich syngas leaves the reformer and enters a water–gas shift reactor. The carbon monoxide contained in the gas is then converted into hydrogen and carbon dioxide using steam in the presence of a ferric-oxide catalytic converter. After the shift conversion, the gas is cooled down to ambient temperature and then purified or treated. The H_2-rich gas is produced and enters into the PSA plant to eliminate the carbon dioxide and any remaining residues, such as CO; this produces hydrogen with a purity of at least 99.9 vol.%. In a large hydrogen-generation plant, the outlet pressure of hydrogen is 30 bars. Depending on the plant design, part of the excess steam could be used to generate electricity, which will increase the demand for water.

The cost of hydrogen separation by reforming is greatly affected by the CO_2-elimination method. If pure CO_2 is used for the further application, then CO_2 can be eliminated after the water–gas shift reaction by washing (e.g., with Rectisol). Almost 90% of CO_2 is eliminated and the reaming captured in a PSA plant. We can add a low-temperature water-shift reactor, which gives a better CO_2 capture rate and good conversion of CO to CO_2. If both pure CO_2 for storage and pure hydrogen for fuel cells are required, then a combination of a CO_2-capture plant (e.g., absorption with Rectisol) and a PSA plant is essential. Only the PSA unit is needed to get pure hydrogen [11].

8.3.2 COST OF CO_2 SEQUESTRATION

Hydrogen separation by the steam methane reforming process produces a considerable amount of CO_2, which is a greenhouse gas; so it becomes necessary to have a CO_2 absorption unit to avoid an environmental hazard. The CO_2 separation unit increases the cost of hydrogen separation by 20%–35% for a large steam methane reforming plant [11].

The economic efficiency of natural gas reforming is considerably dependent on the price of natural gas. The centralized reforming of natural gas, especially steam reforming, is a commercially developed technology. Refineries and chemical industries have a lot of use for this technology. Onsite generation of hydrogen is preferred by these kinds of industries for a large volume of hydrogen. The problem of distribution and storage can be avoided by the generation of hydrogen at the local site for an application such as a fueling station for a fuel cell vehicle. This is known as the decentralized generation of hydrogen. Due to the high cost of CO_2 storage and transportation, capturing pure CO_2 is not preferred for onsite generation of hydrogen.

8.3.3 COAL GASIFICATION

Solid fuel like coal can be used for the generation of hydrogen by pyrolysis and gasification. The classification of gasification methods is based on [12]:

- Mode of heat generation, external or internal
- Nature of contact between oxidation agent and fuel (fixed bed, fluidized bed, and entrained flow)
- The direction of the flow of fuel and the gasification agent (co-current and countercurrent)
- Type of gasification agent (air, oxygen, steam, or a mix)

Most gasification methods present today are classified based on the flow pattern of fuel and oxidant agent. Based on the flow pattern, three groups can be possible: entrained flow (co-current), fluidized bed (countercurrent), and moving bed. The feed input to gasifier can be dry or slurry (i.e., coal and water mixture).

The gasification of solid fuel takes at temperatures of 300°C–2000°C. The process can be divided into the substages of drying, pyrolysis, gasification, and combustion. During the gasification, pyrolysis of the fuel oxidizing agent converts fuel into a combustible gas (syngas). This syngas is then cooled and purified by removing ash particles and sulfur before the gas is fed to the water–gas shift reactor section. The resulting gas mixture is treated to produce hydrogen in a similar way as done in natural gas reforming as shown in Figure 8.3.

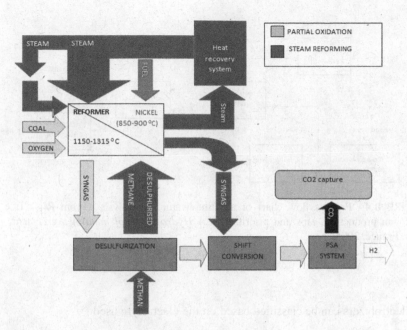

FIGURE 8.3 A block diagram of hydrogen separation from natural gas and coal by the steam reforming process and partial oxidation, respectively.

To avoid N_2 formation, oxygen is used in place of air as an oxidant agent. The high efficiency of the entrained flow gasifier makes it most suitable for hydrogen production using coal gasification. Solids fed into the gasifier have to be very fine and be of uniform quality, making the gasifier technology not appropriate for biomass or waste. The ash is separated as molten slag. External gasifier has a unique quality that can process any rank of coal with a moisture content of 10%.

8.3.4 Hydrogen Separation from Water: Electrolysis

Electrolysis means separating hydrogen from water with the help of a direct electric current. In polymer membrane electrolyzers, the reaction occurring in fuel cells is reversed; that is, water is disintegrated using electricity. To generate a cubic meter of hydrogen, theoretically, 3.54 kWh of energy has to be spent, which is the upper heating value of hydrogen. The water splits into hydrogen and oxygen in an electrolyzer, which consists of two electrodes. Hydrogen and oxygen are formed at cathode and anode, respectively. An electrolyte separates hydrogen and oxygen and only allow ions to flow from the anode to the cathode through the electrolyte. Electrolyzers are always operated with direct current. Upscaling of electrolyzers is very simple because of their modular construction.

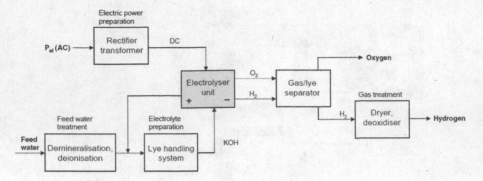

FIGURE 8.4 Process flow chart of alkaline water electrolysis. (From Riis, T. et al., Hydrogen production–gaps and priorities. *IEA Hydrogen Implementing Agreement*, IEA, Paris, France, 2005.)

Electrolyzers can be classified based on the electrolyte used:

- Alkaline water electrolyzers
- Membrane electrolyzers [polymer electrolyte membrane (PEM) electrolyzers]
- High-temperature electrolyzers

Alkaline water electrolysis, the oldest and, therefore, most commonly used technology, is described in more detail in Figure 8.4.

The PEM electrolyzer is mainly a polymer electrolyte membrane fuel cell operating in the reverse condition. At present, the PEM electrolyzer has higher electricity consumption than the alkaline-based electrolyzer; however, the potential for increased energy efficiency in the long term is better. High-temperature electrolysis also offers higher efficiency. At higher temperature, energy consumption increases, and electricity consumption decreases. Hence, high-temperature electrolysis (800°C–1000°C) may offer a good energy balance if high-temperature waste heat is available from other processes [13].

8.4 HYDROGEN STORAGE

The demand for fuel for propelling vehicles is increasing day by day. Today's transport sector is totally dependent upon gasoline. This is a primary fuel for most of the internal combustion engines. The burning of fossil-based fuel produces harmful gas, which leads to climate change and global warming. To eliminate this problem, the hydrogen-based vehicle is becoming impotent. The hydrogen-based vehicle will require a tank to store the fuel for continuous operation over a range of kilometer. Hydrogen storage is a vital issue to be solved to make hydrogen as a commercially alternative fuel for application in transport vehicles.

8.4.1 INITIAL REQUIREMENT

The volumetric storage density of the lightweight gas is low, which leads to the larger size of the storage tank. At ambient temperature and pressure, 5 kg of H_2 would fill a ball of 5 m in diameter, which is roughly comparable to the volume of an inflated hot-air balloon. As a result, the most important technical and economic challenges to be conquered in a practical hydrogen-storage system are the storage density related to the system (together with tank, heat management, and valves), the costs of the system, its safety, a short refueling time, and the ability to supply enough hydrogen during the driving cycle.

8.4.2 DIFFERENT STORAGE METHODS

The physical and chemical properties of hydrogen impose a technical limitation on standard methods of storing H_2 in pure forms, such as a pressurized gas or cryo-liquid. According to the current state of research, five methods are available for storing hydrogen: pressurized hydrogen, liquid hydrogen, storage in solids, hybrid storage systems, and regenerative off-board systems.

8.4.2.1 Pressurized Cylinder Storage

Hydrogen can be stored at high pressure in a cylinder made of composite materials to carry enough fuel for a driving range of 300 to 400 km. The pressurized cylinder has a limitation due to real gas behavior of hydrogen, which leads to no increment in volumetric density as pressure increases. Recent studies revealed that the gravi-metric capacity of a 700-bar system (approximately 4.5 wt% H_2) is less than in a 350-bar system with approximately 6 wt% H_2 [14]. High-pressure vessels consist of an outer shell, which is reinforced by a lightweight and highly stable carbon fibers and a polymer liner inside that acts as a permeation barrier for hydrogen. The outer shell is enclosed by an additional protective shell to prevent mechanical damage. Materials research has been performed to understand degradation effects better in pressurized storage systems and to reduce the risk of failure from fatigue and corrosion. Fatigue may be induced by mechanical stress, resulting from repeatedly changing the load of the vessel and the other components of the storage system. Hydrogen corrosion may occur by physicochemical interaction of the pressurized gas with materials of the pipes and valves, or, to a smaller extent, with the carbon of the vessel.

8.4.2.2 Liquefied Hydrogen Storage

The boiling point of hydrogen is 21 K; hence, to liquefy hydrogen it needs to cool below 21 K, which requires additional work input and increases the cost of hydrogen storage. This is an important technical limitation of storing hydrogen in the liquid form. Overall efficiency is additionally reduced by the so-called boil-off phenomenon. The stored cryogenic liquid starts to evaporate after a certain period of time, owing to unavoidable heat input into the storage vessel, leading to a loss of 2%–3% of evaporated hydrogen per day [15]. This cannot be prevented, even with very effective vacuum insulation and a heat-radiation shield in place. To prevent a high-pressure buildup (the critical temperature of hydrogen is 32 K), the overpressure must be released from the tank, for example, via a catalytic converter. Hydrogen boil-off is considered an issue in terms of refueling frequency, cost, energy efficiency, and safety, particularly for vehicles parked in confined spaces, such as parking garages. Recently, manufactured hydrogen cars with liquid hydrogen tanks are only allowed to be parked in open spaces.

Evaporation loss can be minimized by adopting the system developed by Linde. In the system of Linde, as the temperature of hydrogen increases, it is controlled by liquefying atmospheric air by taking heat from hydrogen. The liquefied air flows through the cooling jacket around the storage tank [16].

8.4.2.3   Solid Storage

Hydrogen is a highly reactive gas that can be attached chemically with the storage medium. Hydrogen can be released by this medium when it is supplied with extra heat. In the case of the fuel cell vehicle, this extra heat can be achieved by waste heat generated by the fuel cell. This reduces the extra equipment needed for heat supply.

Two ways of storing hydrogen by adsorption and absorption over the metals are:

- Chemical absorption (i.e., absorption of hydrogen): This involves the splitting of hydrogen molecules into hydrogen atoms and the chemical bonding of the atoms to a storage medium. Thus, the hydrogen is integrated into the lattice of a metal, an alloy, or a chemical compound.
- Physical adsorption (i.e., adsorption of hydrogen) of molecular hydrogen by weak Van der Waals forces to the inner surface of highly porous storage material. Adsorption has been studied on various nanomaterials, for example, nanocarbons, metal-organic frameworks, and polymers.

Chemical absorption of hydrogen has a high volumetric density because of the strong bonding force as compared to Van der Waals forces. Hydrogen in the gaseous (70 MPa, 300 K) or liquid state consists of H_2 molecules at a mean distance of approximately 0.45 nm or 0.36 nm, respectively. These distances result from repulsive molecular interactions. The minimum H–H separation in ordered binary metal hydrides is 0.21 nm, owing to the repulsive interaction generated by the partially charged hydrogen atoms.

A drawback of storing hydrogen as an atom in the metal matrix is the heat released during the splitting of hydrogen, and heat should be provided when releasing the hydrogen from the metal matrix. It requires an addition thermal management system. Combining the heat recovery system with hydrogen storage will increase the overall efficiency of the system.

Storing hydrogen by physical adsorption needs insulated cryovessels but less heat management because the hydrogen remains in its molecular form, and the enthalpy change between adsorption and releasing of gas is low.

8.4.2.4 Hybrid Storage

Hybrid means a metal hydride storage tank is kept under elevated hydrogen pressure to increase both the gravimetric and volumetric capacity of the system. A pressurized storage vessel can hold a larger volume of hydrogen compared to others and provide a greater driving range. Present-day fuel-cell vehicles typically offer a limited driving range because of the limited amount of fuel carried by their first-generation high-pressure hydrogen storage cylinders.

Storing hydrogen in a metallic alloy at moderate pressures provides a safe and practical method of storing. A Japanese car manufacturer presented the first system in 2005, and with a 1.7 wt% H_2 system capacity, the tank is filled with a usual AB_2 alloy, which has 1.9 wt% storage capacities. The tank volume was 180 L, and the weight was 420 kg [17].

Second-generation systems, which were presented recently, contain an optimized AB_2 alloy based on Mg and Ti and yield 2.2 wt% H_2 in the system (>50 kg/m^3 H_2), which is higher than the volumetric capacity of hydrogen in a 700-bar vessel by 50%. With some 200 kg, the weight of such a system is still comparably high. However, the volumetric density is the more important parameter in this respect, according to Japanese car manufacturers.

8.4.2.5 Hydrogen Safety

Table 8.2 presents some of the most important safety parameters of hydrogen in comparison to the data of methane and n-heptanes. The data show that hydrogen has the lowest ignition energy of the three energy carriers. Moreover, the explosion or detonation range in mixtures with air exhibits the widest spread for hydrogen. Hence, the formation of hydrogen and air mixtures has to be strictly avoided in uncontrolled environments, because there is a high risk of severe incidents, mostly because of the low ignition energy and the wide detonation range. These properties, together with the fact that hydrogen is 15 times lighter than air, are the reasons why an adapted safety strategy is needed to be able to benefit from hydrogen without exposing persons to unnecessary risks.

Research has been performed by fluid dynamics modeling and experimental validation to assess the behavior of hydrogen and air mixtures in various spatial environments, such as private and public garages, tunnels, etc.

TABLE 8.2
Some of the Most Important Safety Parameters of Hydrogen

Properties	Units	Hydrogen	Methane	n-Heptane C_7H_{16}
Lower detonation limit	Vol. %	18.3	6.3	—
Stoichiometric mixture	Vol. %	29.0	9.5	1.9
Minimum ignition energy	mJ	0.02	0.29	0.24
Upper detonation limit	Vol. %	59.0	13.5	—
Upper explosion limit	Vol. %	77	17	6.7
Self-ignition temperature	K	833	868	488

The H_2-release situation, mixing, ignition, flame propagation, and pressure wave expansion can be modeled in three dimensions for various scenarios, and mechanical and thermal loads at various places are obtained in the calculations. As a result of the studies, suggestions can be made for optimized geometries, where the light gas cannot accumulate and is released at low concentrations into the environment.

First studies have shown that critical situations can be avoided if the construction of private and public garages is adapted. It was demonstrated that natural convection may be sufficient to guide rising hydrogen away from a leaking tank to the outer environment, where it is diluted in air.

A slightly tilted roof may channel the hydrogen to an area where it can leave the room, for example. Other combined theoretical and experimental studies have resulted in the recommendation of minimum distances between hydrogen fuel stations and surrounding buildings. Of particular importance is the modeling of the interior in cars. Detailed knowledge of the behavior of hydrogen in voids and the passenger cabin may reduce the number of necessary hydrogen sensors on board and may give hints for the safer construction of the car.

8.5   FUEL CELL- AND HYDROGEN-BASED IC ENGINE

Hydrogen can be directly used to fuel cell- and hydrogen-based internal combustion engines to produce power. A fuel cell is a device that takes hydrogen gas as fuel and produces direct electric current and heat, which can be converted into useful work accordingly. The internal combustion engine is a major component of all vehicles, which can directly feed through hydrogen to produce power. The advantage of using hydrogen in an internal combustion engine is that after combustion of hydrogen, the exhaust contains only nitrogen oxide and water vapor.

REFERENCES

1. Satyapal, S., Petrovic, J., Read, C., Thomas, G., and Ordaz, G., 2007. The US department of energy's national hydrogen storage project: Progress towards meeting hydrogen-powered vehicle requirements. *Catalysis Today, 120*(3–4), pp. 246–256.
2. Hill, P.G., 1988. Combustion, by Irvin Glassman, Academic Press, Orlando, Florida, USA. *The Canadian Journal of Chemical Engineering, 66*(2), pp. 350–351.
3. Heywood, J.B., 1988. *Internal Combustion Engine Fundamentals*. McGraw-Hill, New York.
4. Morley, C., 2005. Gaseq: A chemical equilibrium program for Windows. *Ver. 0.79.*
5. White, C.M., Steeper, R.R., Lutz, A.E., 2006. The hydrogen-fuelled internal combustion engine: A technical review. *International Journal of Hydrogen Energy, 31*(10), pp. 1292–305.
6. da Silva Veras, T., Mozer, T.S., and da Silva César, A., 2017. Hydrogen: Trends, production and characterization of the main process worldwide. *International Journal of Hydrogen Energy*, 42(4), pp. 2018–2033.
7. Chen, H., Cong, T.N., Yang, W., Tan, C., Li, Y., and Ding, Y., 2009. Progress in electrical energy storage system: A critical review. *Progress in Natural Science, 19*(3), 291–312.

8. Baruah, R., Dixit, M., Basarkar, P., Parikh, D., and Bhargav, A., 2015. Advances in ethanol autothermal reforming. *Renewable and Sustainable Energy Reviews, 51,* pp. 1345–1353.
9. Ritter, J.A. and Ebner, A.D., 2007. State-of-the-art adsorption and membrane separation processes for hydrogen production in the chemical and petrochemical industries. *Separation Science and Technology, 42*(6), pp. 1123–1193.
10. Subramani, V., Sharma, P., Zhang, L., and Liu, K., 2010. *Catalytic Steam Reforming Technology for the Production of Hydrogen and Syngas* (pp. 14–126). John Wiley & Sons, Hoboken, NJ.
11. Voldsund, M., Jordal, K., and Anantharaman, R., 2016. Hydrogen production with CO_2 capture. *International Journal of Hydrogen Energy, 41*(9), pp. 4969–4992.
12. Platon, A. and Wang, Y., 2009. Water-gas shift technologies. In *Hydrogen and Syngas Production and Purification Technologies* (pp. 311–328), C. Song, V. Subramani (Eds.), AIChE-Wiley, Hoboken, NJ.
13. Riis, T., Hagen, E.F., Vie, P.J., and Ulleberg, Ø., 2005. Hydrogen production–gaps and priorities. *IEA Hydrogen Implementing Agreement*, IEA, Paris, France.
14. von Colbe, J.B., Ares, J.R., Barale, J., Baricco, M., Buckley, C., Capurso, G., Gallandat, N. et al., 2019. Application of hydrides in hydrogen storage and compression: Achievements, outlook and perspectives. *International Journal of Hydrogen Energy, 44*(15), pp. 7780–7808.
15. Eberle, U., Arnold, G., and Von Helmolt, R., 2006. Hydrogen storage in metal–hydrogen systems and their derivatives. *Journal of Power Sources, 154*(2), pp. 456–460.
16. Eliasson, B., Bossel, U., and Taylor, G., 2002. The future of the hydrogen economy: Bright or bleak? *Fuel Cell World*, pp. 367–382.
17. Ball, M. and Wietschel, M. (Eds.), 2009. *The Hydrogen Economy: Opportunities and Challenges*. Cambridge University Press, Cambridge, UK.

9 Fuel Cell Technology-Polymer Electrolyte Membrane Fuel Cell

Chaitali Morey and Prashant Baredar

CONTENTS

9.1 INTRODUCTION

A fuel cell is a device that converts the chemical energy of fuel into electrical energy directly and produces water vapor as a by-product. Hydrogen is a clean source of energy that is used as fuel in the fuel cell and does not cause pollution [1]. There are different types of fuel cells on the market based on the type of fuel used and electrolyte. Among them, the polymer electrolyte membrane (PEM) fuel cell is the most efficient and promising candidate for applications, such as transportation and stationary and portable sectors. The PEM fuel cell has a low operating temperature, quick startup time, high-power density, excellent dynamic response time, and is lightweight; these features make the PEM fuel cell a good choice for transportation sectors like buses, cars, and trains. The PEM fuel cell uses a polymer electrolyte membrane that is a solid polymer (acidified Teflon) membrane and is also known as Nafion. The polymer nature of the electrolyte provides corrosion resistance to the cell. The maintenance of the cell and its handling becomes very easy due to the solid state of the electrolyte.

PEM fuel cells are highly efficient and clean devices because they consume hydrogen and produce water vapor as a by-product [2]. It has a quiet operation because it has no moving parts and is involved in the energy conversion process. Therefore, it has no frictional and vibration losses. These features increase the overall efficiency of the cell. There are some disadvantages, such as the use of a noble metal as the catalyst, which increases the overall cost of the fuel cell, the life of the fuel cell decreases when using hydrogen-rich gases as fuel, and water and heat management problems [3].

9.2 COMPETING TECHNOLOGIES

Batteries and the internal combustion engine are two competing technologies of fuel cells, and their general structure is shown in Figure 9.1. Both fuel cells and heat engines use the chemical energy of the fuel for producing electrical work. A fuel cell is a single-stage conversion device, whereas heat engines are multi-stage conversion devices. Therefore, heat engines have lower efficiencies than fuel cells due to the multi-step process. The maximum efficiency of the heat engine is approximately is 30%, whereas the fuel cell has an efficiency in the range of 50%–60%. The by-product of a fuel cell is heat and water; on the other hand, the by-product of heat engines are exhaust gases, particulate matter, and heat. The exhaust emitted by the heat engine is the major cause of global warming and air pollution. Due to the dynamic components of the heat engines, wear and tear occur, which produces noise and vibration, rather than a fuel cell, which is static. Due to the quiet operation of the fuel cell, the application in the transport sector increases compares to the heat engine.

Batteries and fuel cells work on the same principle, and they have the same structure. The battery is an energy conversion and storage device, whereas fuel cells are only energy conversion devices. To produce electricity, the battery uses chemical energy, which is stored in the electrode in the form of an electric charge. Therefore, the battery produces electricity as long as the electrode is not depleted, whereas the fuel cell runs as long as fuel is supplied [4]. Hence, the battery has a comparatively shorter life than a fuel cell. The energy density of the fuel cell is higher than the battery. Therefore, fuel cells are lighter in weight than batteries. The number of charge–discharge cycles

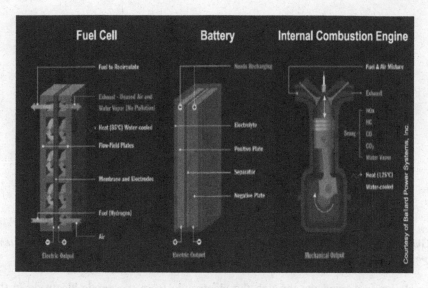

FIGURE 9.1 General structure of the fuel cell, battery, and internal combustion heat engines. (From Omar, Z.S. and Orhan, *Renewable and Sustainable Energy Reviews*, *32*, pp. 810–853, 2014.)

affects the life span of a battery, whereas the fuel cell has no limit of a charge–discharge cycle. The maximum limit of the deep discharge for the battery is 80%, whereas the fuel cell completely supplies its stored energy.

The battery uses its internal components in the energy conversion process, whereas the fuel cell does not involve any consumption of internal components. The recycling of the fuel cell is easy except the catalyst, but recycling and disposal of the battery are difficult and harmful to the environment. The fuel cell is easier to handle than the battery, as it does not require frequent maintenance. Corrosion problem occurs with the battery when not in use, but this is not in the case of a fuel cell [5].

9.3 PRINCIPLE OF OPERATION

Proton exchange membrane is also called a "solid polymer" fuel cell that makes use of an electrolyte, which is made up of a solid polymer film, and it is a form of acidified Teflon. It conducts only hydrogen ions to flow from anode to cathode when hydrogen is supplied at the anode electrode at a specified pressure. The operation of the PEM is shown in Figure 9.2.

The basic reaction involved in the PEM fuel cell is as follows [6]:

At anode:

$$H_2 \rightarrow 2H^+ + 2e^-$$

At cathode:

$$\frac{1}{2}O_2 + 2H^+ + 2e^- \rightarrow H_2O$$

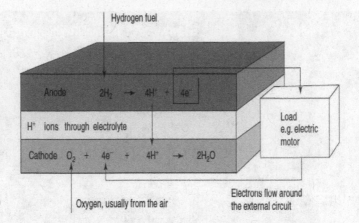

FIGURE 9.2 Operation of the fuel cell. (From Larminie, J. et al., *Fuel Cell Systems Explained, John Wiley & Sons*, Chichester, UK, Vol. 2, pp. 207–225, 2003.)

The H⁺ ion is carried out through the electrolyte from the anode electrode to the cathode electrode, while electrons are forced to move through an external circuit. The electrons and hydrogen ions combine at the cathode with oxygen and produce water. The overall reactions are given as:

$$H_2 + \frac{1}{2}O_2 \rightarrow H_2O$$

9.4 CONSTRUCTION DETAILS

The PEM fuel cell consists of bipolar plates and a membrane electrolyte assembly (MEA). The MEA assembly consists of a gas diffusion layer (GDL) or carbon cloth, catalyst layer, and membrane. The diagram of a PEM fuel cell is shown in Figure 9.3.

The MEA contains an electrolyte, anode, and cathode electrode, which are generally made up of graphite material due to the low electrical resistance. The catalyst

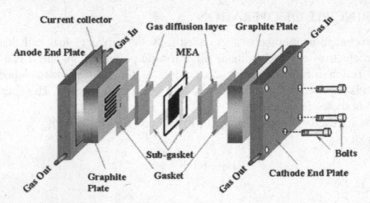

FIGURE 9.3 Construction details of the PEM fuel cell. (From Omar, Z.S. and Orhan, *Renewable and Sustainable Energy Reviews, 32*, pp. 810–853, 2014.)

layer is used over the surface of the electrode, which increases the rate of chemical reactions. The GDL is used after the catalyst layer to diffuse the gases from the anode to the cathode electrode. Additionally, it carries the water product away from the electrode surface to avoid flooding of water at the cathode [7].

Hydrogen fuel and an oxidant are supplied to the opposite side of the MEA. Seals are present at the edges of the electrodes to avoid the problem of leakage. Hydrogen is fed vertically through the bipolar plate at the anode, and the oxygen is fed horizontally at the cathode bipolar plate. Most of PEM fuel cells are connected in series with bipolar plates, and it is connected to the terminals so that maximum power can be extracted [8,9]. The number of connected cells and their size should be designed in such a way that maximum power can be extracted from the stack. Hence, it should be reliable, economical, and have a significant operating life.

9.4.1 MEMBRANE ELECTROLYTE ASSEMBLY

The MEA is a major part of the fuel cell. The MEA contains a solid polymer electrolyte membrane that is placed between the two catalyst layers followed by the GDLs. These layers are merged to the term MEA (membrane electrolyte assembly) and the electrochemical part of the fuel cell. The MEA is as shown in Figure 9.4. Bipolar plates on both sides of the MEA form a complete single cell.

9.4.2 ELECTROLYTE

The chemical reactions take place mainly in the electrolyte of the fuel cell. The electrolyte performs the main functions, such as a gas separator, a good proton conductor, and an electron insulator. The membrane must allow only hydrogen

FIGURE 9.4 Membrane of the fuel cell.

ions to pass through it so that it should have good proton conductivity for hydrogen ions. Hence, the large number of electrons flow through an external circuit, and more output current can be drawn from the fuel cell. It should have considerable mechanical strength and good dimensional stability and lightweight that can be easily manufactured.

The electrolyte mostly used in the PEM fuel cell is polytetrafluoroethylene (PTFE). PTFE is a polymer made up of sulfonated fluoroethylene and is also called as Nafion. The bond between carbon and fluorine is strong, durable, and hydrophobic, which drives the water away from the electrode and thus prevents flooding. The PTFE polymer is sulfonized by replacing the fluorine atom with sulfonic acid HSO_3. This group is ionically bonded, and therefore the side chain ends with SO^-_3. The presence of SO^-_3 and H^+ is a strong pair of ions. The HSO^-_3 bond is highly hydrophilic, which means it attracts water. Thus, Nafion is a compound that does not absorb the water but the sulfonated side chains of Nafion absorbs large quantities of water. The H^+ ions are weakly bonded to the group of SO_3 in these hydrated regions and are therefore more or less mobile. Within this hydrated region, the mobility of the hydrogen ions allows protons to be transferred, provided the hydrated regions are large enough [7].

The main features of the Nafion membranes are:

- Nafion is acidic and when it is hydrated it forms a diluted acid.
- They are highly chemically resistant.
- They absorb large quantities of water.
- H^+ ions are free to move in the hydrated regions and are good proton conductors.

Nafion is one of the commercial membranes produced by Dupont, and the other is produced by Dow Chemical Company. Ballard Power System is one of the manufacturers of fuel cell membranes that have developed their own proprietary membranes.

9.4.3 Catalyst Electrolyte Layer

To increase the rate of the chemical reaction, a catalyst is used on the surface of the electrode. The catalyst present in the layer helps the gases to dissociate the molecules into electrons and proton ions. Platinum is usually used because of its low electrical resistance, better chemical stability, and high catalytic efficiency. Platinum is very costly, and it is a major factor in the fuel cell performance [2]. As the amount of the platinum catalyst exposed to the reactant gases increases, the power density of the fuel cell also increases.

9.4.4 Gas Diffusion Layer

The GDL is applied after the catalyst layer. It is also a critical component after the layer of catalyst in the fuel cell assembly. The GDL is generally made up of carbon cloth or paper, which diffuses the gases through the catalyst from anode to the cathode electrode. It also creates an electrical connection between the bipolar plate and the carbon-supported catalyst [7]. Additionally, it minimizes the flooding of water in the catalyst layer and bipolar plate and forms a layer over the catalyst. Therefore, the GDL should possess the hydrophobic property and increase electronic contact with low resistance and transport the gases to the catalyst layer. The GDL arrangement should be strong enough to resist the pressure in the fuel cell stack, which is important to prevent the leakages [1].

9.4.5 ELECTRODES

Electrode plates are commonly known as bipolar plates. Electrodes have two main functions and they are as follows:

- To distribute the gases from both sides of the membrane.
- Conducts currents as electrons transport from anode to cathode, and it is used to connect the cells in the series to form a fuel cell stack so that maximum voltage can be obtained.

The following are the requirements [7] for the selection of the material of bipolar plates:

1. High electrical conductivity, that is, it must be greater than 10 Siemens/cm.
2. The thermal conductivity valves for the bipolar plate must exceed 20 W/m-K if the heat is to be removed from the surface by the help of cooling fluids. If heat is to be removed from the edges of the bipolar plate, the value of thermal conductivity should be greater than 100 W/m-K.
3. The diffusivity of gas should be less than 10^{-7} mBarL/s-cm.
4. Bipolar plates must be corrosion resistant.
5. Bipolar plates must possess physical properties, such as strength, stiffness, and durability.
6. It should have a low cost because 30% of the total cost is of bipolar plates.

Graphite is used as material for bipolar plates due to its high conductivity, corrosion resistance, high thermal conductivity, and chemical compatibility. The flow field channels are made on the graphite plates for the flow of the gases and are usually done by machining. The machining of the graphite plate is very complex, which increases the cost of the plates. Graphite plates are fragile and are prone to damage during manufacturing and handling, which also increases the cost.

The electrode material must be very thin so that maximum gas and water transport is possible. To maximize the gas flow within the MEA, serpentine design is best suited for the gas channel of the flow field plates. The particular shape of the gas channel is important for stable power generation, better performance of the cell, and most importantly, water management. Flow field plates have to be electrically conductive as a result of the current flow from one cell to another during an electrochemical reaction and then finally to the electrical endplates through which power is obtained.

9.4.6 Humidifiers

The conductivity of H^+ ion depends upon the water content inside the membrane. Hence, dry gases can decrease the life of the membrane considerably. To increase the ion conductivity pre-humidified gases are supplied to the membrane electrode assembly. According to the thermodynamic principle, the temperature of gas decides the specific humidity of gas, and hence high-temperature gas can absorb a large amount of water vapor.

We know that the specific humidity is a strong function of temperature, and hence the humidification temperature should match with the operating temperature of the stack. If the humidification of gas is done above fuel cell stack temperature and the gases come to the operating temperature of the stack, then some of the water vapor gets condensed. Similarly, if the humidification temperature is lower than the stack temperature, the gas will not be saturated as it reaches to stack temperature.

There are two ways by which gases can be humidified. One is by humidifying the gases outside the stack, and these pre-humidified gases are supplied to fuel cell stack. Another is humidifying the gas by using the cooling water flowing inside the cooling plate, and the saturated gas is supplied to the membrane [7].

If the humidification of gases is done by flowing cooling water inside the stack, then the temperature of the gas and stack will be nearly the same. Humidification inside the stack depends upon the coolant, hence the coolant should be in pure form, which imposes a limitation of freezing of cooling water.

Another method consists of using a separate humidifier externally to the fuel cell stack system. This can be of direct-contact type as in case of the inside humidifier used in a stack, or it can be a spray-type humidifier. In the spray type, the humidifier water spray over the hot plate gets vaporized, this vapor is absorbed by gases, and the gases get humidified. For this humidifier, the heat generated from the fuel cell is supplied to the hot plate.

9.5 THERMODYNAMICS OF THE PEM FUEL CELL

Thermodynamics mainly deals with the change in energy of the system during a process. A fuel cell is a control volume system because mass is entering and leaving the system boundary. As we know, the fuel cell converts chemical energy into direct electric current. This current flows out of the control volume and does electrical work. To have an idea about the energy conversion efficiency of fuel, its thermodynamic analysis is important [7].

9.5.1 Reversible Voltage

All practical processes contain a certain amount of irreversibility, which degrades the quality available with energy. If we consider that all the chemical energy is converted into electrical work, then the work obtained is known as reversible, or no-loss work. The energy released during the oxidation of hydrogen is defined by Gibbs free energy.

The Gibbs free energy of formation can be calculated as

$$\Delta G_f = \Delta G_f \text{ of products} - \Delta G_f \text{ of reactants}$$

Consider the simple reaction for the hydrogen/oxygen fuel cell:

$$2H_2 + O_2 \rightarrow 2H_2O$$

which is equivalent to:

$$H_2 + \frac{1}{2}O_2 \rightarrow H_2O$$

One mole of hydrogen reacts with a half mole of oxygen to form one mole of water, and energy is released during the process; hence, Gibbs free energy of formation per unit mole can be calculated as:

$$\Delta g_f = g_f \text{ of products} - g_f \text{ of reactants}$$

Hence,

$$\Delta g_f = \left(g_f \right)H_2O - \left(g_f \right)H_2 - \frac{1}{2}\left(g_f \right)O_2$$

If the fuel cell is working on pure hydrogen and oxygen, then one molecule of hydrogen gives two electrons. According Avogadro's Principle, one mole contains N (6.022×10^{23}) molecules; hence, the total charge produced by one mole of hydrogen is $-2Ne$, where $-e$ is a charge on one electron $(1.602 \times 10^{-19}°C)$.

If E is a voltage of the fuel cell, then electrical work by E to move 2Ne charge is given by

$$\text{Electrical work} = \text{Charge} \times \text{Voltage}$$

$$= \left(-2Ne\right) \times E$$

If the system involves no irreversibility, then all work should be equal to the energy released during the reaction.

$$\Delta g_f = -2Ne \times E_{REV}$$

$$F = \text{Faraday constant} = Ne$$

Thus,

$$E_{REV} = -\frac{\Delta g_f}{2F} \qquad (9.1)$$

If we take z as the number of electrons, then Eq. (9.1) can be written as

$$E_{REV} = -\frac{\Delta g_f}{ZF} \qquad (9.2)$$

9.5.2 Fuel Cell Efficiency

Hydrogen is oxidized in a fuel cell, and energy is released, which is finally converted into electricity. Hence, the efficiency of a fuel cell can be defined as the ratio of electrical work produced to the energy released during the oxidation of fuel (hydrogen).

Oxidation processes in which fuel combines with oxygen and produces heat. This heat is termed as the calorific value of a fuel. This is defined by the enthalpy of formation Δh_f.

Hence, mathematically, the efficiency of the fuel cell is

$$\eta_{fuel\ cell} = \frac{\text{Electrical work produced per mole of fuel}}{\Delta h_f} \qquad (9.3)$$

The maximum electrical work equals to Gibbs free energy of formation, that is, Δg_f when there is no loss of energy due to irreversibility in the system.

$$\eta_{MAX} = \frac{\Delta g_f}{\Delta h_f} \times 100 \qquad (9.4)$$

Theoretically, fuel can have a maximum possible efficiency as described by Eq. (9.4). This is also known as the thermodynamic limit (without violating the law of thermodynamics) of energy conversion that is possible by a fuel cell.

9.5.3 Efficiency in Terms of Cell Voltage

Assuming we have a fuel cell for which $\eta_{MAX} = 100\%$, then Gibbs free energy is equal to its calorific value, and we can write the cell voltage as

$$E = -\frac{\Delta h_f}{2F} \qquad (9.5)$$

Δh_f For hydrogen $= -241.83$ kJ mol^{-1} (for steam as output)
Δh_f For hydrogen $= -285.84$ kJ mol^{-1} (for water as output)

$$E_{MAX} = 1.48V \text{ based on HHV}$$

Or

$$= 1.25V \text{ based on LHV}$$

From the above discussion, we can define the fuel efficiency as the ratio of actual voltage of fuel cell to the maximum possible voltage. Mathematically, it is written as

$$\eta_{\text{fuel cell}} = \frac{V_c}{1.48} \times 100 \left(\text{based on HHV} \right) \tag{9.6}$$

If some amount of fuel gets unused during reaction, then we can define the fuel utilization factor as the ratio of the mass of fuel used in reaction to the total mass of fuel supplied to the system.

$$u_f = \frac{\text{Mass of fuel reacted in cell}}{\text{Mass of fuel input to the cell}} \tag{9.7}$$

Now we can rewrite Eq. (9.6) by considering fuel utilization factor u_f

$$\text{Efficiency } \eta = u_f \frac{V_c}{1.48} \times 100 \tag{9.8}$$

9.5.4 EFFECT OF GAS CONCENTRATION AND PRESSURE

9.5.4.1 Nernst Equations

The performance of the fuel cell is greatly affected by pressure and concentration of gas. As fuel is used in a reaction, the partial pressure of gas decreases, which affects the cell voltage. This variation can be greatly explained by the Nernst equation.

Let a general chemical reaction be

$$P_p + k_K \rightarrow m_M \tag{9.9}$$

where p moles of P reacts with k moles of K to produce m moles of M.

Activity is associated with each reactant and product involved in the chemical reaction. Let the activity associated with P, K, and M denoted by a_p, a_k, and a_m. One should note for an ideal gas activity, a function of the pressure of the gas can be written as

$$\text{Activity } a = \frac{P}{P^0}$$

where P is the pressure or partial pressure of the gas, and P^0 is standard atmospheric pressure, 0.1 MPa.

According to Gibbs free energy equation,

$$\Delta g_f = g_f^0 - RT \ln \left(\frac{a_P^p \times a_K^k}{a_M^m} \right) \tag{9.10}$$

where Δg_f^0 is the change in molar Gibbs free energy of formation at standard atmospheric pressure.

For the fuel cell supplied with hydrogen and oxygen, Eq. (9.10) can be modified as

$$\Delta g_f = \Delta \, g_f^0 - RT \ln\left(\frac{a_{H2}a^{\frac{1}{2}}_{O_2}}{a_{H_2O}} \right) \tag{9.11}$$

From the above equation, it is clear that as the activity of product increases (i.e., $a_{H_2}O$) Δg_f becomes less negative; hence, the energy released during a reaction is decreased. Now cell voltage can be expressed by putting the value of Δg_f in Eq. (9.11).

$$E = -\frac{g_f^0}{2F} + \frac{RT}{2F} \ln\left(\frac{a_{H2}a^{\frac{1}{2}}_{O_2}}{a_{H_2O}} \right)$$

$$= E^0 + \frac{RT}{2F} \ln\left(\frac{a_{H2}a^{\frac{1}{2}}_{O_2}}{a_{H_2O}} \right) \tag{9.12}$$

where E^0 is the electromotive force (EMF) at standard atmospheric pressure.

Equation (9.12) is the governing equation of cell voltage. It gives the relation between cell voltage and activity of reactant and product. This equation is known as the Nernst equation, assuming that the water coming out from cell is in vapor form and acts as ideal gas. Then activity for an ideal gas is given as

$$a_{H_2} = \frac{P_{H_2}}{P^0} \quad a_{O_2} = \frac{P_{O2}}{P^0} \quad a_{H_2O} = \frac{P_{H_2O}}{P^0}$$

Then Eq. (9.12) will become

$$E = E^0 + \frac{RT}{2F} \times \ln \frac{\frac{P_{H_2}}{P^0} \times \left(\frac{P_{O_2}}{P^0} \right)^{\frac{1}{2}}}{\frac{P_{H_2O}}{P^0}}$$

If $P^0 = 1$ atm then

$$E = E^0 + \frac{RT}{2F} \times \ln\left(\frac{P_{H_2} \, P_{O_2}^{\frac{1}{2}}}{P_{H_2O}} \right) \tag{9.13}$$

If the operating system pressure is P, then we can say that

$$P_{H2} = \alpha P, \quad P_{O2} = \beta P, \quad \text{and} \quad P_{H_2O} = \delta P$$

where α, β, and δ are constants depending on the molar masses and concentrations of H_2, O_2, and H_2O.

Equation (9.13) can be written as

$$E = E^0 + \frac{RT}{2F} \ln\left(\frac{\alpha\, \beta^{\frac{1}{2}}}{\delta}\right) \times P^{\frac{1}{2}} \tag{9.14}$$

Equation (9.14) is the modified form of the Nernst equation when the system is operating under total pressure p.

9.5.4.2 Effect of Partial Pressure of Hydrogen

Hydrogen can be used in pure form or as a mixture of gases. Its pressure changes as hydrogen is utilized in a reaction, so the effect of change in pressure can be calculated by the Nernst equation. Equation (9.13) can be written as

$$E = E^0 + \frac{RT}{2F} \ln\left(\frac{P_{O_2}^{\frac{1}{2}}}{P_{H_2O}}\right) + \frac{RT}{2F} \ln\left(P_{H_2}\right) \tag{9.15}$$

Let the hydrogen pressure change from p_1 to p_2, then the difference in voltage can be calculated as

$$\Delta V = \frac{RT}{2F} \ln\left(P_2\right) - \frac{RT}{2F} \ln\left(P_1\right) \tag{9.16}$$

$$\Delta V = \frac{RT}{2F} \ln\left(\frac{P_2}{P_1}\right)$$

9.5.4.3 Fuel and Oxidant Utilization

The fuel cell utilizes fuel and oxidants, and their respective partial pressures decrease, which affect the values of α, β, and δ. In a fuel cell reaction, hydrogen is oxidized; hence, the value of α decreases and δ increases, which affects the cell voltage.

$$\frac{RT}{2F} \ln\left(\frac{\alpha\beta^{\frac{1}{2}}}{\delta}\right) \tag{9.17}$$

The value of α is varied from inlet to an outlet, but due to the high conductivity of the bipolar plate, we cannot have different voltages at different points, which binds current density to decrease [8].

To get a higher system efficiency, fuel utilization should be maximum. But according to the preceding discussion, fuel utilization affects the cell voltage, so it becomes an important parameter to control for optimum performance spatially when fuel is supplied by the reformer.

9.6 APPLICATIONS

Proton exchange membrance fuel is a promising technology for the various applications due to the features of PEM, such as low temperature, quick startup, high-power density, and nearly zero emission [1]. Transportation, stationary, and portable are the main areas of applications of the fuel cell, and they are discussed below.

9.6.1 PORTABLE APPLICATIONS

The power required for the portable device can be supplied by the fuel cell due to its higher energy density and modularity. A portable device such as a laptop, mobile phone, or radio can be run by the small-capacity fuel cell. The emergency devices used in the military for surveillance applications can also be run by fuel cells. The range of power of fuel cells required for portable application is in the range of 5–500 W [2]. The direct methanol fuel cell and PEM fuel cell are the most commercially available ones for portable applications [5,10].

9.6.2 STATIONARY APPLICATIONS

A PEM fuel cell can be used as a stand-alone system, such as auxiliary power units, due to the flexibility and size of the fuel cell when compared to other technologies like solar and wind. Fuel cells combined with a waste recovery plant give an efficiency of about 70%. The stand-alone system can provide electricity to rural areas where grid connectivity is difficult. The mobile tower is the most commercial use of the stand-alone system.

9.6.2.1 Backup Power Supply

Due to the small size and quick-starting facility, it makes fuel cells a backup source for the application where a continuous power supply is necessary. Mobile communication towers and hospitals use diesel-based power generators, which cause pollution; hence, the fuel cell has the opportunity to be used as a backup source for such applications.

9.6.2.2 Stand-Alone Power Supply

A fuel cell can act as a power station for a remote location where the grid supply is not available. Hilly areas and villages are very difficult to obtain a connection with the grid to get a power supply, so we can create a stand-alone system to fulfill their demands.

Presently, remote locations are getting power by either solar or wind-based power plants, but they are dependent upon atmospheric conditions. These disadvantages can be overcome by a fuel-cell-based power plants [11,12].

9.6.2.3 Distributed Power/CHP Generations

Fuel cells produce power and heat simultaneously; hence, combined fuel cell and heat power plants can be used to generate electricity. The conventional power plant has lower efficiency but combined fuel cell and heat power give an efficiency near about 80%. The reduced use of fossil fuel due to some heat supplied by fuel cell stack pollutants coming from the plant is lower in concentration [13,14].

9.6.3 TRANSPORTATION APPLICATIONS

The transportation industry is completely dependent on fossil fuels like gasoline and methane. Seventeen percent of the global greenhouse emission is due to transportation every year [15]. This fossil fuel is the primary source of pollution, so for clean and sustainable development, one needs to shift from conventional to zero-emission vehicle technology. The fuel cell produces no harmful emission, so they can be used for transportation applications. Therefore, fuel cells are the competing technologies to the internal combustion engines and batteries. Fuel cells are lightweight and flexibility in size, which makes them suitable to use in a fuel-powered electric vehicle. The power ranges for applications such as buses, cars, and utility vehicles range from 20–250 kW [16].

9.6.3.1 Light Traction Vehicles

Light traction vehicles such as scooters, motorcycles, and lawn movers can be powered by using fuel cell power electric motors. Forklifts and trucks are material handling equipment that are good candidates for the use of the fuel cell. Forklifts have been successful equipment of the fuel cell and mainly used in the USA. About 2.5 million forklifts were in operation in America, and 1300 fuel-cell-based forklifts are on working in the USA [17,18]. Forklifts are mostly of 5–20 kW PEM fuel cells [2]. Fuel-cell-based scooters [19] and electric vehicles are used in most of the countries such as USA, Japan, and China.

9.6.3.2 Light-Duty Fuel Cell Vehicles

Light-duty fuel cell vehicles are in operation because no moving part are involved in the fuel cell. It's a highly efficient device and has less greenhouse emission as compared to internal combustion engines. Major car manufacturers such as General Motors, Honda, and Toyota, use fuel cells for their main propulsion systems [20]. Light-duty vehicles such as cars can use electricity generated by the fuel cell to run electric motor-based transmissions.

9.6.3.3 Heavy-Duty Fuel Cell Vehicles

Heavy-duty fuel cell vehicles such as heavy trucks, buses, vans, and utility trucks use fuel cells for their electric propulsion systems [2]. In addition, fuel-cell-based buses were considered the best public transport in the transportation sectors. Fuel cell buses produce less emission when using hydrogen as fuel when compared to diesel and combustion-based buses. Fuel-cell-based buses are commonly used in the US, Canada, Europe, China, and Japan. In India, Tata Motors launched a fuel-cell-based luxury bus in 2017.

REFERENCES

1. Wang, Y., Chen, K.S., Mishler, J., Cho, S.C. and Adroher, X.C., 2011. A review of polymer electrolyte membrane fuel cells: Technology, applications, and needs on fundamental research. *Applied Energy, 88*(4), pp. 981–1007.
2. Sharaf, O.Z. and Orhan, M.F., 2014. An overview of fuel cell technology: Fundamentals and applications. *Renewable and Sustainable Energy Reviews, 32*, pp. 810–853.
3. Mekhilef, S., Saidur, R. and Safari, A., 2012. Comparative study of different fuel cell technologies. *Renewable and Sustainable Energy Reviews, 16*(1), pp. 981–989.
4. Wang, C., Nehrir, M.H. and Shaw, S.R., 2005. Dynamic models and model validation for PEM fuel cells using electrical circuits. *IEEE Transactions on Energy Conversion, 20*(2), pp. 442–451.
5. Cowey, K., Green, K.J., Mcpsted, G.O. and Reeve, R., 2004. Portable and military fuel cells. *Current Opinion in Solid State and Materials Science, 8*(5), pp. 367–371.
6. Corrêa, J.M., Farret, F.A., Canha, L.N. and Simoes, M.G., 2004. An electrochemical-based fuel-cell model suitable for electrical engineering automation approach. *IEEE Transactions on Industrial Electronics, 51*(5), pp. 1103–1112.
7. Larminie, J., Dicks, A. and McDonald, M.S., 2003. *Fuel Cell Systems Explained* (Vol. 2, pp. 207–225). Chichester, UK: John Wiley & Sons.
8. Farrington, L., 2003. Fuel for thought on cars of the future. *Scientific Computing World.*
9. Milliken, J., 2011. Hydrogen, *Fuel Cell Technologies Office: Multi-year Research, Development, and Demonstration Plan.* US Department of Energy.
10. Patil, A.S., Dubois, T.G., Sifer, N., Bostic, E., Gardner, K., Quah, M. and Bolton, C., 2004. Portable fuel cell systems for America's army: Technology transition to the field. *Journal of Power Sources, 136*(2), pp. 220–225.
11. Khan, M.J. and Iqbal, M.T., 2005. Pre-feasibility study of stand-alone hybrid energy systems for applications in Newfoundland. *Renewable Energy, 30*(6), pp. 835–854.
12. Wallmark, C. and Alvfors, P., 2003. Technical design and economic evaluation of a stand-alone PEFC system for buildings in Sweden. *Journal of Power Sources, 118*(1–2), pp. 358–366.
13. Briguglio, N., Ferraro, M., Brunaccini, G. and Antonucci, V., 2011. Evaluation of a low temperature fuel cell system for residential CHP. *International Journal of Hydrogen Energy, 36*(13), pp. 8023–8029.
14. Kazempoor, P., Dorer, V. and Weber, A., 2011. Modelling and evaluation of building integrated SOFC systems. *International Journal of Hydrogen Energy, 36*(20), pp. 13241–13249.
15. Plunkett, J.W., 2010. *Plunketts Automobile Industry Almanac 2011.* Houston, TX: Plunkett Research.
16. Lipman, T. and Sperling, D., 2010. Market concepts, competing technologies and cost challenges for automotive and stationary applications. In: W. Vielstich, H. Gasteiger, A. Lamm (Eds.), *Handbook of Fuel Cells: Fundamentals, Technology and Applications.* Chichester, UK: John Wiley & Sons.
17. Wang, C., Mao, Z., Bao, F., Li, X. and Xie, X., 2005. Development and performance of 5 kW proton exchange membrane fuel cell stationary power system. *International Journal of Hydrogen Energy, 30*(9), pp. 1031–1034.
18. Ladewig, B.P. and Lapicque, F., 2009. Analysis of the ripple current in a 5 kW polymer electrolyte membrane fuel cell stack. *Fuel Cells, 9*(2), pp. 157–163.
19. Hwang, J.J. and Zou, M.L., 2010. Development of a proton exchange membrane fuel cell cogeneration system. *Journal of Power Sources, 195*(9), pp. 2579–2585.
20. Wang, Y., Basu, S. and Wang, C.Y., 2008. Modeling two-phase flow in PEM fuel cell channels. *Journal of Power Sources, 179*(2), pp. 603–617.

10 Low Carbon Energy System
Role of Fuel Cell Technology

Madhu Sharma, Debajyoti Bose,
and Tulika Banerjee

CONTENTS

10.1 ENERGY ECONOMY AND TRENDS IN FUEL CONSUMPTION

Energy has become a fundamental part of our civilization; it is used in various forms, such as chemical, electrical, and thermal. It is used for different applications for lighting our homes, cooking meals, and traveling from one place to another. Almost all our activity is dependent on energy, which affects the environment as well as the economy, as well as for future generations, and it also helps to define our civilization [1]. There is a demand for energy that is reliable, affordable, and aims to build a reliable infrastructure that is required to produce and deliver energy. The global demand for energy has risen to greater heights in the last 150 years due to rapid growth in population, industrialization, and technology. In the year 2018, energy demand worldwide grew by 2.3%, its fastest pace this decade. As a result, global energy-related carbon dioxide (CO_2) emissions rose by 1.7% to 33 gigatons (Gt) in 2018, which also included greenhouse gases (GHGs), such as methane (CH_4), N_2O, and fluorinated gases [2]. On the other hand, combustion of fossil fuels creates environmental pollutants, such as SO_x, NO_x, particulate matters (PMs), volatile organic compounds, and toxic heavy metals. Due to the increase in GHGs in the environment, the climate is changing because droughts, torrential heavy rainfall, floods, etc. are increasing [3]. The earth's climate system is extremely complex and is prone to predict future risks. Thus, mitigating the climate risk by dramatically decreasing the GHG emission is a vital aspect these days [4]. There are two forms of energy: renewable and nonrenewable. The renewable form of energy is a path to zero emission [5]. The clean renewable source of energy gets converted into electrical energy in order to be used for different household applications, such as lighting and air conditioning.

10.2   CONVENTIONAL ENERGY SOURCES

Conventional energy sources, or nonrenewable energy sources, are those that are limited in nature because they cannot be created over and over again. This includes nuclear energy, oil, natural gas, coal, etc. Out of these sources, the primary commercial energy source in the world, such as oil, natural gas, and coal, are called fossil fuels.

10.2.1   OIL

Oil is otherwise called crude oil. It is a normally happening combustible fluid found inside layers of shake developments underneath the earth's surface and is a mind-boggling blend, comprising hydrocarbons and other fluid natural mixes. After it is separated, unrefined petroleum is part of different kinds of items in treatment facilities by a procedure called breaking [6]. At the treatment facilities, the smallest and lightest particles ascend to the highest point of the distillation towers, which is the source of propane, butane, and different gases, while the medium molecules gather close to the middle part of the distillation tower, prompting gasoline and aeronautics fuel. Also, the heavier oil-based goods consolidate close to the base of the towers, which are essentially utilized as diesel oil and home-heating oil. The principle results of the handled raw petroleum, for example, gasoline, diesel fuel, aviation or jet fills, home-heating oil, oil for boats, and oil, are utilized in power plants to produce power [7]. Unrefined petroleum is utilized for various applications, for example, a combination of manures, pesticides, and plastics.

Significant air pollution is caused from burning oil by producing different pollutants such as nitrogen oxides (NO_x), sulfur dioxide (SO_2), toxic heavy metals, and volatile organic compounds, which contribute to ground-level ozone [8]. Oil-based power plants likewise strongly affect water, land use, and waste management and require a lot of water for heating and cooling, which can hurt nearby water streams and life in it.

10.2.2 NATURAL GAS

Natural gas is a vaporous type of oil that contains 87% methane, the lightest compound of the hydrocarbons. It is very hard to deal with natural gas when contrasted with fluid oil or solid coal. For moving the petroleum gas, the most advantageous and productive method is transported through pipelines, which are worked to interface the gas fields legitimately to major urban regions [9]. It can likewise be packed in tanks and moved about as liquefied natural gas. Liquefied petroleum gas is discovered broken up in unrefined petroleum. It is separated from liquid oil and put away under strain in substantial metal chambers. These compartments can undoubtedly be transported to homes for utilization in heating and related applications. Natural gas is cleaner when consuming [10]. It is more affordable than oil, so power created from it is more cost-effective. Natural gas can likewise be utilized as a fuel in power generators with ordinary steam boilers like other nonrenewable energy sources.

When contrasted with oil and coal, natural gas represents a more positive effect on the earth. The consuming of natural gas creates about half carbon dioxide as the consuming of coal. It likewise delivers fewer particulate issues, SO_2, and other dangerous outflows. In any case, the utilization of natural gas definitely prompts CH_4 discharges which is an extremely dangerous GHG, like CO_2, adding to worldwide environmental change. Like oil, extraction of natural gas can likewise negatively affect untamed life and wild living spaces. Since natural gas doesn't have any smell and is undetectable, it is important to be blended with a synthetic that gives a solid scent before it is nourished to pipelines and capacity tanks [11]. It can likewise be utilized for different applications, for example, to heat homes and structures, cooking, heating water, drying garments, and other outside activities, for example, grilling, gas lighting, generating heat for a hot tub, and even in a backyard swimming pool.

10.2.3 COAL

Coal is a hard, strong, black rock-like substance that is composed of carbon, hydrogen, oxygen, nitrogen, and changing measures of sulfur [12]. It is mined starting from the earliest stage underground mines and strip mines. In its solid form, most of the coal is transported either via trains or ships. Sometimes, pipelines are utilized to associate coal fields straightforwardly to electric power plants [13]. When transporting the coal through pipelines, the coal is blended with water to frame slurries than can move through the pipelines. Coal is broadly used to create electric power everywhere throughout the world. Coal-terminated generators produce over a portion of the power utilized in the United States [14]. The iron and steel-fabricating industry utilizes a lot of coal, generally as a blend called coke, and a limited quantity of coal is utilized straightforwardly in houses and business structures as a fuel hotspot for heating applications.

With the use of coal, a number of grave environmental problems can occur:

1. Damage to wildlife and soil erosion occurs due to strip mining.
2. Transporting and processing coal could affect water quality for both aquatic and terrestrial life.
3. CH_4 gas is released through the mining of coal; a GHG and contributes drastically to climate change.
4. Similar GHGs, such as a NO_x, are also released during coal production and usage.
5. The presence of mercer in coal also is released as a toxic compound and can have a negative impact on all life forms.

A number of technologies to minimize these negative environmental impacts have been developed, which includes the creation of pressurized fluid and a blast of hot air into the pulverized coal dust to facilitate a floating burning environment [15]. This process can significantly increase the burning efficiency at a lower burning temperature, thus leading to the decreased generation of NO_x.

10.2.4 NUCLEAR ENERGY

Nuclear energy is discharged when a particle is a part, which leads to the release of a large amount of heat and light. The arrival of such a gigantic measure of energy in an uncontrolled manner is risky yet when it is discharged in a controlled manner, this energy can be gathered safely to produce power at high efficiencies [16]. Uranium is utilized to fuel atomic power plants.

Uranium particles are partly separated through a controlled chain reaction in the nuclear reactor. Neutrons shell the particles of uranium-235 (U-235) to activate their split, in this manner framing two new components and moreover two neutrons. These recently framed neutrons at that point keep on bombarding more U-235 particles in order to keep up the chain response, in light of the fact that the number of neutrons increases geometrically. A large amount of heat energy is discharged during this chain reaction. In the center of the reactor, this heat is utilized to heat water along these lines, moving it into a different and radiation-free boiling water framework through a heat exchanger [17]. The steam available in the second set of pipes drives turbines to produce power. Radioactive wastes are available all through the nuclear fuel cycle, for example, in mining and processing of uranium mineral, preparing and creation of uranium, and so forth. In spite of the fact that the measure of solid wastes created in nuclear plants is generally little, radioactive wastes increase health risks that could be incomprehensibly more genuine than any of different sources of electricity. According to studies, approximately 20–25 tons of spent fuel every year is produced by a typical 1000 MW reactor, and the fuel spent is not all around reprocessed, so it can have radioactive dangers for up to 250,000 years.

10.3   RENEWABLE ENERGY SOURCES

Those sources of energy that can be replaced or regenerated from naturally occurring sources are known as renewable sources of energy and include hydro, geothermal, solar, wind, and biomass, which have different characteristics and need to be harvested by using different technologies.

10.3.1   HYDROPOWER

Hydropower facilities are structured and built so that its mechanical energy is converted over into power through turbines. Due to the cycle of water vanishing from the warmth of the Sun and falling back to Earth, it is persistently reestablished by the Sun's energy, it is considered a renewable source of energy. Hydroelectric power plants produce power by exploiting falling water from a specific height. The higher the head and more prominent the flow, the more the power can be created. For producing hydroelectric power, a repository is made by utilizing a dam. Water discharged behind the dam moves through the intake into a pipe called a penstock, where water pushes against cutting edges in a turbine, pivoting the sharp edges of the turbine and creating power [18]. For the measure of power created to be consistent, the speed of the turbine must be steady. The administering doors that open and close as and when required constrains the speed and volume of the water streaming to the turbine.

In various ways, dams can influence the earth. Inside a stream, they can change the water temperature and dimensions of dissolved gases, which thus adds to a variety of related water-quality issues and subsequently are unsafe to the amphibian natural surroundings [19]. With the deposition of sediment, a dangerous concoction or mechanical buildups from upstream sources may likewise be amassed, which could result in the development of exceptionally concentrated toxins.

10.3.2   SOLAR ENERGY

Sun emanates a colossal measure of energy consistently, called solar energy. Just a small amount of the unmistakable brilliant energy that the sun transmits into space ever reaches the earth; however, that sum is adequate. Sun-based energy can either be utilized legitimately as a warm energy source or indirectly as a source of power. Solar energy is considered a sustainable form of energy on the grounds that producing power straightforwardly from daylight does not deplete any of the earth's common resources. The most practical utilization of sunlight-based energy is utilizing it to heat water. Solar water heaters use rooftop boards that contain a system of dark warmth permeable channels. Sun heats the fluid in funnels, which are then siphoned through a heat exchanger to warm household water [20]. These days, solar water heaters are more encouraging than sun-powered photovoltaics, which convert sunlight-based energy into power, since they are more affordable and catch a lot higher extent of the sun's energy.

Both solar photovoltaic and solar thermal collector technologies create zero emissions; however, an emanation of ozone-harming substances and different toxins occurs by their development. Photovoltaic cell assembling produces some perilous materials that should be taken care of cautiously to limit the dangers of presentation to people and the earth [17]. Solar panel construction requires uncommon metals, which prompts mining impacts on untamed life and water quality.

10.3.3   WIND ENERGY

Wind power has turned into the world's quickest developing innovation for the age of power. Flowing of wind happens because of air convection brought about by the Sun's heating impact and the turn of the earth. Every day and seasonal variations in temperature reliably create wind, delivering wind that is never exhausted. In this manner, the wind is considered a renewable source of energy [20]. Wind power plants use large sharp blades to capture the kinetic energy in moving air, which is then moved to turbines, thus creating power. According to the studies, the best site for wind power plants includes normal breeze paces of higher than 16 km/h^{-1}. Although wind power plants do not produce air pollution and do not impact the land, there are some environmental problems caused by them. Three types of environmental impacts can be expected when using wind power plants:

1. On the natural landscape, windmills are viewed as an intrusion.
2. Near the wind plants, wind turbines produce noise that cannot be eliminated completely.
3. Wind turbines are very dangerous for birds. Therefore, to avoid possible bird kills, wind plants are located away from migratory bird routes.

10.3.4   BIOMASS ENERGY

Biomass basically means "organic waste," which is obtained from waste materials, for example, wood, wood squander from assembling exercises, farming squanders and buildups, and civil squanders. It is considered a renewable source of energy since it is obtained from organic waste. They can either be utilized as a solid fuel or can be changed over into fluids or gases to create electric power, heat, or chemicals. Some consume biomass fuel straightforwardly in boilers that supply steam for generators, and most of the biomass power plants consume stumble, farming waste, or wood squander from development and destruction, and this is called direct combustion [21]. Biomass gasification is the conversion of biomass into gases, such as CH_4, to fuel steam generators, burning turbines, consolidated cycle innovations, or energy components.

CH_4 caught from human or animal wastes can be utilized for power generation or for heating fuel. Since biomass advancements use the combustion process to create power, in this manner, they can produce power whenever it is possible [22]. The two most basic sorts of biofuels are bioethanol produced using starches and biodiesel produced using fats or oils. Ethanol, which is one of the significant types of biofuels, is commercially accessible and contributes to sustainable household transportation powers in certain nations.

Besides these advantages, biomass can also be hazardous for the environment. Some of them are listed below:

1. SO_2, which is one of the sources that contributes to acid rain, is emitted from wood in a very small quantity.
2. Due to high nitrogen content in some biomass fuels, there are high NO_x emissions from some biomass power plants after combustion of fuel.
3. Carbon monoxide (CO) is emitted from biomass power plants, sometimes at levels higher than coal plants.
4. The primary GHG responsible for climate change, that is, CO_2, is also produced from biofuels.

The level of CO_2 in the atmosphere will not increase significantly because the cycle of growing, processing, and burning biomass, recycles the CO_2 from the atmosphere.

10.4 GLOBAL ENERGY CRISIS AND SCOPE FOR SUSTAINABLE ENERGY PRODUCTION

As indicated by the United Nations [23], sustainability is characterized as "meeting the needs of the present without compromising the ability of future generations to meet their own needs." Energy sustainability can be accomplished through improved energy conservation and effectiveness. The double challenges of energy and climate, which are the issues of ozone-harming substance outflow and the exhaustion of the nonrenewable energy sources, point to a solitary term, "sustainability." In any case, the current reliance on oil and other petroleum derivatives has expanded outflow of CO_2 in the environment, which is certainly not a feasible action. CO_2 discharge is developing by 22% every decade, which will be a long way past the capacity of the climate to ingest.

According to the laws of thermodynamics, "Energy can be neither created nor destroyed, but can only be transformed from one form to another." Around 80% of worldwide energy is provided by coal, oil, and gas. There are gigantic difficulties in utilizing sustainable power sources. Suitable innovations are still a work in progress, and start-up expenses are still very high. In the event that everybody in creating nations utilized a similar measure of energy as the normal shopper in created nations do, building up the world's vitality utilization would expand more than eightfold somewhere in the range of 2000 and 2050. A few analysts propose that oil wells may be drained inside 70–80 years, natural gas may run out somewhat later, and the present stores of uranium may be satisfactory for just 80–90 years [24]. Nevertheless, the feelings of dread of energy running out may be founded on the present financial models. While there are issues with pollution from coal, the power generation aspect from it is likely to continue at least for the next decade.

10.4.1 ROLE OF HYDROGEN IN SUSTAINABLE ENERGY GENERATION

Hydrogen is the simplest element whose atom comprises just a single proton and one electron. It is the most abundant component known to humankind, which doesn't happen normally as a gas on Earth, and it's constantly combined with different elements. Water (H_2O), for instance, is a mixture of hydrogen and oxygen. Right now, most hydrogen is made along these lines from natural gas [25]. Electrolysis is the process in which an electrical current can be utilized to separate water into the segments of oxygen and hydrogen. Some green growth and microbes, utilizing daylight as their energy source, even radiate hydrogen under specific conditions. Yet an engine that burns pure hydrogen creates no contamination, even if hydrogen is high in energy. Since the 1970s, NASA has utilized liquid hydrogen to move the space transport and different rockets into space [26]. Hydrogen can be utilized in fuel cells to create power, utilizing a chemical reaction as opposed to combustion, delivering just water heat as by-products. It tends to be utilized in vehicles, houses, for convenient power, and in a lot more applications.

Hydrogen is a zero-emission fuel that is presently utilized copiously for sustainable energy generation. It tends to be utilized in fuel cells or internal combustion engines to power vehicles or electric gadgets. It has started to be utilized in commercial fuel cell vehicles, for example, traveler autos, and has been utilized in fuel cell buses for a long time [27]. It is likewise utilized as a fuel for the impetus of the rocket. Hydrogen (H_2) combines with oxygen (O_2) to form water (H_2O) and releases energy in a flame of pure hydrogen gas, consuming in air,

$$2H_2(g) + O_2(g) \rightarrow 2H_2O(g) + \text{Energy} \tag{10.1}$$

The energy discharged empowers hydrogen to go about as a fuel. In an electrochemical cell, that energy can be utilized with moderately high efficiency. Like electricity, hydrogen can be considered as a perfect energy carrier. In the long run, hydrogen will lessen the reliance on foreign oil and the emission of ozone harming substances and various other pollutants. As compared to any other normal fuel by weight, hydrogen has the highest energy content and also has the lowest energy content by volume [28]. Hydrogen is considered an energy carrier. In a structure, energy carriers are utilized to move, store, and convey the energy that can be effectively utilized. Electricity is the most notable case of an energy carrier. Hydrogen as a significant energy carrier, later on, has various preferences. For instance, an enormous volume of hydrogen can be effectively stored in various ways. Where it is hard to utilize electricity, hydrogen acts as a highly efficient and low-polluting fuel that can be utilized for various applications, such as, transportation, heating, and power generation in places. In certain examples, it is less expensive to dispatch hydrogen by pipeline than sending power over long separations by wire (Table 10.1).

TABLE 10.1
Evolution of Fuel Cells from Concept to Applications Over the Years

- 1801—The principle of fuel cells was demonstrated by Humphry Davy.
- 1839—The first fuel cell, that is, the "gas battery" was invented by William Grove.
- 1889—Grove's invention and name of the fuel cell was developed by Charles Langer and Ludwig Mond.
- 1950s—Polymer electric membrane fuel cell gets invented by General Electric.
- 1959—A 5 kW alkaline fuel cell is demonstrated by Francis Bacon.
- 1960s—In space missions, NASA uses its first fuel cell.
- 1970s—The oil crisis prompts the development of alternative energy technologies including phosphoric acid fuel cell (PAFC).
- 1980s—In submarines, US Navy uses fuel cells.
- 1990s—For commercial and industrial locations, large stationary fuel cells are developed.
- 2007—As auxiliary power unit (APU) and for stationary backup power, fuel cells begin to be sold commercially.
- 2008—Honda begins leasing the FCX Clarity fuel cell electric vehicle.
- **2009**—Residential fuel cell.

10.4.2 IMPORTANCE OF FUEL CELLS FOR FUTURE ENERGY ECONOMY

The cost of petroleum product is increasing gradually as a result of the absence of availability. The power framework businesses are rebuilding to sustainable power source-based power generation as a substitute arrangement. By thinking about environmental factors, the fuel-cell-based energy generation is the most appropriate inexhaustible framework than sun-based and wind-energy systems. As of late, fuel cells are rapidly created and commercially accessible with high-, medium-, and low-power range applications [29]. So as to decrease the expense of fuel cells, scientists have been engaged to improve the unwavering quality and efficiency of fuel-cell-based power systems. The history of the fuel cell is subsequently discussed.

An electrochemical cell that converts the chemical energy of a fuel (frequently hydrogen) and an oxidizing agent (regularly oxygen) into electricity through a chemical reaction is known as a fuel cell. Fuel cells can create power persistently as long as fuel and oxygen are provided to it, which is then utilized in various areas, including transportation, material handling, stationary, portable, and emergency backup power applications. In many power plants and traveler vehicles, as of now, fuel cells have few advantages over conventional combustion-based technologies. They work at much higher efficiencies than combustion engines and convert the chemical energy of the fuel into electrical energy with efficiencies of up to 60%, and they also have lower emissions than combustion engines [30]. Hydrogen fuel cells discharge just water as a by-product, so there is no emission of carbon dioxide and air pollutants that create smog and cause medical issues at the point of operation. Additionally, fuel cells are quiet during operation because they have fewer moving parts.

10.4.3 Components of the Fuel Cell

Fuel cells consist of an anode, a cathode, and an electrolyte that allows positively charged hydrogen ions or protons to move between the two sides of the fuel cell [31]. A catalyst causes the fuel to undergo oxidation reactions at the anode that generates protons (positively charged hydrogen ions) and electrons. After the reaction, the protons flow from the anode to the cathode through the electrolyte. At the same time, electrons are drawn from the anode to the cathode through an external circuit, producing direct current electricity. Another catalyst, at the cathode, causes hydrogen ions, electrons, and oxygen to react, forming water. Further, it satisfies the essential capacity of generating power, which could be utilized to control something as straightforward as a light or a whole city [32]. A basic synthetic response, which happens inside an energy unit, is in charge of the number of electrons, which would eventually return back to the cell, so as to finish the electric circuit.

Anode and Cathode. Materials that have high electron conductivity and zero proton conductivity in the form of porous catalyst (porous catalyst or carbon). Usually, carbon-based electrodes are used, with platinum as a catalyst on the cathode.

At Anode:

$$2H_2 \rightarrow 4H^+ + 4e^- \qquad (10.2)$$

At Cathode:

$$O_2 + 4H^+ + 4e^- \rightarrow 2H_2O \qquad (10.3)$$

Overall Reaction:

$$2H_2 + O_2 \rightarrow 2H_2O \qquad (10.4)$$

Electrolyte. High proton conductivity and zero electron conductivity are the required feature of any electrolyte in such systems. In the whole process, the electrolyte assumes a critical task in keeping up only the correct extent of particles to go between the anode and the cathode [33]. Other free ionic migrations could upset the progressing electrochemical response in the cell.

Catalyst. Mostly platinum is used as a catalyst for the hydrogen oxidation reaction occurring at the anode and the oxygen reduction reaction at the cathode [34]. Usually, the platinum catalyst takes the form of small particles on the surface of electrodes, that is, the carbon particles that act as supports.

10.4.4 TYPES OF FUEL CELLS AND APPLICATION AREA

1. Alkaline Fuel Cell (Figure 10.1)

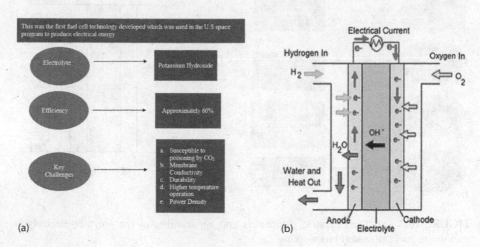

(a) (b)

FIGURE 10.1 (a) The operating parameters and applicability of the alkaline fuel cell, and (b) its working.

2. Solid Oxide Fuel Cell (Figure 10.2)

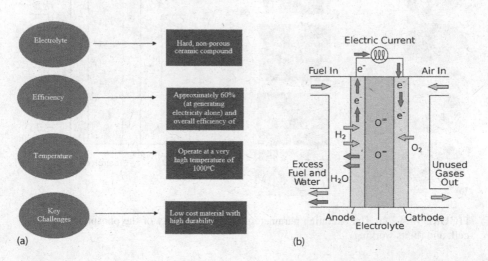

(a) (b)

FIGURE 10.2 (a) The operating parameters and applicability of the solid oxide fuel cell, and (b) its working.

3. Polymer Electrolyte Membrane Fuel Cell (Figure 10.3)

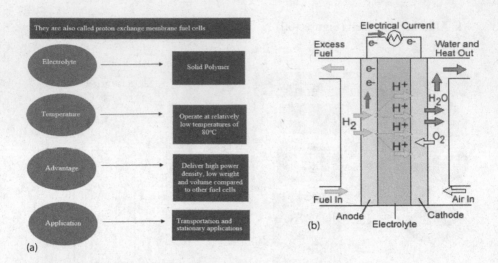

(a)

(b)

FIGURE 10.3   (a) The operating parameters and applicability of the polymer electrolyte membrane fuel cell, and (b) its working.

4. Phosphoric Acid Fuel Cell (Figure 10.4)

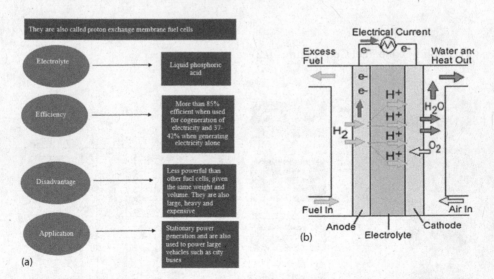

(a)

(b)

FIGURE 10.4   (a) The operating parameters and applicability of the phosphoric acid fuel cell, and (b) its working.

5. Molten Carbonate Fuel Cell (Figure 10.5)

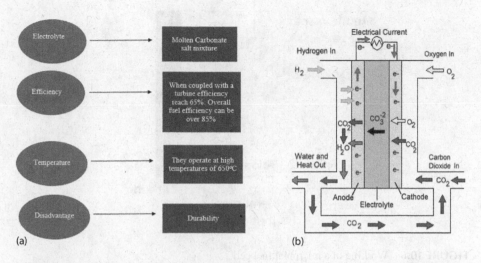

(a)   (b)

FIGURE 10.5   (a) The operating parameters and applicability of the molten carbonate fuel cell, and (b) its working.

6. Microbial Fuel Cell

It is a bioelectrochemical device that harnesses the power of respiring microbes to convert organic substrates directly into electrical energy [35]. It has the following characteristics:

- Transforming chemical energy into electricity using oxidation-reduction reactions.
- It consists of an anode, a cathode, a proton, or cation exchange membrane and an electrical circuit. The anode respiring bacteria cling to the anode of the microbial fuel cell (MFC).
- These bacteria strip electrons from organic waste in the course of their metabolic activity.
- Electricity is then produced in the process in addition to CO_2 and water when electrons flow through a circuit to the cathode.
- From anode hydroxide, or OH^-, ions are transported into the surrounding electrolyte (Figure 10.6).

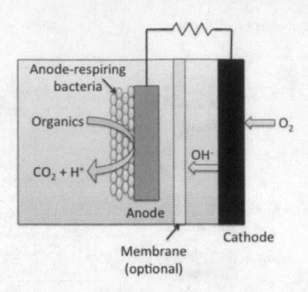

FIGURE 10.6 Working of a microbial fuel cell.

10.5   ISSUES WITH THE HYDROGEN ECONOMY AND FUEL CELLS

In 2018, over one million electric vehicles were sold worldwide. The number of Nissan Leaf, Tesla, and hydrogen fuel cell vehicles such as the Honda FCX Clarity is now in circulation of over one million. While there are many brands of e-vehicles to choose from, there are only two choices when it comes to powering these vehicles, either batteries or fuel cells. Both produce electricity to drive electric motors, eliminating the pollution and inefficiency of the fossil-fuel-powered internal combustion engine. The energy needed by the vehicles can be produced from zero-carbon sources, such as wind and solar, hence both are being pursued by car manufacturers and researchers as the possible future of electric vehicles; however, a great debate is waged across several communities over which power source to choose [36]. For example, Elon Musk (of Tesla) has claimed that hydrogen fuel cells are more of a marketing ploy for auto makers, and not a long-term strategy; in contrast, Japan has announced its ambitious plan to become the first hydrogen society by 2022.

Although hydrogen is the clear winner on the face of such automobiles, it falls behind when it comes to the end-to-end production process. For instance, fully charging a Tesla Model 3 battery with a 75 kWh battery costs between $10 and $12 with a rated range of 500 km, which makes it around 2–2.4 cents per kilometer. Now with a fuel cell vehicle, say a Toyota Mirai, the hydrogen from filling stations costs around $80–$85 to fill a 5 kg tank, which has a range of 480 km, which comes to around 17.7 cents per kilometer. This is eight times the price as compared to batteries, and here lies the problem [30]. Hydrogen production is an energy-intensive process. To understand the economic viability, we need to see the production process. Before any hydrogen vehicle can hit the road, the hydrogen needs to be available. Although it is the most abundant element in the universe, it is not a readily available source of energy. Hydrogen is usually stored in water, hydrocarbons such as methane, and other organic matter. The process of extracting it from these compounds is one of the main challenges of using hydrogen as an energy storage mechanism. Primarily, hydrogen is produced in refineries by a process called steam reforming, where steam reacts with hydrocarbons at high pressure and temperature. While hydrogen fuel cells do not create pollution, the process of steam reforming does; so if we were to assume a future scenario where negative carbon emissions are to be considered, this method will not be taken into account. Another method to produce hydrogen is electrolysis—separating the hydrogen from water using an electric current. While the electricity needed in this process can be taken from renewable energy sources, it requires more energy input than steam reforming.

Additionally, the transport and storage of hydrogen is a concern, as it has a high gas density. That means, in order to utilize it effectively as a gas, it has to be stored in highly pressurized containers, or at cryogenic temperatures for the liquid, both of which are expensive processes. Although we can be critical about hydrogen, we must remember comparing the current battery electric vehicle costs per kilometer with the current hydrogen fuel cell electric vehicle is at present not entirely justifiable, because they are at different places in their development. While the cost per kilometer for hydrogen is more, projected costs for larger stations with higher utilization are much lower [37]. Most of the costs of delivered H_2 occur at the refueling station from equipment located there, especially compressors. Work is being done to avoid this cost, for instance, searching for low-pressure storage methods that do not require refrigeration.

Further, in considering battery electric vehicles, one must consider what a scaled-up system looks like, one in which so many vehicles are being charged that it disrupts the grid architecture, especially near the ends of the power network. For example, if a single vehicle requires 25 kW when charging, a street with 40 homes could require an additional megawatt of capacity since it is not unreasonable to think people come home and plug in at comparable times. Also, as you move upstream in the distribution network, these loads aggregate. While current over capacity handles the existing electric vehicle charging, future scenarios must consider significant changes in the power grid, and this should be factored into the cost assessment. One must take great care to ensure that the two things being considered to have comparable system boundaries. The example of not factoring in grid upgrade costs is an example.

In hindsight, it is critical to optimize with respect to the right parameter. Energy is not the limiting factor, capital is. It is important to realize that the broad problem being analyzed is the electrification of not just transportation (which is about one-third of our emissions) but electrification of industry and heating. Achieving this with renewables faces challenges with energy storage. When you consider hydrogen in this context, it may offer grid services that facilitate the build-out of high levels of renewable power, and the cost of those services (voltage regulation, for example, as well as storage) should be compared in a hydrogen scenario vs. a battery scenario.

10.6 WORLDWIDE IMPLEMENTATION OF FUEL-CELL-BASED SYSTEMS: PERSPECTIVES AND OUTLOOK

Hydrogen advancements have encountered cycles of excessive expectations pursued by disillusions. In any case, a developing assortment of evidence proposes these advances form an appealing alternative for the deep decarbonization of worldwide energy systems and that ongoing upgrades in their expense and performance point toward economic viability too. The US Department of Energy is working intimately with its national research laboratories, colleges, and industry accomplices to beat basic specialized boundaries to fuel cell development (Figures 10.7 and 10.8, Table 10.2).

The best example for fuel cell technology so far is Japan, which has kept with it on residential fuel cells and pushed the business on hydrogen generation, shipment, refueling infrastructure, and vehicle deployment; Korea, which turned into the main market for fuel cell power generation by including fuel cells as an inexhaustible innovation; California, which keeps on financing infrastructure development and activity to push businesses to develop heavier vehicles; Germany, making efforts for renewables has made choices around green hydrogen; and China, where fuel

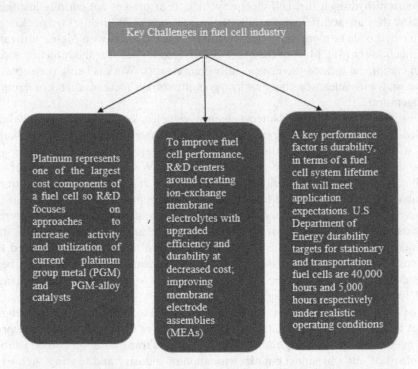

FIGURE 10.7   Representation of key challenges faced by the fuel cell industry.

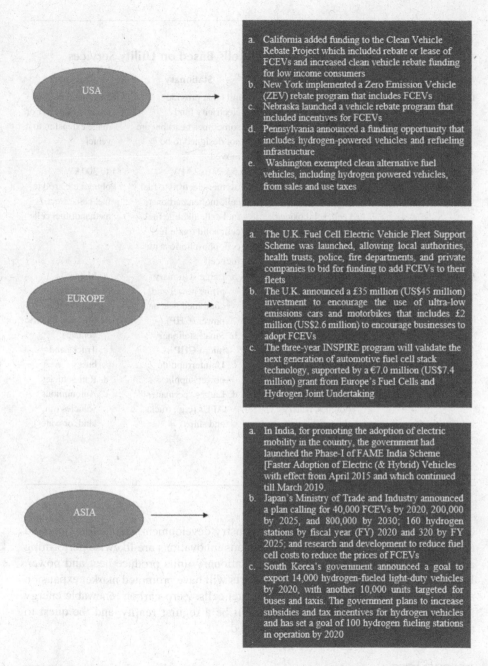

FIGURE 10.8 List of policies that supported fuel cell electric vehicles and hydrogen infrastructure development since 2016.

TABLE 10.2

Applications of Different Types of Fuel Cells Based on Utility Services

Application Type	Portable	Stationary	Transport
Definition	Units that are built into, or charge up, products that are designed to be moved, including small APUs	Units that provide electricity (and sometimes heat) but are not designed to be moved	Units that provide propulsive power or range extension to a vehicle
Typical potier range	1 W to 20 kW	0.5 kW to 2 MW	1 to 300 kW
Typical technology	Polymer electrolyte fuel cell, direct methanol fuel cell, solid oxide fuel cell	Polymer electrolyte fuel cell, molten carbonate fuel cell, alkaline fuel cell, solid oxide fuel cell, phosphoric acid fuel cell	Polymer electrolyte fuel cell, direct methanol fuel cell
Example	a. Small "movable" APUs (campervans, boats, lighting) b. Military applications (portable soldier-borne power, skid-mounted generators) c. Portable products (torches, battery chargers), small personal electronics (mp3 player, cameras)	a. Large stationary prime power and combined heat and power (CHP) b. Small stationary micro-CHP c. Uninterruptible power supplies d. Larger "permanent" APUs (e.g., trucks and ships)	a. Materials handling vehicles b. Fuel cell electric vehicles c. Trucks and buses d. Rail vehicles e. Autonomous vehicles (air, land, or water)

cells fit into its five-year plan high-tech industry development, smog reduction, and reduction in carbon intensity [38]. The various innovations are likewise performing admirably—transports and vehicles run, stationary units produce heat and power. In this way, by 2050, it is expected that efforts will have prompted market expansion and zero-emission transport, powered by fuel cells. Zero-carbon renewable energy from fuel cell generators and hydrogen will be a regular reality, and the quest to achieve a greener world will be accomplished.

REFERENCES

1. Dunn, B., Kamath, H. and Tarascon, J.M., 2011. Electrical energy storage for the grid: A battery of choices. *Science*, *334*(6058), pp. 928–935.
2. Geng, Y., Ji, W., Wang, Z., Lin, B. and Zhu, Y., 2019. A review of operating performance in green buildings: Energy use, indoor environmental quality and occupant satisfaction. *Energy and Buildings*, *183*, pp. 500–514.

3. Urry, J., 2015. Climate change and society. In *Why the social sciences matter*, edited by Cooper, C.L. and Michie, J. (pp. 45–59). Palgrave Macmillan, London.
4. Bose, D., Kandpal, V., Dhawan, H., Vijay, P. and Gopinath, M., 2018. Energy recovery with microbial fuel cells: Bioremediation and bioelectricity. In *Wastebioremediation*, edited by Varjani, S.J., Gnansounou, E., Gurunathan, B., Pant, D., and Zakaria, Z.A. (pp. 7–33). Springer, Singapore.
5. Bose, D. and Bose, A., 2017. Graphene-based microbial fuel cell studies with starch in sub-Himalayan soils. *Indonesian Journal of Electrical Engineering and Informatics (IJEEI)*, *5*(1), pp. 16–21.
6. Al-Awadhi, H., Al-Mailem, D., Dashti, N., Khanafer, M. and Radwan, S., 2012. Indigenous hydrocarbon-utilizing bacterioflora in oil-polluted habitats in Kuwait, two decades after the greatest man-made oil spill. *Archives of Microbiology*, *194*(8), pp. 689–705.
7. Bose, D., 2015. Design parameters for a hydro desulfurization (HDS) unit for petroleum naphtha at 3500 barrels per day. *World Scientific News*, *9*, pp. 99–111.
8. Ahmadov, R., McKeen, S., Trainer, M., Banta, R., Brewer, A., Brown, S., Edwards, P.M. et al., 2015. Understanding high wintertime ozone pollution events in an oil-and natural gas-producing region of the western US. *Atmospheric Chemistry and Physics*, *15*(1), pp. 411–429.
9. Cesur, R., Tekin, E. and Ulker, A., 2016. Air pollution and infant mortality: Evidence from the expansion of natural gas infrastructure. *The Economic Journal*, *127*(600), pp. 330–362.
10. Rahman, F.A., Aziz, M.M.A., Saidur, R., Bakar, W.A.W.A., Hainin, M.R., Putrajaya, R. and Hassan, N.A., 2017. Pollution to solution: Capture and sequestration of carbon dioxide (CO_2) and its utilization as a renewable energy source for a sustainable future. *Renewable and Sustainable Energy Reviews*, *71*, pp. 112–126.
11. Perera, F.P., 2016. Multiple threats to child health from fossil fuel combustion: Impacts of air pollution and climate change. *Environmental Health Perspectives*, *125*(2), pp. 141–148.
12. Vassilev, S.V., Kitano, K., Takeda, S. and Tsurue, T., 1995. Influence of mineral and chemical composition of coal ashes on their fusibility. *Fuel Processing Technology*, *45*(1), pp. 27–51.
13. Whitehurst, D.D., Mitchell, T.O. and Farcasiu, M., 1980. *Coal liquefaction: The chemistry and technology of thermal processes.* Academic Press, Inc., New York, 390 p.
14. Rowe, C.L., Hopkins, W.A. and Congdon, J.D., 2002. Ecotoxicological implications of aquatic disposal of coal combustion residues in the United States: A review. *Environmental Monitoring and Assessment*, *80*(3), pp. 207–276.
15. Flores, R.M., Rice, C.A., Stricker, G.D., Warden, A. and Ellis, M.S., 2008. Methanogenic pathways of coal-bed gas in the Powder River Basin, United States: The geologic factor. *International Journal of Coal Geology*, *76*(1–2), pp. 52–75.
16. Zinkle, S.J. and Was, G.S., 2013. Materials challenges in nuclear energy. *Acta Materialia*, *61*(3), pp. 735–758.
17. Whicker, F.W. and Schultz, V., 1982. *Radioecology: Nuclear energy and the environment* (Vol. 2). CRC Press, Boca Raton, FL.
18. Deane, J.P., Gallachóir, B.Ó. and McKeogh, E.J., 2010. Techno-economic review of existing and new pumped hydro energy storage plant. *Renewable and Sustainable Energy Reviews*, *14*(4), pp. 1293–1302.
19. Yang, C.J. and Jackson, R.B., 2011. Opportunities and barriers to pumped-hydro energy storage in the United States. *Renewable and Sustainable Energy Reviews*, *15*(1), pp. 839–844.
20. Boyle, G., 2004. *Renewable energy*, edited by Boyle, G. (p. 456). Oxford University Press, Oxford, May 2004.

21. Kapoor, L., Bose, D. and Mekala, A., 2020. Biomass pyrolysis in a twin-screw reactor to produce green fuels. *Biofuels*, *11*(1), pp. 101–107.

22. Kapoor, L., Mekala, A. and Bose, D., 2016, November. Auger reactor for biomass fast pyrolysis: Design and operation. In *2016 21st Century Energy Needs-Materials, Systems and Applications (ICTFCEN)* (pp. 1–6). IEEE.

23. Kates, R.W., Clark, W.C., Corell, R., Hall, J.M., Jaeger, C.C., Lowe, I., McCarthy, J.J. et al., 2001. Sustainability science. *Science*, *292*(5517), pp. 641–642.

24. Clark, W.C. and Dickson, N.M., 2003. Sustainability science: The emerging research program. *Proceedings of the National Academy of Sciences*, *100*(14), pp. 8059–8061.

25. Bose, D., 2015. Fuel cells: The fuel for tomorrow. *Journal of Energy Research and Environmental Technology*, *2*(2), pp. 71–75.

26. Farooque, M. and Maru, H.C., 2001. Fuel cells-the clean and efficient power generators. *Proceedings of the IEEE*, *89*(12), pp. 1819–1829.

27. Cook, B., 2002. Introduction to fuel cells and hydrogen technology. *Engineering Science & Education Journal*, *11*(6), pp. 205–216.

28. Momirlan, M. and Veziroglu, T.N., 2005. The properties of hydrogen as fuel tomorrow in sustainable energy system for a cleaner planet. *International Journal of Hydrogen Energy*, *30*(7), pp. 795–802.

29. Sieniutycz, S. and Jezowski, J., 2018. *Energy optimization in process systems and fuel cells*. Elsevier, Amsterdam.

30. Cano, Z.P., Banham, D., Ye, S., Hintennach, A., Lu, J., Fowler, M. and Chen, Z., 2018. Batteries and fuel cells for emerging electric vehicle markets. *Nature Energy*, *3*(4), p. 279.

31. Bose, D., Gopinath, M. and Vijay, P., 2018. Sustainable power generation from wastewater sources using Microbial Fuel Cell. *Biofuels, Bioproducts and Biorefining*, *12*(4), pp. 559–576.

32. Edwards, P.P., Kuznetsov, V.L., David, W.I. and Brandon, N.P., 2008. Hydrogen and fuel cells: Towards a sustainable energy future. *Energy Policy*, *36*(12), pp. 4356–4362.

33. Bashyam, R. and Zelenay, P., 2011. A class of non-precious metal composite catalysts for fuel cells. In *Materials for sustainable energy: A collection of peer-reviewed research and review articles from Nature Publishing Group*, edited by Dusastre, V. (pp. 247–250). Nature Publishing Group, London.

34. Bose, D., Sridharan, S., Dhawan, H., Vijay, P. and Gopinath, M., 2019. Biomass derived activated carbon cathode performance for sustainable power generation from microbial fuel cells. *Fuel*, *236*, pp. 325–337.

35. Bose, D., Dhawan, H., Kandpal, V., Vijay, P. and Gopinath, M., 2018. Bioelectricity generation from sewage and wastewater treatment using two-chambered microbial fuel cell. *International Journal of Energy Research*, *42*(14), pp. 4335–4344.

36. Whittingham, M.S., Savinell, R.F. and Zawodzinski, T., 2004. Introduction: Batteries and fuel cells. *Chemical Reviews*, *104*, 4243–4244.

37. Gröger, O., Gasteiger, H.A. and Suchsland, J.P., 2015. Electromobility: Batteries or fuel cells? *Journal of the Electrochemical Society*, *162*(14), pp. A2605–A2622.

38. Das, V., Padmanaban, S., Venkitusamy, K., Selvamuthukumaran, R., Blaabjerg, F. and Siano, P., 2017. Recent advances and challenges of fuel cell based power system architectures and control–A review. *Renewable and Sustainable Energy Reviews*, *73*, pp. 10–18.

11 Solid Oxide Fuel Cells
Opportunities for a Clean Energy Future

Anand Singh and Prashant Baredar

CONTENTS

11.1 INTRODUCTION

Solid oxide fuel cells (SOFCs) are electrochemical gadgets that convert the compound vitality of a fuel and oxidant legitimately into electrical vitality. Since SOFCs produce power through an electrochemical response and not through a burning procedure, they are considerably more proficient and naturally more kindhearted than traditional electric power age forms. Their inborn attributes make them exceptionally reasonable to address the ecological, environmental change, and water concerns related to non-renewable energy-source-based electric power age. The SOFC works at exceptionally high temperatures, the most noteworthy of all the power device types at around 800°C to 1000°C. They can have efficiencies of over 60% when changing over fuel to power. If the warmth they delivered is additionally tackled, their general effectiveness in changing over fuel to vitality can be over 80%. SOFCs utilize a strong clay electrolyte, for example, zirconium oxide balanced out with yttrium oxide, rather than a fluid or layer. Their high working temperature implies that powers can be changed inside the energy component itself, disposing of the requirement for outer transforming and enabling the units to be utilized with an assortment of hydrocarbon energizes (Li et al. 2012). They are likewise generally impervious to little amounts of sulfur in the fuel, contrasted with different kinds of power devices, and can consequently be utilized with coal gas. Figure 11.1. shows the schematic diagram of (a) proton ion and (b) oxygen ion transport processes in an SOFC. A further favorable position of the high working temperature is that the response energy is improved, evacuating the requirement for a metal impetus. There are a few weaknesses to the high temperature: these cells take more time to

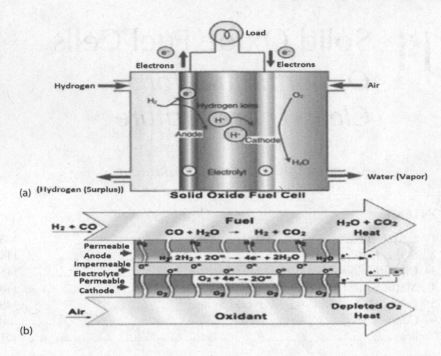

FIGURE 11.1 Schematic diagram of (a) proton ion and (b) oxygen ion transport processes in an SOFC.

fire up and achieve working temperature; they should be developed of strong, heat-safe materials; and they should be protected to forestall heat misfortune. There are three distinctive SOFC geometries of SOFC: planar, coplanar, and miniaturized scale rounded (Lu et al. 2017). In the planar structure, segments are amassed in level stacks where the air and hydrogen customarily stream; however, the unit through diverts worked into the anode and cathode. In the cylindrical structure, the air is provided within an all-inclusive strong oxide tube (which is fixed toward one side) while fuel streams around the outside of the cylinder. The cylinder itself frames the cathode, and the cell parts are built in layers around the cylinder. SOFCs are utilized widely in huge and little stationary power age: planar sorts discover application in, for instance, Bloom Energy's 100 kW off-matrix control generators and SOFCs with produce of a combine of kilowatts are being tried for smaller cogeneration applications, for example, residential combined heat and power (CHP). Smaller-scale rounded SOFCs with yield in the watt range are additionally being produced for little compact chargers (Zhang et al. 2014).

11.2 SOFC HISTORY

SOFC is a group of electrochemical gadgets that produce power by advancing a redox response over an ionically conductive layer. Despite the fact that energy components were first announced in 1839 by Sir William Grove, it was not until 1961, when NASA started Project Gemini, that they found their first pragmatic application (Barelli et al. 2013). Power modules are normally named regarding two key attributes: the portable particle and the electrolyte material, with the working temperature additionally being utilized to subclassify sometimes. SOFCs are named after their particle leading, earthenware oxide electrolyte, and their history is attached to a portion of the extraordinary names in science and designing. Faraday's initial examinations of conduction in earthenware production during the 1830s drove him to arrange conductors into two classifications, even though the careful instrument for these two methods of conduction was obscure. It was not until some other time, during the 1890s, when Walther Nernst watched the altogether expanded conductivity of blended oxides over their unadulterated constituents that the primary innovative ramifications of particle conduction in solids were considered. Albeit at last not a business achievement, due to some extent to its mind-boggling expense, the "Nernst Glower" was about twice as proficient as the carbon fiber lights of the day (Yang et al. 2014). The gadget comprised an earthenware oxide bar made of yttria-doped zirconia (regularly alluded to as the "Nernst Mass") which, after preheating to around 1000°C, would start to direct under a burden; this prompted the temperature expanding further, making the pole gleam. Despite the fact that the framework was a disappointment because of high Ohmic misfortunes, it impelled another flood of examination concerning leading blended oxides. Over the accompanying 30 years, an efficient examination concerning ion-conducting anode materials so as to discover structures that had both the mechanical and electrochemical properties required for a strong power device. By 1970, the reception of electro-earthenware production for a wide scope of other mechanically pertinent applications, for example, sensors (e.g., lambda sensors that are generally utilized today to gauge the air/fuel proportion in motor fumes gases) and oxygen detachment films, prompted key advances in materials handling and the materials store network. Other related advances, for instance, in the semiconductor business, brought about procedures rising, for example, electrochemical vapor testimony. This took into consideration a lot slenderer layers of high immaculateness material to be saved, which could not just lessen Ohmic misfortunes, yet in addition, opened the likelihood of utilizing materials recently regarded excessively expensive. Following the first and second oil emergencies of the 1970s, which aggregately prompted a 10-overlay increment in the cost of oil, governments from fuel-producing countries started to put all the more vigorously in the innovative work of elective vitality advances. Since the mid-1990s, a succession of SOFC organizations, transcendently from the United States, Western Europe, and Japan, have risen going for putting up a scope of SOFC designs for sale to the public. Regular to these applications is the need for the gadgets to work for broadened periods (5 to 10 years) without requiring huge upkeep or substitution. It is likewise basic for the cells, stacks, and frameworks to have the option to withstand the inescapable shutdown occasions, which represents a specific issue for SOFCs because of their high working temperature and fragile clay segments. Best in class, SOFC gadgets would already be able to accomplish electrical efficiencies of above half and joined warmth and power frameworks that exist with all-out efficiencies in an overabundance of 90%. These two measurements are extremely noteworthy all alone, yet in blend with the absence of NOx/SOx or particulates in the fumes stream and the low commotion/

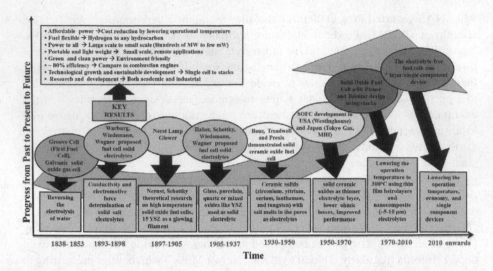

FIGURE 11.2 Timeline graph representing the progress of SOFC over the last 17 decades.

vibration of these frameworks, the intrigue of SOFC gadgets is clear. In any case, SOFCs won't most likely completely convey on their potential until the debasement issues key to lifetime is settled, which is the subject of this section (Al-Sulaiman et al. 2010). Figure 11.2 shows the timeline graph representing the progress of SOFCs over the last 17 decades.

11.3 CLASSIFICATION OF FUEL CELLS

An assortment of power modules is in various phases of advancement. They can be ordered by utilization of different classifications, contingent upon the mix of kinds of fuel and oxidant, regardless of whether the energy component is handled outside (outer improving) or inside (inner transforming) the cell. The types of electrolyte temperature of activity, regardless of whether the reactants are boosted to the cell by inner or external manifolds (Fong and Lee 2014). The most normal order of power modules is by the sort of electrolyte utilized in the cell that appears in Table 11.1.

11.4 MATHEMATICAL MODELING OF FUEL CELL

The fuel cell is a static device that converts the chemical energy of a fuel (hydrogen) and an oxidant (air or oxygen) into electrical energy. A simplified mathematical model is shown in Figure 11.3. It represents a particular fuel cell stack operating at nominal conditions of temperature and pressure. The parameters of the equivalent circuit can be modified based on the polarization curve. A diode is used to prevent the flow of negative current into the stack (Singh et al. 2017).

The principal components that influence the polarization bend are cathode weight, reactant halfway weight, cell temperature, and film moistness (Ji et al. 2017).

TABLE 11.1
Types of Fuel Cells

Fuel Cell →	AFC	PEMFC	PAFC	MCFC	SOFC
Electrolyte	Aqueous potassium hydroxide	Sulfonated organic polymer	Phosphoric acid	Molten lithium/sodium potassium carbonate	Yttria stabilized zirconia
Operating temperature	60°C–90°C	70°C–100°C	150°C–220°C	600°C–700°C	650°C–1000°C
Charge carrier	OH^-	H^+	H^+	CO_3^{2-}	O^{2-}
Anode	Nickel or precious metal	Platinum	Platinum	Nickel/Chromium oxide	Nickel/Yttria-stabilized zirconia
Cathode	Platinum (Pt) or Lithiated NiO	Platinum	Platinum	Nickel oxide (NiO)	Strontium-doped lanthanum manganite
Co-generation heat	None	Low quality	Acceptable for many applications	High	High
Electrical efficiency	60	40–45	40–45	50–60	50–60
Fuel sources	H_2	H_2	H_2	H_2, CO Natural gas	H_2, CO Natural gas
Application	• Military • Space	• Backup power • Portable power • Distributed generation • Transportation • Specialty vehicles	• Distributed generation	• Electric utility • Distributed generation	• Auxiliary power • Electric utility • Distributed generation

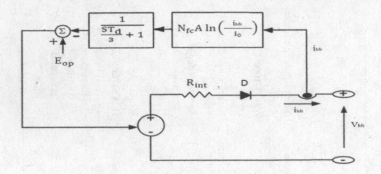

FIGURE 11.3 Electrical equivalent circuit diagram of a fuel cell.

The polarization bend is processed by utilizing the tafel condition, which subtracts the different voltage misfortunes from the open circuit DC voltage.

$$V_{hfc} = E_{op} - \left\{ N_{FC} \times A \times \ln\left(\frac{i_{hfc}}{i_0}\right) \times \frac{1}{\frac{ST_d}{3}+1} \right\} - \left(R_{int} \times i_{hfc}\right) \qquad (11.1)$$

where i_{hfc} is the fuel cell current (A), V_{hfc} is the fuel cell voltage (V), E_{oc} is the open circuit voltage (V), N_{fc} is the number of cells, A is the Tafel slope (V), i_0 is the exchange current (A), ST_d is the response time at 95% of the final value (s), and R_{int} is the internal resistance (ohm).

$$E_{op} = K_c \times E_{Nernst} \qquad (11.2)$$

where K_c is the voltage constant at nominal condition, and E_{Nernst} is the Nernst voltage (V).

$$i_0 = \frac{z \times F \times k \times \left(P_{H_2} + P_{O_2}\right)}{R \times h} \times e^{\left(\frac{-\Delta G}{R \times T}\right)} \qquad (11.3)$$

$$A = \frac{R \times T}{z \times \alpha \times F} \qquad (11.4)$$

where R is the gas constant [8.3145 J/(mol K)], F is the Faraday's constant [96485 A s/mol], z is the number of moving electrons (z = 2), a is the charge transfer coefficient, PP_{H_2} is the partial pressure of hydrogen inside the stack (atm), PP_{O_2} is the partial pressure of oxygen inside the stack (atm), k is the Boltzmann's constant $\left(1.38 \times \frac{10^{-23} \, J}{K}\right)$, h is the Planck's constant $\left(6.626 \times 10^{-34} \, J \, s\right)$, ΔG is the activation energy barrier (J), and T is the temperature of operation (K).

$$UtH_2 = \frac{60000 \times R \times T \times i_{hfc}}{z \times F \times P_{Sfuel} \times V_{FRuel} \times i\%} \qquad (11.5)$$

$$UtO_2 = \frac{60000 \times R \times T \times i_{hfc}}{z \times F \times P_{Sair} \times V_{FRair} \times j\%}$$
(11.6)

where P_{Sfuel} is the supply pressure of fuel (atm), P_{Sair} is the supply pressure of air (atm), V_{FRfuel} is the fuel flow rate (l/min), V_{FRair} is the airflow rate (l/min), i% is the percentage of hydrogen in the fuel (%), and j% is the percentage of oxygen in the oxidant (%).

11.5 POTENTIAL OF A SOFC-BASED HYBRID SYSTEM

Power age and related natural effects have turned out to be significant issues in the world. Today, most electrical power is given by customary power age advances that generally depend on non-renewable energy source burning. Be that as it may, these will incite two genuine dangers: an unnatural weather change and air contamination. To stay away from this circumstance, vitality frameworks must be refreshed, and another power age framework ought to be grown right away. Distributed generation (DG) framework is increasingly being considered worldwide as an elective age source (Hao et al. 2017). Among the few DG sources, the energy component is the most appropriate mixture framework with sustainable and non-inexhaustible power sources. In this recognition, the model of power module dependent on the sort of the co-age framework utilized is basic for the solid and safe activity of the crossover DG framework. The expanding infiltration of DG has a few specialized ramifications and opens significant inquiry as to the conventional ways to deal with tasks and improvement of intensity framework. Among the few energy units, the SOFC is the most encouraging for crossbreed activity because of their high temperature working, control coordinating, and remote application (Kazempoor et al. 2011). These energy components can be utilized to repay the power for the current inexhaustible and non-sustainable power sources. The molten carbonate fuel cells (MCFCs) SOFC can be completely used with the gas turbine age framework in using the fume's warmth and remaining fuel, which is changed over to extra-electrical vitality through the gas turbine. The microturbines are the development of gas innovation and are rising crossbreed DG sources with energy units. The mixture task of SOFC and another renewable source (solar PV, wind, biomass) framework can improve the power quality, framework dependability, and in general the execution of the framework. Among the few half-breed DG frameworks, the crossbreed SOFC and other renewable energies (solar PV, wind, biomass) are most appropriate because of the intensity coordinating, load sharing, and remote application (Mahato et al. 2015). This paper displays a writing survey on crossbreed activity of power devices with customary and non-traditional vitality assets in the utility interconnected mode. From this writing audit, the half and half SOFC and (Solar PV, Wind, Biomass) framework is the most effective and conservative DG power source in network-associated activity. This likewise incorporates the powerful hardware and its application in the utility-associated half-breed DG framework (Yu et al. 2018). A crossover control framework comprises a blend of at least two power age innovations to utilize their working attributes. This can expand the efficiencies higher than that could be achieved from

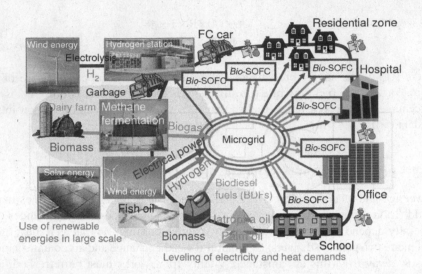

FIGURE 11.4 Microgrid system using Bio-SOFC as a major distributed power source.

a solitary power source. A crossover framework could be intended to work either in the detached mode or in the utility interconnected mode, through a power electronic interface. The advantages of the partial and semi DG innovation over the person of DG framework is, they can increment in the general framework proficiency and unwavering quality, less or no outflow and lessening of in general cost (Gao et al. 2011) (Figure 11.4).

11.6 CONCLUSIONS

The world is plentiful in sustainable power source assets that can satisfy its vitality needs. In any case, at present, all national governments are as yet attempting to overcome its vitality emergency by methods for traditional vitality innovations. A viable and supportable long-haul arrangement must be embraced as far as sustainable power source advancements. It is with this lament that these assets have not been gathered, because of social, monetary, and bureaucratic boundaries. To fortify the sustainable power source advances in the nation, the general population and private segments ought to enormously dedicate interests in sustainable power source advances for making a practical future. SOFC innovation is one of the most encouraging sustainable power source advances because of its compatibilities with a few sustainable power source assets and being without ignition. What's more, the assorted variety of SOFC innovations make it an appropriate possibility to legitimize our future vitality requests and support the advancement of nations. The fuel utilized in SOFC innovation is exceptionally compelling contrasted with traditional energizes. Later on, we will expect that the expense of SOFCs will be to a great extent decreased by the advancement of improved fuel stockpiling systems. Subsequently, SOFC half and half frameworks will establish a more splendid point of view in all nations.

REFERENCES

Al-Sulaiman FA, Dincer I, Hamdullahpur F. Energy analysis of a trigeneration plant based on solid oxide fuel cell and organic Rankine cycle. *Int J Hydrogen Energy* 2010;35(10):5104–13.

Barelli L, Bidini G, Gallorini F, Ottaviano PA. Design optimization of a SOFC-based CHP system through dynamic analysis. *Int J Hydrogen Energy* 2013;38:354–69.

Fong KF, Lee CK. Investigation on zero grid-electricity design strategies of solid oxide fuel cell trigeneration system for high-rise building in hot and humid climate. *Appl Energy* 2014;114:426–33.

Gao Z, Raza R, Zhu B, Mao Z. Development of methanol fueled low-temperature solid oxide fuel cells. *Int J Energy Res* 2011;35:690–6.

Hao SJ, Wang C, Liu TL, Mao ZM, Mao ZQ, Wang JL. Fabrication of nanoscale yttria stabilized zirconia for solid oxide fuel cell. *Int J Hydrogen Energy* 2017;42:29949–59.

Ji Y, Deng YB, Liu ZY, Zhou T, Wu YH, Qian SZ. Optimal control-based inverse determination of electrode distribution for electroosmotic micromixer. *Micromachines* 2017;8:247–53.

Kazempoor P, Dorer V, Weber A. Modeling and evaluation of building integrated SOFC systems. *Int J Hydrogen Energy* 2011;36:13241–9.

Li CX, Yun LL, Zhang Y, Li JC, Guo LJ. Microstructure performance and stability of Ni/ Al_2O_3 cermet supported SOFC operating with coal-based syngas produced using supercritical water. *Int J Hydrogen Energy* 2012;37:13001–6.

Lu YZ, Li JJ, Souamy L, Wang J, Zhang YM, Zhu B. Model analysis on hydrogen production by hybrid system of SOEC and solar energy. *Eng Lett* 2017;25(4):382–8.

Mahato N, Banerjee A, Gupta A, Omar S, Balani K. Progress in material selection for solid oxide fuel cell technology: A review. *Prog Mater Sci* 2015;72:141–337.

Singh A, Baredar P, Gupta B. Techno-economic feasibility analysis of hydrogen fuel cell and solar photovoltaic hybrid renewable energy system for academic research building. *Energy Convers Manag* 2017;145:398–414.

Yang WY, Zhao YR, Liso V, Brandon N. Optimal design and operation of a syngas-fuelled SOFC micro-CHP system for residential applications in different climate zones in China. *Energy Build* 2014;80:613–22.

Yu SC, Zhang GP, Chen H, Guo LC. A novel post-treatment to calcium cobaltite cathode for solid oxide fuel cells. *Int J Hydrogen Energy* 2018;43:2436–42.

Zhang XQ, Wang Y, Liu T, Chen JC. Theoretical basis and performance optimization analysis of a solid oxide fuel cell gas turbine hybrid system with fuel reforming. *Energy Convers Manag* 2014;86:1102–9.

12 Comprehensive Study of District Heating (DH) in the UK
Techno-Economic Aspects, Policy Support, and Trends

Abdur Rehman Mazhar, Shuli Liu, and Ashish Shukla

CONTENTS

12.1 INTRODUCTION

The sustainability of the heating sector, especially in the UK, where the winter temperatures are around subzero [1], will play an integral part in the development of resilient cities. Approximately 44% of the primary energy consumption in the UK is used for heating, while nearly 50% of heating uses natural gas [2]. The UK building sector contributes up to 29% of the total carbon dioxide emissions, most of which are derived from heating [3]. To fully implement sustainable energy policies, it is critical to revamp the heating sector, especially for non-industrial buildings. A reduction of greenhouse gas emissions, by promoting sustainable energy technologies, has been the ultimate goal for the future of the UK [4]. District heating (DH) provides a flexible interconnected topology to heat urban centers in a sustainable cost-effective manner and is considered as the best substitution for these large metropolitans based on the experience of several European states [5]. It has a long association, especially with skyscrapers and dense urban environments both in Europe and North America [6]. Over the recent years, there have been a series of consistent government efforts to facilitate widespread use of DH, with the most recent policies defined in the fourth Carbon Plan and the Future of Heating. Several simulation tools and research conducted by the UK government indicate that heat networks being developed could supply up to 20% of the domestic heat demand by 2030 as compared to only 2% at the moment [2]. At the same time, financial incentives and research grants to local authorities are in place to compete with present-day heating technologies.

In DH, both large-scale centralized and smaller decentralized sources supply hot water, which is transmitted to consumers via insulated water pipes. Depending on the geographical location and the environmental conditions, the same network could be used for district cooling in summer. These systems are flexible since they can incorporate any fuel including renewables, waste industrial energy, and most importantly combined heat and power (CHP) systems, which are available in a range of sizes [7]. DH systems are also more efficient from the viewpoint of exergy since low-grade heat is utilized and there are lower losses [8]. The most sustainable DH networks are found in Scandinavia [9]. The most recent fourth-generation DH systems ensure renewable energies will be the dominating heat source [10], in a less expensive system also having the possibility of integrating into an overall smart-energy system. These low-temperature grids will make it possible to integrate even non-conventional sources, which is a breakthrough for the overall transformation of domestic heating. Several pilot schemes have already been implemented successfully, which will eventually pave the way for the commercialization of these fourth-generation grids in the coming years. With DH, the need for individual gas suppliers, boilers, flue-gas-treating technologies, and storage of potentially hazardous fuel is eliminated from built environments. In densely populated residential areas, DH not only makes more economic sense but is also safer and reliable [5].

Research has been carried out considering various aspects of DH within the UK, such as, the policy-based aspects of DH focusing on the social and legislative criteria [11–13]; comprehensive studies focusing on the financial aspects and environmental effects [3,14,15]; and studies focusing on calculating heat demands along with the technical aspects of DH [16–18]. However, with the recent development of fourth-generation DH [10], all these studies conducted in the past take a whole new dimension in terms of practicality and feasibility. Meanwhile, there is a lack of a holistic study in the literature, which links the governance, legislation, and regulations, with the technical, social, and economic variables of DH, focusing on the UK. This study is linking the current trends with the future goals within the UK. The objective of this paper is to comprehensively investigate the development and the associated constraints of expanding DH within the UK. Therefore, the following topics have been analyzed to address the research objective, with respect to the UK:

- The basic characteristics of current DH and domestic heating systems
- Legislations, technological frameworks, and policies to enhance expansion
- Financial aspects and incentives
- A critical analysis of the obstacles and underlining reasons for the underdevelopment, despite continual government efforts over the past decades
- An understanding of the development tools used to assist in the planning and development of DH

12.2   CURRENT STATUS OF DH IN THE UK

12.2.1   Ranking of UK in Terms of Worldwide DH Networks

District heating in the UK is almost non-existent when compared to networks over the world or more particularly in neighboring European states. Based on a case study on Italy, an exergy analysis was done for the entire energy mix, using a method to compare social-economic aspects with the second law of thermodynamics. It was proved that DH could save substantial amounts of energy in the entire system, especially in the residential sector, irrespective of the regional constraints [19]. In spite of having similar weather conditions and demand patterns, the UK has lagged behind considerably compared to its neighbors. Based on a recent study [9], the world's DH networks are categorized, where the network in the UK is classified as being in the emerging stages. A comparison of some statistics [20], of the UK DH grids and other

TABLE 12.1
European DH Stats for 2013

		Denmark	Finland	France	Germany	Iceland	Poland	UK
Energy Sales	TJ	105,563	114,160	86,112	254,839	28,181	248,693	—
Sales	mEuro	2945	1861	1634	5701	145	3083	437.5
Citizens Served	%	63	50	7	12	92	53	2
Length	km	29,000	13,850	3725	20,219	—	20,139	361
Number	—		˙394	400	501	3372	48	317	2000
Capacity	MWth		23,270	21,230	49,691	2290	56,521	335
CHP Share	%	73	73	23	81	—	57	80

European counterparts are presented in Table 12.1. The networks in Scandinavia are much more developed and sustainable while those in eastern European states, for example, Poland, are much more comprehensive but not that efficient [21]. On the other hand, Germany has the fastest growing sustainable network due to its policies and aggressive approach toward a sustainable transition [15].

An important point to note is that although the number of DH grids in the UK is second only to Germany, this is in fact not a measure of usage since many small networks are present as compared to Denmark where large interconnected grids exist. Energy sales values in the UK are not available since most networks are operated privately or by small associations with minimal government interference.

12.2.2 The UK Heating Sector and Associated Fuels

According to historians, the first DH network in the UK was initiated in 1742 to distribute steam for heating purposes [22]. In 1791, a scientist from Halifax, Yorkshire, received a patent for heating a set of buildings using an array of steam-carrying pipes. There were many notable DH grids till the mid-1950s, and it was considered by that time that the UK would be a world leader in this technology. A series of bad experiences, government negligence toward this technology, the availability of cheaper natural gas, and the complete liberalization of the energy market are some reasons for its downfall ever since.

While Denmark and most western European countries spent the past 40 years implementing a sustainable heat grid, the UK has spent the same time developing its gas grid. Shared with Norway, vast amounts of gas reserves were discovered and developed in the North Sea since the 1970s. Consequently, the UK had access to this relatively cheaper fuel and was a net exporter of energy from much of the 1980s till the last decade. However, it was in 2004 that the UK first became a net importer of fuels, and since then, the urge to develop sustainable fuels has increased. About 80% of households in the UK are connected to the gas grid, and 95% of them use gas boilers, mostly being of the condensation type [2], resulting in the UK becoming a world leader in boiler technology. Gas boilers are also dominant in non-domestic use; however, they are equipped with better control systems, for more optimized heating [12]. The UK heat-generation sector consumes more energy compared to the other two major sectors, electricity generation and transport.

Another important aspect, especially within the UK housing stock, is the importance of the indoor environment and fuel poverty. About 15% of households in the UK are affected by surface condensation and mold growth [5]. This is attributed to insufficient heating, poor ventilation, and excessive moisture production in the building. About 11% of households within the UK are suffering from fuel poverty with the main reasons being low incomes, poor efficiencies of heating systems, and rising fuel costs [23]. The current fossil-fuel-based boiler systems, in typical houses, increase fuel poverty, causing insufficient heating with no control over the indoor environmental conditions. On the contrary, cheaper, sustainable, and well-controlled DH is the answer to this long British problem. In an experimental study in Sheffield [5], the heating system was changed from an electric underfloor heating to a DH installation. Results showed much less mold growth and improved indoor conditions.

12.2.3 DEVELOPMENT OF DH IN THE UK

As informed in Section 12.2.2, some municipality-sized heating networks in the UK were developed in the early 1960–1970s, predominantly in urban cities, which are still existent today, although they are not very efficient [9]. These initially built schemes were neither expanded nor refurbished over the years and have completely lost attractiveness. More than 85% of the grids are outdated because they were established before 1990. Many of these schemes were operated by housing scheme operators and were on a small scale that was not up to the necessary merit [5]. These schemes were not engineered per international standards; the construction techniques were poor and, most importantly, individual components were not maintained. Consequently, the operational costs were high, much to the dismay of the customers, while the reliability was low. Most of these schemes used fossil fuels including oil which made them even more non-eco-friendly compared to conventional domestic gas boilers. During this decade of the 1960–1970s, unlike electricity, heating tariffs were completely unregulated with no legal pricing mechanism. Customers were dependent on the operators of these small grids in terms of supply and operational charges [24]. Some of these schemes have been abandoned while others destroyed because they completely lost their attractiveness to customers. In the UK, most operational DH grids are operated by:

- Local authority-led schemes for commercial buildings.
- Housing schemes and private sector firms.
- Standalone schemes for a specific organization, schools, hospitals, universities, etc.

Approximately 2000 (registered and unregistered) DH networks provide heat to 210,000 dwellings and 1700 commercial buildings across the UK. In total, there are about 27 million households and 1.8 million commercial buildings, corresponding to 2% of the total demand, which shows the non-existence of DH [25]. Based on size [25], the networks in the UK could be categorized according to Table 12.2:

TABLE 12.2

Distribution of UK DH Grids in Terms of Size

Network Size	Domestic and Non-Domestic Buildings Served	% of Total DH Grids	No. of Networks
Small	<100 and <3	75	1280
Medium	100–500 and 3–10	20	315
Large	>500 and >10	5	75

FIGURE 12.1 The spread of DH networks across the UK. (From Department of Energy and Climate Change UK, The Future of Heating: Meeting the Challenge, 2013.)

A large number of small grids exist throughout the country, which are not inter-linked. Figure 12.1 visualizes how these small networks are spread out in clusters, especially in the south of the country. It could be seen that the north of the UK has less DH, even though it is colder, while middle England and areas surrounding London have more DH. This is due to the higher population density toward urban centers and in the south of the country. More than half of the DH schemes are based in London alone, with most of them not interconnected existing at very small scales. Several notable schemes in Birmingham, Nottingham, and the central UK have emerged over the past few years.

12.2.4 CURRENT HEAT SOURCES AND STORAGE FOR DH

To achieve the targets of the carbon plan, it is essential that the heat sources of DH should be renewable and sustainable. At the moment, most of the heat used in DH networks is derived from CHP plants with a small fraction from renewable sources. At the same time, for such intermittent renewable heat sources, it is crucial to develop enough energy storage capacity to decouple demand and supply. These two characteristics will be vital for every DH network of the future.

Most large networks in the UK are using CHP plants, while a majority of the smaller grids are using boilers or renewable sources. Nearly 90% of all the grids use gas as the primary source of energy [2]. Still, less than 0.1% of households in the UK are connected to a DH-CHP heat source [26] because most of the CHP stations in the UK supply heat for industrial purposes [27]. Less than 10% of the heat supplied to DH grids comes from renewable sources. According to figures in 2010 [28], the main sources of renewable heat supplied to DH grids are biomass (90%), solar thermal heating (8%), and heat pumps (2%). Biomass is mostly in the form of wood chips or straw [29]. Heat pumps operated by renewable sources accounted for the remaining 2% (963 MW) of renewable heat sources because they are currently a less mature technology. Ground source heat pumps (GSHP) and air source heat pumps (ASHP) are the most common subtypes in operation [30].

In terms of energy storage, the UK is also amongst the lowest performing nations in Europe [31]. At the moment, the only major energy storage option is that of stocking fossil fuels. Heat storage in the form of hot water storage is mostly decentralized and only at building levels. About 14 million homes in the UK have hot water storage capabilities, giving a maximum storage capacity of about 80 GWh [32]. However, these small-scale decentralized hot water storage systems could be used in future DH grids in conjunction with centralized sources. Several of the localized DH schemes also have hot water storage capacities but mostly up to a maximum of 5000 m³.

12.2.5 EXAMPLES OF DH IN PROMINENT CITIES

Approximately, 55% of the DH grids of the UK are based in London with 65% of them being small non-interconnected ones. However, several projects for building new and expanding existing DH networks in cities like Leeds, Manchester, Newcastle, Nottingham, and Sheffield are being supported by the DECC at the moment [2]. Some of the most notable DH grids present in the UK are as follows:

- Nottingham has one of the largest DH networks in the UK. It comprises of a 65 km network of pipelines serving about 4600 homes and over 100 businesses, which accounts for roughly 3.5% of the city's heat consumption [2].
- The Southampton DH grid is a tri-generation scheme utilizing renewable sources, including biomass. It has about 14 km worth of pipelines, with heat being provided by CHP plants and backup boilers, in a total of about 27 MW. This grid serves only the city center and a few buildings in its vicinity [33].

- The DH grid in Stratford, London, is about 16 km in length providing about 91 MW of heat from biomass-fired boilers and CHP plants. It was established for the London Olympics in 2012 [34]. It is envisioned that by 2030, 25% of the heat demand in London would be met by interconnecting the current DH grids supplied with decentralized renewable sources of heat [35].
- The DH grid of Sheffield has expanded rapidly over the recent years and is believed to be the largest in the UK [17]. The heat source is an energy from waste (EfW) plant, partially operated by the Sheffield City Council. Approximately 28 t/h of non-recyclable waste is combusted. An approximate estimate of the carbon savings of this facility is about 21,000 t/a, which has made it the most sustainable network within the UK.
- The CHP-DH grid of Leicester was among the best in the country in the early 1990s [22]. It supplied heat to several commercial buildings, serving thousands of customers. Thus, in 1990, Leicester was chosen as the "Green City" of the UK and since then has played an active role in promoting the cause of sustainable heating.

12.3   POLICIES, REGULATIONS, AND OUTLOOK

12.3.1   SUSTAINABILITY AND THE EU DIRECTIVE

The issue of greenhouse gas emissions and the human impact on climate is expected to reach the point of no return where the global average temperature would rise by more than 2°C [1]. Countries representing more than 80% of these global emissions have pledged domestic targets under the Copenhagen Accord and the Cancun agreements, to curb carbon dioxide generation in the years to come. The important requirements for most energy systems in the future will be [36]:

- To ensure a sustainable system to meet the end user demand (secured energy);
- To have an optimized system having the least operating cost (competitiveness); and
- A low-carbon system (sustainable development).

Being at the forefront of these summits and being a trendsetter in sustainable energy technologies, the EU has developed comprehensive policies and roadmaps for such a future energy system, including a focus on the heating sector. In spite of sustainable heating technologies being present, proper governance and planning are essential for infrastructure development at urban levels [37]. In 2009, the European Parliament presented the daring Renewable Energy Directive [38]. As part of this directive, by 2020, 20% of the final energy consumed within the EU would be from renewable sources, which would eventually increase to about 30% by 2030. As an expansion of this directive, the "Roadmap 2050" [39] was published. This is a general paper defining the outlines of such a typical energy system in 2050. With this focus on sustainability and a low-carbon future, the energy performance of buildings [40] was introduced, and consequently, the Energy Efficiency Plan [41] was put forward. The overall theme of these policies in terms of heating is to have a sustainable and carbon-free system by 2050, which promotes DH, especially in urban buildings. According to statistics published by the European Technology Platform, the potential of DH in Europe is huge. It is expected to supply at least 25% of the heat consumed by 2020, which may increase to about 50% by 2030 [42].

On an individual basis, most of the countries within the EU, including the UK, have also set similar legislation toward the pathway of such a system. A balance between policies by the central government and local initiatives by city councils is the key ingredient for rampant urban infrastructure change [37]. An overview of the central government's policies in terms of the expansion of DH is presented in this section.

12.3.2   UK POLICIES ON DH DEVELOPMENT

Although the UK has the necessary resources for the development of renewable energy, exploitation has been extremely low, as it is ranked 25th out of the 27 EU member states, with only 4.4% of the primary energy demands met by renewable sources [28]. Accordingly, 201 TWh of heat was generated from renewable sources in 2013, corresponding to only about 2.8% of the total heat demand. As indicated in Section 12.2, domestic heating consumes a major chunk of the UK energy demand coming mostly from unsustainable natural gas. It is clear that the integration of renewables in the domestic heating sector is essential to lower carbon emissions and develop a sustainable sector. DH has the potential to integrate these low-grade renewable sources and to minimize carbon emissions within this sector. Based on the directives of the EU, the British government formulated its National Action Plan, with a strong support amongst all political factions; the UK parliament initiated the Climate Change Act in 2008. The target was to reduce greenhouse gas emissions by 80% of the 1990 values in the year 2050 and to provide energy security by reducing the reliance on imported fossil fuels. Since 2008, a series of carbon budgets have been implemented to drive progress and to formulate legislation in each sector, to achieve this overall target. The fourth of these carbon budgets is the most recent, aiming to achieve targets for the years 2023–2027. Consequently, including the carbon budget (within the carbon plan), there are three specific policy papers that are directly concerned with the heating sector, especially DH:

* The Carbon Plan [4];
* The Energy Efficiency Strategy [43]; and
* The Future of Heating [2].

A summary of the main points of each policy eventually progressing toward the implementation of DH is summarized in Figure 12.2.

12.3.2.1   The Implications of the UK Carbon Plan

This fourth "Carbon Budget" was set into motion in 2011 and introduces policies with an overall aim of reducing emissions by 50% (1950 MtCO$_2$e) of 1990

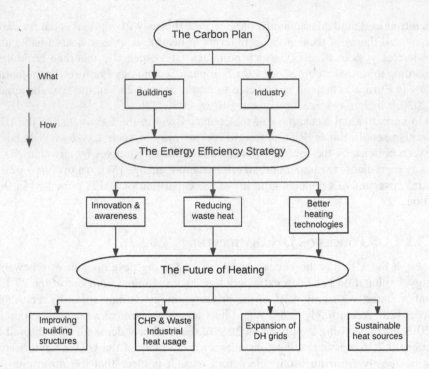

FIGURE 12.2   UK policy toward the expansion of DH.

levels by 2030. Although this policy paper does not specify targets and pathways
for individual technologies, it gives a general target figure for the major sectors
involved. It directly influences the heating sector as follows [4]:

1. *Buildings:* It is envisioned that by 2050 all buildings, both domestic and
   non-domestic, in the UK would be net-zero energy buildings. The government plans to achieve this target by improving the energy efficiency
   of buildings and by supplying the energy demand from low-carbon sustainable sources, where the use of DH becomes inevitable. It is targeted
   to reduce buildings emissions by 24%–39% of the 2009 base values by
   2030.
2. *Industry:* Emissions reductions are expected to be achieved by introducing
   higher energy efficiency standards, including waste heat recovery, replacing
   fossil fuels with sustainable biofuels, and by using carbon capture and storage (CCS) techniques. By 2030, emissions from industrial activities should
   be 20%–24% lower than the 2009 levels. Waste heat recovery using DH is
   considered to play a pivotal role in large-scale future heating, both industri-
   ally and non-industrially [44].

12.3.2.2   The Implications of the Energy Efficiency Strategy

To elaborate on the specific targets, the government issued the "Energy Efficiency
Strategy." The aim is to address possibilities to enhance efficiency measures in all

the sectors defined in the carbon plan. It is estimated that about 196 TWh of energy could be saved by more efficient practices by 2020. The need for efficient heating is addressed by three vital strategies [43] and they are as follows:

12.3.2.2.1 Innovation and Awareness

There are three major innovative areas likely to transform UK dwellings into low carbon structures [45]. These include advanced retrofitting techniques using better construction materials, more efficient technologies, and decentralized microgeneration strategies for energy. The last two areas point toward the development of DH grids. In conjunction, a range of innovative technologies, including better insulations, efficient lighting, smart appliances, efficient regulatory mechanisms, and passive heating/cooling technologies, would transform buildings to be more sustainable [46].

To realize the complete benefits of DH, it must be coupled with public awareness. In an experimental study [16], the energy performance of a low-carbon housing society comprising 25 houses using DH fueled by biomass (wood-chip boilers) was conducted for a time period of one year. A significant observation in this study was that although the overall carbon footprint of the system decreased, there was a wide disparity in the actual heat demand of the consumers, even with those having similar characteristics, for example, floor size and occupants. The concluding factor is that despite having sustainable heat-generation sources, to reduce the overall carbon footprint, the heat demand is also to be minimized, which is strongly linked to consumer behavior and maturity. Cheng et al. [47], developed the Domestic Energy and Carbon Model to predict the future energy consumption and carbon dioxide production of the UK residential sector. The model was validated with national statistics with only marginal errors, making it a valuable tool for holistic consumption predictions of all household utilities. Results show that 85% of the variance in energy consumptions is dependent on the dwelling type and socioeconomic conditions of the dwelling, making public awareness an essential part.

12.3.2.2.2 Reducing Waste Heat

At a domestic level, this can be done through demand-side management techniques. It includes installing smart meters so that consumers can make better decisions, with real-time data to efficiently consume energy. DH is the best way to reduce waste heat since low-temperature grids can incorporate waste heat sources and simultaneously have smart meters with the capability to be interlinked to other energy utilities. At the same time, a DH grid is the only heating technology capable of having a two-way flow of heat in which heat generated from decentralized renewable sources or from customers can be sold back to the grid. Another addition to this clause is the introduction of incentives from the UK government for better energy utilization. The government has introduced some policies including the Green Deal, Renewable Heat Incentives (RHI), and Feed-In Tariffs (FIT) for net-metering of renewable energies. The Green Deal is a popular policy in which all consumers can increase the energy efficiency of their respective properties with no upfront cost; instead, it is paid in monthly installments in the energy bill.

In an interesting study aimed to show that a huge potential also exists in commercial buildings, Davies et al. [35] highlights the potential of reducing waste heat in commercial data centers within London. Normally such centers have a range of servers and IT facilities that need to be cooled at a continuous rate. The cooling of these IT facilities with the inclusion of this heat in a low-temperature DH is analyzed, with the utilization of heat pumps to boost the temperatures. A 3.5 MW data center could save £1 million and 4000 tons of CO_2 annually.

12.3.2.2.3 Better Heating Sources

At the moment, fossil fuels are the main source of energy in the heating sector, and as stated earlier, 90% of the DH grids are using fossil fuels as well. In order to transform this sector with sustainable heating sources, it is envisioned that heat pumps for low-density areas and DH for high-density areas, based on renewable sources, would replace the current gas network [48]. According to a recent study [46], limiting gas usage not by mass electrification of the heating sector but by renewable growth is the key to a sustainable UK heating system, where the use of DH grids becomes desirable once again.

Unfortunately, at the moment the only major renewable source for DH is biomass, which of course is likely to increase in the future. In general, biomass sources, heat pumps, and solar thermal systems are mature sustainable technologies in the UK and are considered to be an integral part of the future heat mix [33]. It is estimated that if the current total potential of biomass was utilized in the heating sector, about 31% of the total heat demand could be supplied [49]. It is important to note that for biomass to be economically feasible, both consumption and production must be located close together, which is favorable when using DH networks on a localized level [50]. About 80% of the biomass produced has a potential consumption within a radius of 25 km, while 95% can be consumed within 40 km. It is estimated that in the future about 31 Mt of biomass will be cultivated, producing 366 PJ of heat, most of which would be used in DH. It is quite interesting to note that in case DH is not developed, this potential of biomass would also be lost [51]. Similarly, it is envisioned that by 2020 at least 1.2 TWh worth of decentralized solar thermal heating will be installed in the UK, ideal for decentralized low-temperature fourth-generation DH grids [28]. GSHP and ASHP from renewable sources are also estimated to produce about 19.7 TWh of thermal energy by 2020.

12.3.2.3 The Implications of Future of Heating

Due to the wide range of heating applications in industries and domestic consumers, the heating sector is one of the most complex and difficult to decarbonize, especially in an industrialized economy of the UK [12]. However, a recent study shows that if the cost and performance of DH are comparative to the current system, the general public is willing to shift to a more sustainable option, in spite of the difficulties [12]. Based on the fundamentals established in the previously discussed policy papers, the UK government further expanded on this objective based on the EU directive [28], with the "Future of Heating" report. Effective measures for improved heating techniques are broken down into three sublevels and they are as follows:

12.3.2.3.1 CHP and Waste Industrial Heat Usage

As mentioned in the carbon plan, the government plans on a reduction of 70% of emissions from the industrial sector. Out of the three defined mechanisms to do so, two are directly linked to the expansion of DH:

- The inclusion of more efficient processes and the promotion of biofuels have paved the way for the expansion of CHP plants [52]. As compared to electricity-only power plants, the increase in overall efficiency is about 20%–30% for CHP plants [53]. Along with DH for domestic consumers, the transmission of heat from CHP plants to clustered industries is normally done via heating grids as well [27]. In 2011, the government introduced specialized tariffs for non-domestic heating systems using renewable sources. The RHI is the first of its kind and even provides financial assistance to non-domestic users [28]. There are also different schemes to promote the utilization of CHP systems including the CHP Quality Assurance, CHP Focus, and the Carbon Price Floor schemes. It is envisioned that by 2020 approximately 18 GW_{el} of CHP plants will be connected to the grid.
- Industrial waste heat must be utilized in order to increase the overall energy efficiency of the system [54]. The UK has a huge potential not only in industrial plants but also in thermally fired power stations [44]. Additionally, the concept of EfW is also becoming quite popular both in the UK and Europe as well, especially for heat production. The launch of the Green Investment Bank provides direct funding to energy efficient projects for non-domestic improvements, including industrial waste heat recovery.

Although high-grade thermal heat is easy to recover within the industrial sector itself, the utilization of low-grade heat is only possible via DH [55]. There exists an extremely high potential for such recovery in the UK, especially in small and medium enterprises (SMEs). These industries consume 45% of the total industrial energy because their processes and technologies are inefficient due to a lack of financial muscle, with which the waste energy could be recovered and reused. Compared with larger industries in which at least 8% of easy energy saving potential exists, these SMEs have a potential of 20% [55].

12.3.2.3.2 Expansion of Heating Grids

Expanding CHPs and utilizing waste industrial heat pave the way toward the expansion of DH [53]. To facilitate this, the government set up of the Heat Networks Delivery Unit (HNDU), within the Department of Energy and Climate Change (DECC) to work closely with local authorities and specialized firms. Through this scheme, initial funding to local authorities for the setup of heating grids would be provided via the Green Investment Bank. At the moment, HNDU supports 122 projects in 91 municipalities. In response to this positivity by the government, energy companies, consultants, experts, manufacturers, and a range of industrial individuals helped set up the Combined Heat and Power Association and the UK District Energy Association. At least 14%–20% of the domestic heat demand by 2030 could be supplied by DH and 40% by 2050 [56] as in Figure 12.3.

Several schemes are already in place to expand DH, including the "Building Regulations Act," "Zero Carbon Homes Policy," "EU ETS Policy," etc. [2]. The Low Carbon Pioneer Cities program is specifically targeted for larger cities to decentralize their heat supply through the development of DH. At the moment, the DECC is funding up to £1 million for projects in Manchester, Leeds, Nottingham, Sheffield,

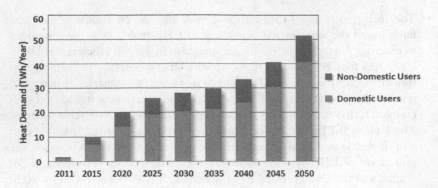

FIGURE 12.3 Future heat demand by DH. (From Department of Energy and Climate Change UK, The Future of Heating: Meeting the Challenge, 2013.)

and Newcastle. With a further £7 million available to councils to carry out feasibility studies, the intensity of the government's approach is quite serious. As an example of such incentives, the recently developed DH network in Coventry using an EfW plant was initially funded by the DECC for its feasibility analysis [57].

12.3.2.3.3 Improving Building Structures and Having Sustainable Heat Sources

It is expected that the overall heating demand would reduce due to improved building structures and technologies. At the same time, it is envisioned that this demand will be met by sustainable sources despite this decline.

In a study of the future [15], the space heating and hot water demand for four EU nations including the UK was calculated. Using heat data from the residential sectors from 1970 to 2005, along with using an index decomposition, econometric models, and a co-integration analysis technique, an algorithm was developed to predict the demand up till 2050. Results show that monetary policies along with simple technical improvements will reduce the final energy consumption in the residential sector by about 1% on an annual basis. However, according to studies by the government, after the implementation of the energy efficiency strategy, it is estimated that the demand for space heating and hot water would be steady at about 500 TWh annually. Different refurbishment technologies are commercially mature in the market and are ranked hierarchically according to cost and effectiveness [45].

Once again in urban centers, DH is considered vital for future domestic heating, especially when integrating renewable heat sources. At the same time, DH has the capability to efficiently and reliably integrate itself in improved building structures, to supply low carbon heat in these retrofitted buildings. Although future heat demands for domestic buildings would decrease, it is still economically justified, especially in urban centers when the current gas boilers are outdated in the future [48]. In areas of low densities, several different heating technologies are to be introduced, which are economically feasible. The Strategic Framework presented in the Future of Heating outlines that electrically driven heat pumps, biomass systems, solar thermal devices, micro CHP systems, fuel cells, and electric storage heaters would be the dominant commercially available heating technologies for off-grid buildings. Most modeling software utilized by the DECC suggest a high share of electrically driven pumps (both air and ground) would be commonly used in hybrid with other technologies [56].

The promotion of these non-DH sustainable heating technologies has been a part of government policies [2]. The FIT scheme, established by the government, focuses on the growth of micro CHP plants. Under this scheme, the government guarantees a buying rate from electricity produced by micro CHP units. Some public funds, including the Enhanced Capital Allowance, Rural Communities Renewable Energy Fund, and the Renewable Heat Premium Payments, are also supporting the development of small-scale efficient heating technologies.

12.3.3 OUTLOOK OF THERMAL ENERGY STORAGE

It is widely agreed that for future energy systems, the land below the earth would have immense value for thermal storages [58]. Thermal storage is expected to play a key role in two circumstances. The first in conjunction with DH and second with the usage of heat pumps, as it is a replacement for electrical storage since it is comparatively cheaper.

Thermal storage linked with DH ensures a reliable supply at any time of the day, lowers operating costs, and improves the resilience of the overall energy system. At the moment, most thermal storage capabilities in the UK are decentralized, on an individual basis. DH will change this with the overall aim of flattening peak demand both on a daily and seasonal basis. At the same time, storage costs are inversely proportional to size, as presented in the next section, making large-scale storage favorable. However, the exact potential to be realized in the future is dependent on an array of factors, including the size of the DH networks, their interconnectivity, peak loads, load-shifting required, thermal structures of buildings, and electrification of the heat supply. Nevertheless, it is estimated that a storage capacity of 48 MWh having a volume of $2060m^3$ is required by a DH grid serving 4000 customers with a peak load of 4 kW per customer [59]. As presented earlier, the government plans to serve 40% of the total heat demand by DH in 2050. Approximately 40% of the 27 million households in the UK served by a DH system would mean about 10.8 million customers. Assuming each grid serves about 4000 customers, this would mean an addition of 2700 grids having a potential thermal storage of about 130 GWh. The Low Carbon Innovation group [31] proposes, that the development of energy storage technologies has the potential to save £2–10 billion till 2050 and would thus have a market value of £3–26 billion.

With increased fluctuating renewable sources to provide electricity, the most efficient way to store excess electricity is in the form of heat since electrical storage is much more expensive [31]. To do this, heat pumps combined with heat storage would be ideal to expand thermal storages [48]. Ongoing research has helped in the evolution of latent heat storage technologies from sensible technologies, especially phase change materials (PCMs) being a promising development in the future [32]. It is also expected that in case PCMs replace the current hot water sensible storage systems, the required storage volumes would shrink by two-thirds [59].

12.4 FINANCIAL IMPLICATIONS FOR DH IN THE UK

Up until now, the prime reason for a low penetration of DH in the UK is its relatively high cost compared with the conventional gas heating system. Nevertheless, the value of the entire UK DH grid rose from about £350 million in 2010 to about £530 million in 2015, providing heat to 2% of the total demand [2]. According to the government's plans, if 40% of the demand is met by DH by 2050, this would account for an estimated value of about £10.6 billion.

The carbon abatement costs of upgrading the housing stock of the UK, from a holistic perspective, using the famous software UK MARKAL, were calculated [3]. According to the marginal abatement cost curve derived from this model, the decarbonization of the electricity and heat sector by 2030 would cost about £50/tCO$_2$. This cost is reflected in an increased share of renewables in the electricity sector, an increase in wood-fired and pellet boilers, an expansion of the overall DH grid, and the introduction of heat pumps.

Based on data collected from existing DH grids within the UK [25], the characteristics of typical networks are in Table 12.3.

12.4.1 DISTRIBUTION

Currently, most houses in Britain have a gas boiler in conjunction with centrally heated wet radiators and hot water storage tanks. In case DH is implemented, the gas boilers are to be replaced by substations to exchange heat with the DH grid. Additional heat meters with adjustments to the pressure and/or temperature are to be

TABLE 12.3
Statistics of Typical Heat Networks in the UK

Parameter	Value
Distribution losses in % of the heat generated	6%–28%
Thermal storage capacity per MWh of the network in m³/MWh	0.016
Operating temperature of supply in °C	84–88
Operating temperature of return in °C	54–62
Cost of heat meters per domestic building in £	491–668
Cost of heat meters per non-domestic building in £	2878–3433
Cost of heat network to building connections in £ per building	738–1326
Cost of the substation in £/KW	16–53
Cost of transmission heat network in £/million	422–1472
Cost of distribution heat network in £/million	94–244
Cost of thermal storage associated with the heat network in £/m³	843–962
Overall operational costs for the entire network in £/MWh	30–36
Heat rates charged to the customer in p/KWh	4.64–9.68

done along with other minor considerations to the indoor centrally heated network of the house. In a less expensive scenario, a direct connection between the building and the DH grid could be made in which case only an upfront heat meter would be required. Alternatively, there are several possible arrangements on the positioning of the substation and meter, which can considerably reduce upfront costs for individual customers [21].

The cost of DH is highly dependent on the scale of the distribution grid. The higher the heat load and density of buildings the more economical it gets in comparison to other technologies [60]. According to Table 12.3, an optimistic average capital cost for a consumer installing a DH connection is somewhere between £1, 229 and £1, 994, while in comparison a domestic gas boiler costs somewhere around £1, 800 to £2, 500 [61]. On the other hand, the costs of a unit KWh of heat from DH is in the range of 4.64–9.68 pence [62], while a unit of heat from a gas boiler at the moment costs around 4.77 p/KWh. It is clear that from the viewpoint of the customers, the figures are quite comparable for both the capital and operational costs. The only difference is in the comparison of the already existent gas infrastructure and the non-existent DH infrastructure, which requires intensive amounts of capital investment. At the moment, there is also not a regularized tariff structure for DH operators, which makes the billing to customers non-transparent, and in some cases, it doesn't even guarantee the investor the rates of returns desired [63].

12.4.2 TRANSMISSION

The experience in setting up a modern sustainable DH grid is almost non-existent in the UK, and a relatively higher capital cost is expected when laying out the network [63]. In Scandinavia and central European countries, the number of multi-family dwellings is considerably higher as compared to the UK, where most are single-family houses [60]. This makes it even more costly to set up transmission lines since the heat densities are spread out. This cost can broadly be subdivided into the civil works required and the mechanical equipment to be installed. Typically, both these costs are almost similar to each other in value. A heat network to supply about 270,000 homes would roughly cost about £1.5 billion [61]. In a recent study carried out [14], it was concluded that the excavation costs in the UK are relatively higher compared to other European states and consumes the highest share of capital investment costs. To minimize this cost, it is proposed to use twin pipes wherever applicable since their performance-to-price ratio is the best.

As an estimate, the capital costs of installing a DH grid in the UK are 20% higher than many central European states [61]. At the same time, no standardized procedure is in place both for the designing of the network and the non-technical formalities. This slows down the entire development process and makes it financially riskier for potential investors. Such high investment is only favored by investors when the rate of return is high. Some studies suggest that if authority figures reduce the discount rate to about 3.5% from the normal 10%, only then will the full potential of setting up DH be realized [61].

TABLE 12.4

Characteristics and Costs of Heat Sources

	Capital Costs	Operation and Maintenance/a
Small gas engine CHP—500 KWe	£864/KWe	£80/KWe
Large gas engine CHP—2 MWe	£657/KWe	£48/KWe
Small combined cycle gas turbines (CCGT)—50 MWe	£805/KWe	£32/KWe
Medium CCGT—90 MWe	£759/KWe	£32/KWe
Gas DH boiler—100 KWth	£60/KWth	£3/KWth
Biomass DH boiler—100 KWth	£615/KWth	£15/KWth
Small biomass CHP—100 KWe	£4000/KWe	£180/KWe
Medium biomass CHP—8 MWe	£3500/KWe	£80/KWe
Large biomass CHP—30 MWe	£1780/KWe	£80/KWe
EfW: Anaerobic digestion—1 MWe	£7745/KWe	£775/KWe
EfW: Combustion—24 MWe	£8750/KWe	£517/KWe
Industrial waste heat—30 MWe	10%–15% of electrical transmission lines	

12.4.3 GENERATION

The liberalization of the energy market in the UK has been a major hurdle in the growth of renewable technologies, over the recent years. Critics believe that the stakeholders have monopolized the utility market and government neutrality in this aspect has caused a lag in growth [13]. If benefit costs of carbon reductions are realistically considered, renewable technologies combined with DH could even compete with present-day fossil fuel systems. Presently, regulations do not give substantial advantages to low-carbon heating technologies, particularly in terms of carbon savings [64]. According to government projections, about £5 billion are expected to be invested in natural-gas-fueled CHP plants by 2020 [2]. On the other hand, the cheapest and most promising option to supply heat in DH is from waste industrial heat. The characteristics and costs [18,61] of some common heat sources used in conjunction with DH are presented in Table 12.4.

Thermal energy storage is also considered integral to DH grids, since they act as a buffer, being both a source and supply. Small storage tanks in the range of 300 m³ cost about £390/m³, while large-scale underground storage of 75,000 m³ is at about £25/m³ [59].

12.5 ANALYSIS OF UNDERDEVELOPMENT OF DH IN THE UK

Analysts believe that fluctuating energy prices along with the financial recession of 2008 has led to some of the goals and targets of the carbon budget to slip [36]. There have been uncertainties in the government's priorities over the past, following

TABLE 12.5

Basic Differences in DH Setups of Scandinavia and the UK

Scandinavia	United Kingdom
Need driven to minimize fossil fuel imports as no national energy reserves—driven by public demand	Change driven to transform into a low-carbon sustainable system with large gas reserves—driven as a precautionary measure
DH is less capital intensive due to past experience	DH is relatively more expensive due to a lack of experience and unfamiliar organizational system
Experienced and skilled manpower in DH	Generally inexperienced and unmotivated manpower
Direct financial involvement by central and local governments	Financial involvement by the government only to the extent of grants and initial funding
Strong city councils with the motive of sustainable transformation to the benefit of the people	Weak city councils having private firms running DH with the aim of profit
Direct involvement by central governments to promote renewable technologies	No favoritism and technology-neutral attitude of the central government
Liberalized markets with considerable government tariffs and involvement	Completely monopolized and liberalized energy market
Long-term visionary policies	Short-term impulsive policies
Regulation of heat tariffs and financial aspects on both the investor and customers end	Tariffs and financial aspects decided solely by private firms
Systematic development of grids followed by heat sources	Development of the entire setup as a whole, making it economically unfavorable

the development of shale gas reserves and the recent expansion of nuclear power plants, which make the realization of these renewable goals unlikely [65]. Although the policy and legislation are in place, there has always been a conflict between these goals, considering past historical trends. Several ambitious attempts in the past have failed, and critics believe this attempt is no different [24]. The basic differences between the system of Scandinavian leaders in DH and the UK are in Table 12.5.

Due to the diversity of local and national factors, the system in one country cannot be completely replicated to produce successful results in another [66]. However, the well-established DH system in Scandinavia can be mimicked to a certain extent or at least provide a source of direction for the UK.

12.5.1 FINANCIAL AND ECONOMIC REASONS

As a consequence of the 1973 oil crisis, several pro CHP-DH pressure groups have been advocating the case [22]. Since then, all stakeholders have agreed upon its importance and its inclusion in the energy mix; however, it has always been neglected when taking actions. The primary reason being that since then, it has not been an economically viable option compared to the already existing infrastructure of heating mechanisms. The high capital expenditure compared to the low revenue earned is a major factor intimidating investors. Critics consider many of the government policies to be short-term and unrealistic for longer sustainable targets. Like many European nations, a direct financial involvement is necessary instead of only legislative and supportive policies [22]. Funding is limited to feasibility studies and financial incentives but should cover more of the initial capital cost [5]. Sources of private financing, low-carbon technologies, through local citizens is an interesting option suitable for a liberalized economy like the UK [67]. Although government intervention in the form of quota schemes, soft loans, tax incentives, etc., are needed, such a financial mechanism reduces the risk on individual investors and generates the necessary momentum in the society in favor of DH, direly required in the UK as was studied in East Europe [68].

Until now, most of the DH schemes in the UK have had long-term heat contracts with a single supplier [69]. The heat tariff is regularized in such a manner that a minimum return on investment is guaranteed to the investor. Although this is encouraging from the point of view of investment as a comparative assessment, this is not how DH in Scandinavia evolved [20]. Initially, in these countries, heat prices had more regulation either by a non-profit operation, cost-based tariffs, market-based tariffs compared to alternative energy utilities, or through government subsidies. These mechanisms assure that a balanced law exists both for the investor and customers.

The classical method by which DH evolved in Scandinavia was by first developing the heating grid followed by a heating source [5]. As the heat load is gradually built up, the initially installed inexpensive large-scale boiler could be replaced by a more sustainable source. This makes it easier, in terms of the economic prospects, to justify such a large investment because the potential is easier to predict. After the establishment of a consistent heat load, the case to install a sustainable source and even expanding the network is made stronger.

The UK has one of the most liberalized energy markets in the world [70]. Although it is beneficial to a great extent, in recent years this system has been one of the greatest hurdles in promoting renewable and sustainable technologies [71]. A handful of large corporations control the generation and supply of most utilities. Even with the ongoing unpredictability in fuel prices, the major players of this setup still haven't paved the way for the expansion of renewables [28]. This setup has made government intervention extremely difficult, and consequently, in spite of passing renewable-friendly regulations, most have been overlooked [24]. The role of the government in such a market should be redefined, and an upper hand must be established to ensure the current setup doesn't jeopardize the future setup. One suggested way is to regulate certain aspects of the market and redefine the concept of liberalization, while another approach is to introduce competition to diversify the supply sources [70]. It is clear that policy and security must be the drivers instead of cost [46]. In an interesting development, The Labour Party announced in 2013 that if it forms the next government, it would freeze energy prices for 20 months in order to eliminate this monopoly and reset the energy market [13].

Based on a case study, a combination of a policies, subsidies, and direct funding is essential for the cultivation of DH in new setups [68].

12.5.2 SOCIAL AND POLITICAL REASONS

12.5.2.1 Social Barriers

Relative to other European counterparts, the public demand for the requirement of DH in the UK has been non-existent in the past [22]. Rocco [72] pinpoints the fact that the UK was one of the first nations to be industrialized in Europe, with an economy based on fossil fuels. Since it has a huge housing stock developed on old standards and a relatively longstanding mind-set, the transformation to a sustainable system will take time because it is a tangled mixture of social, economic, political, and cultural hurdles. A case study of a city Stoke-on-Trent, a former industrial hub in the south of the UK, is performed to conclude that although the transformation is slow and hard, with the right policies and financial support, indeed there is the possibility to change.

At the same time, there is a lack of knowledge about DH in the general public, especially with respect to the operation, bill charging, benefits, and awareness of the services offered, particularly due to the immaturity of this technology [12]. To make it easier for future users to cope with this technology, a public awareness scheme should be developed [13].

The development of human skills and technical expertise is an area where there has been a complete lack of concern by both the government and the private sector. There is also a lack of skill and institutional flexibility in many city councils since such infrastructure projects require simultaneously interconnected expertise between different departments like engineering, procurement, construction, planning, finance, legal services, and contracts.

In an interesting study [71], to bridge the gap between the public and policymakers, the role of "intermediaries" is analyzed. These are actors who facilitate the introduction of new technologies to the benefit of all stakeholders involved. Based on a case study in Newcastle, the results show that "intermediaries" delivered several critical functions that paved the way for the implementation of a successful DH grid.

12.5.2.2 Political Barriers

Compared to other European states, DH in the UK has not been operated by strong city-administrated energy firms. Although not completely under the control of city councils, in Scandinavia these utility firms are commercial companies with most their ownership by the specific city [20]. In the UK, DH operations have mostly been privatized [5]. This causes DH to face strong competition in the UK, compared to gas and electricity utility firms, which eventually decay its growth. This is rather unfair since DH has a non-existent infrastructure and competes against these much-developed utilities, which causes the operating firms to monopolize the overall energy mix to their benefit. However, in many European states, the city councils have much stronger administrative responsibilities and financial muscle [24]. These city utility firms, formulate their own policies and operate the overall energy mix to the advantage of the locality instead of monetary gain [8]. This ensures that the advantage is always toward the city and not the firms. At the same time, this strategy helps develop local manpower, which is another main deficiency in the UK. The operational policies and technical standards of these firms follow more stringent rules, compared to the liberalized firms. With the Local Government Act of 2000 and Sustainable Communities Act of 2007, local authorities have started to gain functional and financial autonomy, but this is limited compared to Scandinavian counterparts [11].

There is also a certain lack of communication between the central and local governments, both in terms of funding and political motives, especially further up north of the UK. In a recent case study [69], of the Scottish city of Aberdeen, the city council was given more power and autonomy to eventually establish a reliable and independent DH grid. Consequently, Aberdeen has a comparatively better DH grid compared to its neighboring cities. This proved the fact that not only are technical and economic considerations the driving forces behind development in the UK, but also social and political criteria are equally important. The development of an independent energy services company Aberdeen Heat and Power was created, similar to the Scandinavian model. Aberdeen's city council, having more financial autonomy than regular councils, has developed three gas-fired CHP stations along with a thorough network of about 14 km of heat pipes, supplying about 34 MWh/pa of heat to commercial and residential buildings. Carbon savings are 45% compared to the conventional system before this DH grid. The total expenditure on this system was about £8 million.

For the case of the UK, government intervention has become inevitable, if the long-term sustainable objectives are to be achieved [11]. So far, the government has been technologically neutral, by not favoring renewable technologies over nonrenewable ones [13]. This might be because it may disrupt the liberalized energy market, or there may be pressure from the fossil-fuel industry, especially gas-based monopolies. In the long term, this policy is criticized because the government needs to set priorities instead of being diplomatic [26].

Many critics believe that most policies and regulations, in the UK, are short-term and do not support the longer vision of a renewable-based heating sector [11]. According to a study [11], the development of infrastructures is interconnected between a range of social and technical drivers. It is not only the supply side of a grid that should be adjusted but the demand side as well. Infrastructure transformations are gradual, only developing with the correct long-term structural regulations and governance. Based on this study, there are four stages of an infrastructure lifecycle. The development of DH in the UK is classified to be in the first stage defined as the "stagnation and inertia" stage, where long-term investments and regulations are required. More successful demonstrations of DH technologies are needed in the UK before investors are comfortable with developing secure financial models for such systems [36]. A summary of all barriers is given in Table 12.6:

TABLE 12.6
Barriers in the Deployment of DH in the UK

Financial barriers	• Cost intensive as compared to conventional heating systems.
	• No standardization in terms of the supply chain resulting in more than conventional infrastructure costs.
	• Benefits of carbon savings not properly incorporated for heating technologies, making it impossible to compete with the gas network.
	• Uncertainty in fuel pricing of sustainable heat sources.
	• Less revenue and more uncertainty for investors over the long term.
Project barriers	• Accessing capital with a reasonable discount rate.
	• Not enough concentrated heat demand density as compared to other European counterparts.
	• Lack of experienced developers and operators of DH.
	• Uncertainty of consistent demand and customers along with uncertain heat sources and the associated fuel prices.
	• Unpredictabilities of renewable heat supply from sources.
	• Access to land for excavation.
Policy barriers	• Non-existent tariff framework.
	• Lack of awareness amongst consumers and the public sector.
	• No standardized contracts or legal practices in this area.
	• Low public sector involvement and motivation.
	• High lead times due to inefficient coordination of different departments and authorities involved.
	• Local councils not powerful enough.
	• Lack of technical human skills.
	• Monopoly by other utility providers, that is, gas and electricity.
	• Technology-neutral policies of government are hurting renewable development, especially biomass harnessing in DH.
	• Uncertain long-term policies based on past experience.
	• Lack of initiative to minimize waste heat by industrial players.

12.6 PLANNING TOOLS IN THE UK

The UK is new to this field of DH planning, with only a handful of developmental tools and assessment models [73]. Most of these tools incorporate both the usage of industrial waste heat along with potential CHP stations, as these are the most beneficial points of a DH scheme in terms of exergy [52]. Some tools have been used simply for policy studying, others for the planning or expansion of DH, while some are only for the financial analysis for future investors. The tools presented in this section focus on development/planning and not operation/design.

Most commonly, there are two steps in computer-based tools used for DH [74]. Initially, data should be collected. This can either be done via questionnaires, thermal imaging techniques to sense heat variations in different areas, remote sensing techniques to gather data via satellites, or simply by calculations [75]. In the second step, the data is analyzed, which is normally done using geographical information system (GIS) techniques or simple calculation-based energy modeling software. In reality, a combination of both GIS and modeling programs are used, for this second phase. GIS tools to be used with DH normally collect, analyze, and digitally display data on spatial maps. On the other hand, modeling software normally collects data and uses mathematical relationships along with optimization software to analyze the data and present the desired output, which can be incorporated in any visual form [74]. An overview of some major tools, within the UK, is presented in this section.

12.6.1 HEAT ROADMAP EUROPE

The purpose of this EU-funded project was to match wasted heat sources relative to heat demands, in order to identify regions capable of hosting large-scale DH within 27 European Union (EU 27) nations, including the UK [76].

A GIS-based software ArcMap 10.1, having a bottom-up perspective, was used to gather and map data, while a modeling software having a top-down perspective was used to produce alternative projections and possibilities of the future DH grids. The data were clustered within square kilometers in the entire EU 27 boundaries to form a high-density heat demand map. Excess heat ratios were computed based on the waste heat available and heat demand. Based on these ratios using data from the GIS tool, a modeling tool was used to develop strategic heat synergy regions where the setup of DH was considered favorable. This study established that there are at least 139 possible sites having excess heat and sufficient demand, while 82 are quantified as being techno-economically feasible. Most of these regions lie in London, Southampton, Bristol, Liverpool, and Edinburgh as shown in Figure 12.4.

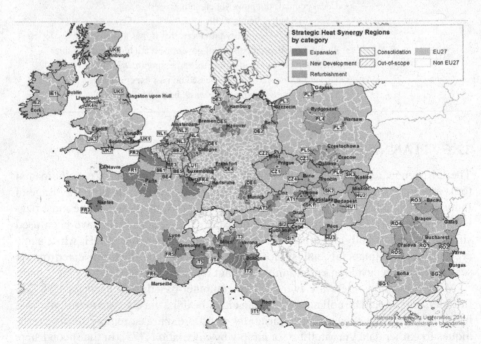

FIGURE 12.4 Feasible locations for potential setups using the energy systems modeling tool. (From Persson, U. et al., *Energy Policy*, 74, 663–681, 2014.)

However, since the approach of this project was on a regional level instead of a national one and focused only on industrial waste heat, the amount of detail and planning is lacking. This resulted in the use of this study only for policy development and motivating national governments to form a stronger case.

12.6.2   UK MARKAL Expansion

The UK MARKAL is a famous national energy system modeling tool, particularly used by the government of the UK for the analysis of future policies [77]. However, because it models the energy system as a whole, it is not very beneficial, especially for the heating sector. A team of researchers from the UCL Energy Research Group analyzed the shortcomings of the version 3.26 and expanded the heating subsector to gain a better insight into future developments. This extended model enhances both demand and production technologies, including DH, used in the base version.

The objective of this bottom-up, dynamic, linear optimizer is to match supply and demand throughout the different subsectors of the economy, in the most cost-effective manner. The combination of technologies and the various scenarios are presented as an output for further sensitivity analysis, to be used in policy-making.

The advantage of this tool is that different combinations of building types with heating technologies can be assessed and compared. Heating trends across different residential sectors pointed to the fact that urban setups would utilize DH while non-dense rural setups would use electrical heat pumps or micro CHP systems. This tool only provides an insight of the future distribution of technologies based on the least expensive combination. It does not incorporate the waste heat utilization from industries nor the technical feasibility of associating different demand and supply combinations. Hence, so far it is only being used as a policy-making tool and not practical for planning and detailed assessment purposes.

12.6.3   UK in Demand for Waste Heat Utilization

UK in Demand is a research center funded by the government, consisting of six participating institutes with the aim focused on the government's strategy in energy efficiency. An insightful study answers in detail the question of whether there is a potential to minimize heat losses in the industry and link it up with heating grids, in the UK [78,79]. Although the annual production of waste heat in industries is around 10–20 TWh, the estimated demand of grid-connected houses could be 65 TWh, and several factors must be assessed before a consensus on the viability of using this heat could be established. A range of possible scenarios is analyzed to develop a consensus on the feasibility to use industrial waste heat (excluding power plants) for both domestic and non-domestic buildings.

The initial step is to map the production of waste heat and the demand of DH in different regions of Britain. According to this study, if a region has a demand greater than 3 MW/km^2, it is deemed economical to be used with DH; hence, values greater than this are mapped in the demand portfolio. Several factors including the heat values, distances, temperatures, heat losses, and seasonal variations of the production and demand are assessed.

Heat demands and sources connected
(3MW/km², 32km range criteria)

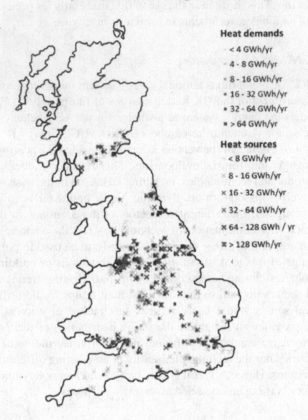

Heat demands

< 4 GWh/yr

4 - 8 GWh/yr

8 - 16 GWh/yr

16 - 32 GWh/yr

32 - 64 GWh/yr

> 64 GWh/yr

Heat sources

< 8 GWh/yr

8 - 16 GWh/yr

16 - 32 GWh/yr

32 - 64 GWh/yr

64 - 128 GWh / yr

> 128 GWh/yr

FIGURE 12.5 Layout of heat sources and sinks in the UK. (From Cooper, S. J.G. et al., *J. Energy Inst.*, 1–13, 2015.)

A layout of the distribution of heat sources and sinks in the UK for a specific scenario are in Figure 12.5.

Based on this mapping technique, it is concluded that the demand for non-domestic buildings is clustered while that of domestic buildings is unevenly distributed. Due to this reason, although the overall demand for domestic heating is greater, the potential to utilize waste heat is shared equally among both building types. Another important observation was that the heat demand varies during the season while industrial waste production is constant; this normally results in oversizing the potential. Generally, with the baseline scenario of a distance of 16 km and a heat density of 3 MW/km², the most optimistic potential of waste heat usage is 6.82 TWh, and the most pessimistic potential is 4.02 TWh. With a set of controlled criteria, which may not completely depict real-world scenarios, this study proves that a high potential for waste heat usage exists in the UK. A more powerful computing algorithm along with enhanced spatial techniques would give a much more realistic and practical picture.

12.6.4 AEA INDUSTRIAL HEAT RECOVERY IN SCOTLAND

This project was commissioned by the government of Scotland and delivered by a firm called AEA [80]. The aim was to develop a technical and financial feasibility study of converting four thermal power plants into CHP stations connected with DH. As a scenario for the future, considering the impact of CCS in conjunction with the grid is also considered. About 40 TWh/a of heat is lost in thermal power plants in Scotland, while about 64 TWh/a worth of gas is consumed for space heating alone. Instead of disposing this waste heat to the environment, a DH grid could benefit consumers and reduce carbon emissions from this low-quality heat source.

The aim was to assess the feasibility of converting these power plants to CHP stations and matching this source with potential heat loads within a 30 km radius. Initially, a GIS mapping tool, namely, Network Analyst (ArcGIS) was used to detail all the sources and sinks. Following that, a screening tool was used to assess and select technically feasible sinks in the system and compute the associated parameters. Finally, spreadsheets were used to calculate the different financial scenarios to get an insight into the economic viability of these feasible options.

The technical screening tool proved that all four potential sites were feasible to convert to CHP systems, with a range of different configurations. The technical tool also showed that despite the inclusion of costly CCS technologies, the operation of the DH grid would still be feasible. Contrary to this, the financial model showed that returns on the investments were quite low, making the schemes non-practical. Financial incentives from the government are required to make it economically feasible.

Although the study shows the huge potential available in Scotland, it is mostly used for policy-making purposes, as a case study. The fact that there are a lot of assumptions and considering that fossil-fueled heat sources are not sustainable, the study is not really practical, especially for a future renewable energy system. It covers the financial aspects in quite a lot of detail while the real technical picture is a bit blurry due to the numerous assumptions and the guess-work involved.

12.6.5 LEEDS HEAT PLANNING TOOL

The Centre for Integrated Energy Research at the University of Leeds has developed this free tool to identify possible DH sites within England and Wales, primarily used for pre-planning of DH [23]. Unlike most tools, it is based on an analysis of both the techno-economic and social criteria in Microsoft Excel. This is done with spatial mapping techniques based on input data from social, economic, geographic, technical, and housing backgrounds. The main target groups for this tool are local city councils. With this tool, potential sites are assessed and an evidence base in favor of a viable site is established to gain political support. With the prime objective of meeting social requirements of the city council, for example, by eradicating fuel poverty, further assessment is done. This is a unique tool because it also uses social criteria, since these criteria are the prime reason DH has not integrated into the UK heating sector.

Based on inputs from the user to a specific region, the tool gives a potential district—a score indicating how likely a DH grid would suit it. Weightings between −5 and +5 are assigned to the different criteria used in the assessment of that area. In the first stage of this tool, the score of a specific area is calculated in Excel. In the second stage, it is output in a CSV file, which can be linked with GIS software to provide a better visual representation.

Based on this tool, an analysis was done on a case study in the city of Leeds to identify promising sites for future DH grids. Using a sensitivity analysis technique, the sites having commercial, social, and technical favoring could easily be distinguished. Although this tool considers all aspects of planning a DH grid and is quite useful especially for the pre-planning stages, it lacks a complete technical picture. The annual energy balances, the technical characteristics of a potential setup, and details on waste heat utilization potentials are some features lacking. However, it is an easy-go tool that can be used for a less detailed preliminary assessment of a potential site.

12.6.6 University of Manchester Tool

This tool is only concerned with the possibility of using potential waste heat sources to connect to an existing DH grid [81]. It is a programming model without any graphical interface, only computing and analyzing variables based on different scenarios. Potential waste heat sources are put under economic and engineering constraints to assess their usability, with results published for a case study on a site in the Manchester area. The tool has two important criteria for positively selecting waste heat sources:

- Distance relative to the heat demand
- Part load performance of the already installed plants in the DH network

The emphasis is to provide an economic comparison of the two options of using a potential waste heat site in an existing DH grid or not to. An important consequence learned from this study is that in an already existing DH grid the inclusion of waste heat sources results in partial loads for the already installed plants, resulting in increased operational costs. Hence, this is another aspect of the feasibility of integrating these waste heat sources. The tool analyzes both the scenarios by matching demand and supply in the least expensive manner, with the help of an optimizer. Linear models are used to describe the performance of all the equipment involved. A hierarchical algorithm has been set in place to govern the operation and behavior of this equipment involved.

Although this model can be extended to the integration of distributed renewable sources, the fact that it only assesses existing DH grids make it of limited practical use, especially in the UK, where DH is already scarce. On the other hand, the focus is more on the economic feasibility of integration rather than having an overall approach. In another study conducted by the University of Manchester [14], a comprehensive model for the development of the design of a heating network, the operation according to specific criteria, and the calculation of the associated costs is developed.

It is used as a feasibility tool, using Visual Basic programming, to be used in the initial stages of DH planning and can perform a sensitivity analysis as well. Unlike similar software, the strength of this tool is that it incorporates low-temperature flow conditions that can be compared to general conditions and analyzed in detail. Initially, the heat demand of users is modeled in this tool for the testing area after which the heat network is designed, with the capability of modeling each component of the network, including pipes, substations, and heat exchangers. The only drawback of this tool is that it is aimed at community-level DH, on a relatively small scale.

12.6.7 SHEFFIELD DH EXPANSION TOOL

The purpose of this tool, developed by authorities in Sheffield, was to create a GIS map for the analysis of an expansion of the already existing DH grid, within the boundaries of this city [75]. After the collection of data, a study was performed to generate detailed heat maps to identify potential heat sources and sinks to form a consensus on the expansion. Sheffield has an EfW facility that produces about 120 GWh of heat annually, serving about 140 buildings of all types. This district energy network is claimed to be the largest in the UK and is looked upon especially due to the low carbon heat source. For assessing the feasibility of expanding this network, a GIS tool called ESRI ArcGIS was used. However, the role of the GIS was limited to gathering data and formulating multi-layered heat maps, while most planning was done manually.

Based on case studies of Barnsley and Sheffield, deductions were drawn about the outcomes of this tool. Using mathematical modeling and advanced monitoring techniques, Barnsley has developed a series of mini biomass-fueled DH grids having extremely low carbon emissions along with being quite successful, especially as a case study for further development in UK cities. In an extension of this study [17], the environmental, social aspects along with other factors causing hindrance in the expansion of DH in the UK are analyzed, with regards to the Barnsley case study [26].

12.6.8 TECHNOLOGIES AND URBAN RESOURCE NETWORKS

This program was developed by a research team at the Imperial College of London and Newcastle University [82]. As an eventual improvement for the Vantage Point software used by the Newcastle City Council, this new tool was launched. Although it is used to model the entire energy system of a region, specific subsectors can also be analyzed in detail, including DH. This computational model inputs energy demands for various times of the year and uses mixed integer linear programming techniques to find the mix of technologies to supply this demand, having the least costs. The timescale of this model is flexible, enabling it to be analyzed for both long-term and short-term planning. The inputs are controlled by the user, which helps in performing a sensitivity analysis to further study specific scenarios.

The computing time and depth of detail are quite high, resulting in an accurate output. However, the biggest drawback of such detailed custom-made model is that it is only applicable to a specific region. Currently, this model is only for the Newcastle area and would have to be altered for use by other city councils. It also lacks social factors, and considering the high amount of input data required, it is quite difficult to update, especially the financial variables.

12.6.9 NATIONAL HEAT MAPS

In 2012, the DECC launched a national heat map of England using Google Maps, accurate to street level [23]. It displays annual heat demands, which can be aggregated in clusters of square kilometers. Buildings by sector can also be differentiated. It also displays the existing sources of heat, including only CHP plants for the moment. This tool is being expanded to display heat maps along with possible pipeline routes to be used after the initial feasibility assessment stage. In the future, the data set of this tool will be used in combination of modeling and GIS software, which gives it a promising future in terms of practicality.

This heat map does not incorporate seasonal variations and uses only annual averages, for now. It also does not predict future heat loads as a result of energy efficient measures. Many critics believe that the strategic use of these maps has been limited, due to these reasons [69]. Having identified heat centers, there is no enforcement strategy that the building owners will comply with and use the DH network. Considering this point, the Scottish government has developed a heat mapping toolkit in which users can develop their own heat maps, with their custom requirements. Currently, this toolkit is in the form of a free online manual along with online datasets. Several maps developed by local authorities based on the guidelines of this toolkit are also available online.

12.6.10 ENERGY PATH—DEVELOPED BY ENERGY TECHNOLOGIES INSTITUTE

The Energy Technologies Institute is a research center that links firms with the government for promoting sustainable technologies. Their latest tool with regards to DH is the Energy Path [83]. Several industrial partners are onboard this project, which develops pathways for all aspects of future energy networks, including DH. The development of this project is to be done in two phases. In the first phase, software tools are being developed for every important location within the UK. The input data, parameters, and the constraints are region-specific. The feasibilities of different locations are predicted in this initial phase. In the second phase, the output of the first phase are designed and planned upon. This is again done with the help of additional software tools to practically demonstrate the functionality of researching and designing a future DH system. The Energy Path program has been split into different areas dealing with consumer, technology, economic, and supply chain aspects of DH. The latest version uses both GIS and modeling techniques to analyze and formulate a consensus on the data. The Energy Path Networks tool is in the development stages at the moment. However, it is believed to be more concerned with the financial and business aspects to lure future investors in developing DH.

There is a wide variety of practical and innovative tools focusing on different aspects of planning and development. However, there is a dire need for a holistic tool incorporating all aspects of planning for the entire country. A common consensus by many investors and housing authorities was established in the Vanguard's network meeting [23], that a comprehensive tool is yet to be established in the UK. Ideally, it would have holistic capabilities applicable throughout the country be transparent, modular, and open source to suit both designers and policy-makers [73].

12.7 CONCLUSIONS

Although about 2000 DH schemes have been built in the UK, most of these are extremely small and unsustainable, making the UK lag considerably in this technology compared to its European neighbors. A series of government policies and initiatives have been set into motion under the umbrella of the Carbon Plan, through which the UK plans to supply 40% of its approximate 500 TWh of heat demand in 2050. Keeping in view past hurdles and loopholes in the UK's organizational setup along with the fact that DH development costs are 20% higher than most of Europe, this target seems quite steep, at least for the time being. However, the development of several innovative simulators and research tools are paving the way to future development and is a promising contribution by the research community.

In conclusion, the state-of-the-art DH is expected to develop comprehensively over the future. The concept of fourth-generation DH may take lead to transform the current infrastructure to a simpler and low-cost one [10]. The holistic picture presented in this study shows that the potential for such transformation is high in the UK.

ACKNOWLEDGMENTS

The research presented here is a result of the activities of the Low Carbon Technologies research group at the School of Energy, Construction, and Environment at Coventry University, UK. We would like thank our colleagues, associated staff, and the university for their support and cooperation. The work was funded by the EPSRC project (EP/K002716/1) "An Intelligent Digital Household Network to Transform Low Carbon Lifestyles."

REFERENCES

1. Jenkins GM, Perry M, Prior J. *The Climate of the UK and Recent Trends*. Met Office Hadley Centre, Exeter, UK: 2009.
2. Department of Energy and Climate Change UK. The Future of Heating: Meeting the Challenge. 2013.
3. Kesicki F. Costs and potentials of reducing CO_2 emissions in the UK domestic stock from a systems perspective. *Energy Build* 2012;51:203–11. doi:10.1016/j.enbuild.2012.05.013.
4. Department of Energy and Climate Change UK. The Carbon Plan: Delivering our low carbon future. London: 2011.
5. Nelson J, Amos J, Hutchinson D, Denman M. Prospects for city-scale combined heat and power in the UK. *Appl Energy* 1996;53:119–48. doi:10.1016/0306-2619(95)00058-5.
6. Gallo E. Skyscrapers and District Heating, an inter-related History 1876–1933. *Costr Hist* 2003;1.

7. Matson C. District heating: A real alternative. *Reinf Plast* 2016;17:32–5. doi:10.1016/j.ref.2015.11.003.

8. Lucia U. Econophysics and bio-chemical engineering thermodynamics: The exergetic analysis of a municipality. *Physica A* 2016;462:421–30. doi:10.1016/j.physa.2016.06.119.

9. Wiltshire R. *Advanced District Heating and Cooling (DHC) Systems*. Cambridge, Woodhead Publishing Ltd: 2015.

10. Lund H, Werner S, Wiltshire R, Svendsen S, Thorsen JE, Hvelplund F, et al. 4th Generation District Heating (4GDH). Integrating smart thermal grids into future sustainable energy systems. *Energy* 2014;68:1–11. doi:10.1016/j.energy.2014.02.089.

11. Bolton R, Foxon TJ. Infrastructure transformation as a socio-technical process—Implications for the governance of energy distribution networks in the UK. *Technol Forecast Soc Change* 2015;90:538–50. doi:10.1016/j.techfore.2014.02.017.

12. Chaudry M, Abeysekera M, Hosseini SHR, Jenkins N, Wu J. Uncertainties in decarbonising heat in the UK. *Energy Policy* 2015;87:623–40. doi:10.1016/j.enpol.2015.07.019.

13. Hawkey DJC. District heating in the UK: Prospects for a third national programme. *Sci Technol Stud* 2014;27:68–89.

14. Ahmed A, Mancarella P. Strategic techno-economic assessment of heat network options for distributed energy systems in the UK. *Energy* 2014;75:182–93. doi:10.1016/j.energy.2014.07.011.

15. Broin EO, Nassen J, Johnsson F. The influence of price and non-price effects on demand for heating in the EU residential sector. *Energy* 2015;81:146–58. doi:10.1016/j.energy.2014.12.003.

16. Gill ZM, Tierney MJ, Pegg IM, Allan N. Measured energy and water performance of an aspiring low energy/carbon affordable housing site in the UK. *Energy Build* 2011;43:117–25. doi:10.1016/j.enbuild.2010.08.025.

17. Finney KN, Chen Q, Sharifi VN, Swithenbank J, Nolan A, White S, et al. Developments to an existing city-wide district energy network: Part II—Analysis of environmental and economic impacts. *Energy Convers Manag* 2012;62:176–84. doi:10.1016/j.enconman.2012.03.005.

18. Pirouti M, Bagdanavicius A, Ekanayake J, Wu J, Jenkins N. Energy consumption and economic analyses of a district heating network. *Energy* 2013;57:149–59. doi:10.1016/j.energy.2013.01.065.

19. Lucia U, Grisolia G. Unavailability percentage as energy planning and economic choice parameter. *Renew Sustain Energy Rev* 2017;75:197–204. doi:10.1016/j.rser.2016.10.064.

20. EuroHeat & Power. Guidelines for District Heating Substations & Countrywise Survey. Brussels: 2008.

21. Lygnerud K. Challenges for business change in district heating. *Energy Sustain Soc* 2018;8. doi:10.1186/s13705-018-0161-4.

22. Babus'Haq RF, Probert SD. Combined heat-and-power implementation in the UK: Past, present and prospective developments. *Appl Energy* 1996;53:47–76. doi:10.1016/0306-2619(95)00054-2.

23. Bale C, Bush R, Taylor P. Spatial mapping tools for district heating (DH): Helping local authorities tackle fuel poverty. *Final Report* 2014:1–51.

24. Hawkey DJC. District heating in the UK: A Technological Innovation Systems analysis. *Environ Innov Soc Transit* 2012;5:19–32. doi:10.1016/j.eist.2012.10.005.

25. Department of Energy and Climate Change UK. Summary evidence on District Heating Networks in the UK. London: 2013.

26. Finney KN, Zhou J, Chen Q, Zhang X, Chan C, Sharifi VN, et al. Modelling and mapping sustainable heating for cities. *Appl Therm Eng* 2013;53:246–55. doi:10.1016/j.applthermaleng.2012.04.009.

27. Kelly KA, McManus MC, Hammond GP. An energy and carbon life cycle assessment of industrial CHP (combined heat and power) in the context of a low carbon UK. *Energy* 2014;77:812–21. doi:10.1016/j.energy.2014.09.051.

28. Connor PM, Xie L, Lowes R, Britton J, Richardson T. The development of renewable heating policy in the United Kingdom. *Renew Energy* 2015;75:733–44. doi:10.1016/j.renene.2014.10.056.

29. Grant J, Horne R, Mortimer N, Hetherington R. A comparative assessment of the energy and carbon balance of utilizing straw. *Energy* 1996;21:77–86.

30. Lucia U, Simonetti M, Chiesa G, Grisolia G. Ground-source pump system for heating and cooling: Review and thermodynamic approach. *Renew Sustain Energy Rev* 2017;70:867–74. doi:10.1016/j.rser.2016.11.268.

31. Taylor PG, Bolton R, Stone D, Upham P. Developing pathways for energy storage in the UK using a coevolutionary framework. *Energy Policy* 2013;63:230–43. doi:10.1016/j.enpol.2013.08.070.

32. Xie Y, Gilmour MS, Yuan Y, Jin H, Wu H. A review on house design with energy saving system in the UK. *Renew Sustain Energy Rev* 2017;71:29–52. doi:10.1016/j.rser.2017.01.004.

33. Wright DG, Dey PK, Brammer J. A barrier and techno-economic analysis of small-scale bCHP (biomass combined heat and power) schemes in the UK. *Energy* 2014;71:332–45. doi:10.1016/j.energy.2014.04.079.

34. International Energy Agency. CHP-DHC Country Scorecard. Paris: 2009.

35. Davies GF, Maidment GG, Tozer RM. Using data centres for combined heating and cooling: An investigation for London. *Appl Therm Eng* 2016;94:296–304. doi:10.1016/j.applthermaleng.2015.09.111.

36. Watson J, Gross R, Ketsopoulou I, Winskel M. The impact of uncertainties on the UK's medium-term climate change targets. *Energy Policy* 2015;87:685–95. doi:10.1016/j.enpol.2015.02.030.

37. McCormick K, Anderberg S, Coenen L, Neij L. Advancing sustainable urban transformation. *J Clean Prod* 2013;50:1–11. doi:10.1016/j.jclepro.2013.01.003.

38. European Union. Renewable Energy Directive. Brussels: 2009.

39. European Union. Roadmap 2050. Brussels: 2012. doi:10.2833/10759.

40. European Union. Energy Performance of Buildings. Brussels: 2010. doi:10.3000/17252555.L_2010.153.eng.

41. European Union. Energy Efficiency Plan 2011. Brussels: 2011.

42. AEBIOM. EU Handbook District Heating Markets. Brussels: 2012.

43. Department of Energy and Climate Change UK. The Energy Efficiency Strategy: The Energy Efficiency Opportunity in the UK. London: 2012.

44. Swithenbank J, Finney KN, Chen Q, Yang YB, Nolan A, Sharifi VN. Waste heat usage. *Appl Therm Eng* 2012;60:430–40. doi:10.1016/j.applthermaleng.2012.10.038.

45. Xing Y, Hewitt N, Griffiths P. Zero carbon buildings refurbishment—A Hierarchical pathway. *Renew Sustain Energy Rev* 2011;15:3229–36. doi:10.1016/j.rser.2011.04.020.

46. Eyre N, Baruah P. Uncertainties in future energy demand in UK residential heating. *Energy Policy* 2015;87:641–53. doi:10.1016/j.enpol.2014.12.030.

47. Cheng V, Steemers K. Modelling domestic energy consumption at district scale: A tool to support national and local energy policies. *Environ Model Softw* 2011;26:1186–98. doi:10.1016/j.envsoft.2011.04.005.

48. Quiggin D, Buswell R. The implications of heat electrification on national electrical supply-demand balance under published 2050 energy scenarios. *Energy* 2016;98:253–70. doi:10.1016/j.energy.2015.11.060.

49. Jablonski S, Pantaleo A, Bauen A, Pearson P, Panoutsou C, Slade R. The potential demand for bioenergy in residential heating applications (bio-heat) in the UK based on a market segment analysis. *Biomass Bioenergy* 2008;32:635–53. doi:10.1016/j.biombioe.2007.12.013.

50. Thomas A, Bond A, Hiscock K. A GIS based assessment of bioenergy potential in England within existing energy systems. *Biomass Bioenergy* 2013;55:107–21. doi:10.1016/j.biombioe.2013.01.010.

51. Levidow L, Papaioannou T. UK priorities for decarbonisation through biomass. *Sci Public Policy* 2015;43:1–16. doi:10.1093/scipol/scv016.

52. Mollenhauer E, Christidis A, Tsatsaronis G. Evaluation of an energy- and exergy-based generic modeling approach of combined heat and power plants. *Int J Energy Environ Eng* 2016;7:167–76. doi:10.1007/s40095-016-0204-6.

53. Stankeviciute L, Riekkola AK. Assessing the development of combined heat and power generation in the EU. *Int J Energy Sect Manag* 2011;2020:1–7. doi:10.1108/IJESM-08-2012-0004.

54. Perry S, Klemes J, Bulatov I. Integrating waste and renewable energy to reduce the carbon footprint of locally integrated energy sectors. *Energy* 2008;33:1489–97. doi:10.1016/j.energy.2008.03.008.

55. Ammar Y, Joyce S, Norman R, Wang Y, Roskilly AP. Low grade thermal energy sources and uses from the process industry in the UK. *Appl Energy* 2012;89:3–20. doi:10.1016/j.apenergy.2011.06.003.

56. Weber C, Shah N. Optimisation based design of a district energy system for an eco-town in the United Kingdom. *Energy* 2011;36:1292–308. doi:10.1016/j.energy.2010.11.014.

57. Anna Bright. Implementation steps to large scale district heating. Birmingham: 2014.

58. Evans D, Stephenson M, Shaw R. The present and future use of "land" below ground. *Land Use Policy* 2009;26:302–16. doi:10.1016/j.landusepol.2009.09.015.

59. Eames P, Loveday D, Haines V, Romanos P. The Future Role of Thermal Energy Storage in the UK Energy System: An Assessment of the Technical Feasibility and Factors Influencing Adoption. London: 2014.

60. Persson U, Werner S. Heat distribution and the future competitiveness of district heating. *Appl Energy* 2011;88:568–76. doi:10.1016/j.apenergy.2010.09.020.

61. Davies G, Woods P. *The Potential and Costs of District Heating Networks*. Oxford: 2009.

62. Department of Energy and Climate Change UK. Assessment of the Costs, Performance, and Characteristics of UK Heat Networks. London: 2015.

63. Hawkey D, Webb J, Winskel M. Organisation and governance of urban energy systems: District heating and cooling in the UK. *J Clean Prod* 2013;50:22–31. doi:10.1016/j.jclepro.2012.11.018.

64. Kelly S, Pollitt M. An assessment of the present and future opportunities for combined heat and power with district heating (CHP-DH) in the United Kingdom. *Energy Policy* 2010;38:6936–45. doi:10.1016/j.enpol.2010.07.010.

65. Shackley S, Green K. A conceptual framework for exploring transitions to decarbonised energy systems in the United Kingdom. *Energy* 2007;32:221–36. doi:10.1016/j.energy.2006.04.010.

66. Heiskanen E, Matschoss K. Understanding the uneven diffusion of building-scale renewable energy systems: A review of household, local and country level factors in diverse European countries. *Renew Sustain Energy Rev* 2017;75:580–91. doi:10.1016/j.rser.2016.11.027.

67. Curtin J, Mcinerney C, Gallachóir BÓ. Financial incentives to mobilise local citizens as investors in low-carbon technologies: A systematic literature review. *Renew Sustain Energy Rev* 2017;75:534–47. doi:10.1016/j.rser.2016.11.020.

68. Ziemele J, Cilinskis E, Zogla G, Gravelsins A, Blumberga A, Blumberga D. Impact of economical mechanisms on CO_2 emissions from non-ETS district heating in Latvia using system dynamic approach. *Int J Energy Environ Eng* 2017;9:1–11. doi:10.1007/s40095-017-0241-9.

69. Webb J. Improvising innovation in UK urban district heating: The convergence of social and environmental agendas in Aberdeen. *Energy Policy* 2015;78:265–72. doi:10.1016/j.enpol.2014.12.003.

70. Jamasb T, Pollitt M. Security of supply and regulation of energy networks. *Energy Policy* 2008;36:4584–9. doi:10.1016/j.enpol.2008.09.007.

71. Bush RE, Bale CSE. The role of intermediaries in the transition to district heating. *Energy Procedia* 2017;116:490–9. doi:10.1016/j.egypro.2017.05.098.

72. Rocco R. Policy Frameworks for Energy Transition in England: Challenges in a former industrial city. *J Settlements Spat Plan* 2016:41.

73. Sousa G, Jones BM, Mirzaei PA, Robinson D. A review and critique of UK housing stock energy models, modelling approaches and data sources. *Energy Build* 2017;151:66–80. doi:10.1016/j.enbuild.2017.06.043.

74. Delmastro C, Mutani G, Schranz L. The evaluation of buildings energy consumption and the optimization of district heating networks: A GIS-based model. *Int J Energy Environ Eng* 2016;7:343–51. doi:10.1007/s40095-015-0161-5.

75. Finney KN, Sharifi VN, Swithenbank J, Nolan A, White S, Ogden S. Developments to an existing city-wide district energy network—Part I: Identification of potential expansions using heat mapping. *Energy Convers Manag* 2012;62:165–75. doi:10.1016/j.enconman.2012.03.006.

76. Persson U, Möller B, Werner S. Heat Roadmap Europe: Identifying strategic heat synergy regions. *Energy Policy* 2014;74:663–81. doi:10.1016/j.enpol.2014.07.015.

77. Dodds PE. Integrating housing stock and energy system models as a strategy to improve heat decarbonisation assessments. *Appl Energy* 2014;132:358–69. doi:10.1016/j.apenergy.2014.06.079.

78. McKenna RC, Norman JB. Spatial modelling of industrial heat loads and recovery potentials in the UK. *Energy Policy* 2010;38:5878–91. doi:10.1016/j.enpol.2010.05.042.

79. Cooper SJG, Hammond GP, Norman JB. Potential for use of heat rejected from industry in district heating networks, GB perspective. *J Energy Inst* 2015:1–13. doi:10.1016/j.joei.2015.01.010.

80. McNaught C, Williams E, Stambaugh J, Kiff B. A study into the recovery of heat from power generation in Scotland. Glengarnock: 2011.

81. Kapil A, Bulatov I, Smith R, Kim J-K. Process integration of low grade heat in process industry with district heating networks. *Energy* 2012;44:11–9. doi:10.1016/j.energy.2011.12.015.

82. Keirstead J, Calderon C. Capturing spatial effects, technology interactions, and uncertainty in urban energy and carbon models: Retrofitting newcastle as a case-study. *Energy Policy* 2012;46:253–67. doi:10.1016/j.enpol.2012.03.058.

83. Energy Technologies Institute. Smart Systems and Heat Manual. Birmingham: 2015.

13 Potential of the Thermal Energy Storage System in Peak Shaving

Chinnasamy Veerakumar and Appukuttan Sreekumar

CONTENTS

13.1 INTRODUCTION

The fossil fuel reserves are depleting drastically, and the energy demand is increasing simultaneously. This requires the need for more attention toward the utilization of renewable energy sources and adopting various energy-saving measures in order to reduce the energy demand. The energy demand is not constant, and it varies with time. Peak energy demand is a major issue to be focused on while dealing with energy demand and supply. It is well known that buildings contribute a major role in global energy consumption by cooling, heating, and lighting requirements. By implementing smart-energy management measures in buildings, the peak energy demand can be balanced. Several techniques are conventionally adopted to address the peak energy demand, and peak shaving is one among them. This chapter discusses the overview of energy demand and energy consumption by buildings. The various strategies for peak shaving and how thermal energy storage helps in addressing the problem of peak demand are also discussed.

13.2   ENERGY DEMAND

Energy demand is defined as the energy consumed by all human activities. It influences the amount of energy used, the location and type of fuel used, and the characteristics of the energy system and technology. According to the International Energy Agency, energy demand increased globally by 2.3% in 2018 (International Energy Agency IEA, 2018). In the present decade, the fastest growth of energy demand is due to increased heating and cooling needs and a robust global economy. The demand for heating and cooling contributes one-fifth of an increase in global energy demand due to an increase in summer and winter temperatures. A cold snap increases the heating demand, and a hot summer increases the demand for cooling. Three countries, namely China, India, and the United States, are together responsible for a 70% increase in the global energy demand (International Energy Agency IEA, 2011).

The important sectors of the global economy contributed to the major part of global energy demand. The energy transition depends on the method of energy storage and consumption in the different sectors of the global economy. The primary energy consumption by the end-use sector is shown in Figure 13.1. The industrial sector around the world currently consumes almost half of the global energy. The building sector consumes 29%, and the transportation sector accounts for the remaining 21%. In the developing countries, there is a significant increase in energy demand for lighting, space heating, and cooling requirements, which expands due to growth in the economy and lifestyle of people. The growth of annual demand and sector-wise contribution is shown in Figure 13.2.

13.2.1   Base Load

Base load is the minimum level of energy demand required over a period of 24 hours. It is needed to provide power to components that keep running at all times (also referred to as continuous load).

13.2.2   Peak Load

Peak load is the highest amount of energy that a consumer draws from the grid in a specific time period. These peaking demands are often for only shorter durations.

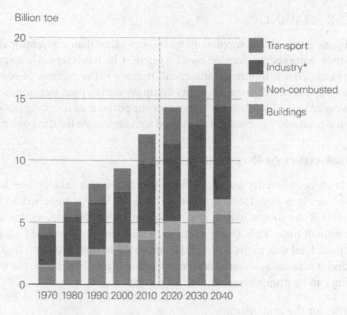

FIGURE 13.1 Primary energy consumption by end-use sector.

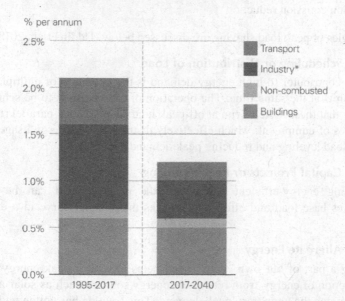

FIGURE 13.2 Annual demand growth and sector contribution.

In mathematical terms, peak demand could be understood as the difference between the base demand and the highest demand. Understanding peak load is essential for any commercial energy management strategy because it is used to determine a part of your energy bill. In some places, the cost of electricity during peak load is higher than the normal consumption period.

13.3   PEAK SHAVING

During the peak demand, the supplies from sources other than conventional supplies are used to reduce the energy demand on the system, which is referred to as peak shaving (Uddin et al., 2018). In other words, peak shaving is the process of reducing the amount of energy purchased from the utility company during peak demand hours. Peak shaving is similar to load leveling but maybe for the purpose of reducing peak demand rather than the economy of operation. It is used to compensate for the power shortage.

13.3.1   IMPORTANCE OF PEAK LOAD SHAVING

Peak load is an occasionally occurring sensitive factor that takes place for a short duration of time in a day. Conventionally, the peak load is matched to the additional capacity of the source, which is not economical or efficient, as it is used for a shorter duration of time. This also leads to high fuel consumption and CO_2 emission. Therefore, peak load shaving is a preferable approach to match sudden high demand and other disadvantages associated with capacity addition. In general, the benefits of peak shaving can be grouped into three categories:

- Benefits for the grid operator
- Benefits for end-user
- Carbon emission reduction

The strategies of peak load shaving are discussed below (Uddin et al., 2018).

13.3.1.1   Scheduling or Distribution of Load

The major contributor to high energy demand is the operation of multiple energy-intensive units at the same time. The operation of the system is to be scheduled in such a way that these systems run at off-peak hours. This avoids parallel running of all the types of equipment, which effectively distributes the energy usage, thereby providing load leveling and reducing peak demand.

13.3.1.2   Capital Projects or Peak Shrinking

By installing energy-efficient equipment, the peak demand can be reduced. This reduces base load and effectively shrinks the demand curve in a downward direction.

13.3.1.3   Alternate Energy

Generating a part of our own energy need helps to reduce peak energy demand. The production of energy from renewable energy sources, such as solar and wind, is used to match the temporary peak demand. This solution has gotten wide acceptability nowadays due to technological advancements.

13.3.1.4   Energy Storage

Energy storage provides an alternative option for reducing peak energy demand. During a low-energy consumption period, the energy is stored, and when there is high energy consumption or peak energy demand, this stored energy is used to match the increased load.

13.3.2 Peak Shaving Using the Energy Storage System

Peak shaving can be achieved by integrating energy storage systems to the load and providing an energy supply at the time of high demand for a short duration of time. The different types of commonly adopted energy storage technologies are:

1. Mechanical energy storage
2. Electrochemical energy storage
3. Thermal energy storage

Among the different types of energy storage techniques, electrochemical energy storage is the most commonly used technology for temporary energy storage. In electrochemical energy storage, batteries are used to store electrical energy, which is available in surplus or during the off-peak time, and stored energy will be used during peak time.

13.4 ENERGY CONSUMPTION IN BUILDINGS

Buildings are responsible for about 40% of the global energy consumption, and air-conditioning contributes to a major part of it (Cui and Riffat, 2011; Khadiran et al., 2016). Due to increased living standards, the air-conditioning requirement is drastically increasing. The growth of energy use in buildings is presented in Figure 13.3. The final energy consumption in buildings categorized by fuels is presented in Figure 13.4.

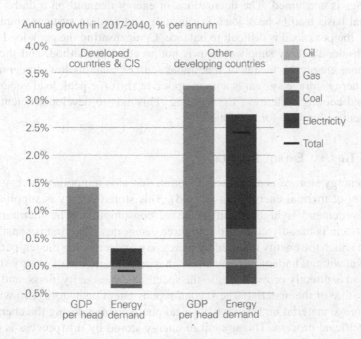

FIGURE 13.3 Growth of energy use in buildings.

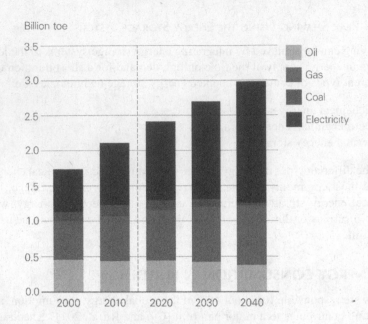

FIGURE 13.4   Final energy consumption in buildings by fuels.

13.4.1   NEED FOR THERMAL ENERGY STORAGE IN BUILDINGS

Heating or cooling is the major requirement for which most of the final energy in buildings is consumed. The fluctuations in energy demand on a daily, weekly, or seasonal basis lead to peak load. For heating and cooling systems employed in buildings, the peak load is difficult to balance. Compensating the peak load with an overdimensioned energy supply system is not an efficient method, and the supply of additional energy by means of renewable sources remains low. Integrating the thermal energy storage system is a good option to shift the peak load generated by heating and cooling systems of the building. This aids to develop efficient energy management technology for buildings.

13.4.2   THERMAL ENERGY STORAGE

Thermal energy storage is an attractive option that aids temporary energy storage in the form of thermal energy (hot or cold). This stored energy is supplied when the energy demand is higher than the usual consumption. The thermal energy storage system is broadly classified into three categories. The first is sensible heat storage in which the energy is stored by increasing or decreasing the temperature of the material without undergoing any phase change. The amount of energy stored by this method is directly proportional to the specific heat capacity, mass, and change in temperature of the material. The second type is latent heat storage in which the energy storage material undergoes a physical phase change during the energy storage and retrieval process. The amount of energy stored by this process is directly proportional to the latent heat of melting/freezing and mass of the energy storage

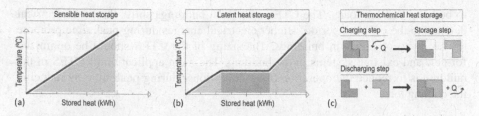

FIGURE 13.5 Thermal energy storage methods: (a) sensible, (b) latent, and (c) thermochemical storage. (From Lizana, J. et al., *Renew. Sustain. Energy Rev.*, 82, 3705–3749, 2018.)

material. In the third method, the energy is stored by reversible chemical reactions. The different types of thermal energy storage technologies are discussed in detail by many researchers (Hahne n.d.; Li et al. 2012; Sharif et al. 2015). The amount of energy stored is directly proportional to the reaction potential of the chemical reaction. The three types of thermal energy storage are clearly illustrated in Figure 13.5.

Out of the three methods of energy storage, latent heat thermal energy storage possesses several advantageous over others, which are listed below (Dheep and Sreekumar 2015):

- The amount of energy stored is comparatively higher (5 to 14 times higher than sensible heat storage).
- Isothermal heat storage and retrieval—almost at a constant temperature or with less variation in temperature.
- Less space requirement—more energy can be stored with less volume.
- Wide range of applications based on the phase change temperature of energy storage material.

In this chapter, latent heat thermal energy storage, which uses phase change material (PCM) as an energy storage medium, is discussed further.

13.4.3 FREE COOLING

Free cooling is a process of storing cold energy available in the ambient air with low temperature and utilizing the stored energy when it is required. It is also termed as night ventilation if the nighttime cold air is used to charge the energy storage medium. This method consumes less energy or no energy. The free cooling or night ventilation system will be efficient if the diurnal temperature difference is 10°C or more (Osterman et al., 2012, 2015; Zalba et al., 2004).

13.4.4 ADVANTAGES OF COLD THERMAL ENERGY STORAGE (CTES) IN BUILDINGS

In a building, cooling load is the largest contributor to peak energy demand. During summer, the electrical utilities experience their peak electrical demand mainly due to building cooling loads. The CTES system in buildings shifts the peak electrical demand to the off-peak period. Higher electrical charges during peak time promote the necessity of CTES in buildings. The sizing of the CTES should be optimized for new and existing systems in the building. The main application of CTES in the building is to reduce the operation of air-conditioner during peak time by reducing the cooling load.

13.5 METHODS OF INTEGRATING THERMAL ENERGY STORAGE WITH BUILDINGS

The thermal energy storage can be integrated into the building structures in different ways. Encapsulation of PCM, incorporating PCM in building materials, and various experimental designing parameters are important and discussed in this section.

13.5.1 PCM ENCAPSULATION

The PCM cannot be directly applied in the application, and it has to be encapsulated in containment in order to prevent direct exposure to the environment. Also, PCM, which is the energy storage medium, undergoes physical phase change during energy storage and retrieval. It should not be leaking during the phase change process. Hence, the PCM is packed in an enclosure according to the requirement. Based on the size, it is classified as nano, micro, and macro encapsulation, which is selected as per necessity. The encapsulation materials generally used are metals, polymers, etc. It is made in differential geometry, and a suitable model is selected for the respective applications. The PCM filled in blocks, pouches, spherical capsules, etc. in macro-scale made of metallic or polymeric film is referred to as micro-encapsulation (Figure 13.6a–f) (Raj and Velraj 2010).

13.5.2 PCM USED IN WALLS AND CEILINGS

Encapsulated PCM is used as wall materials, and nano and micro encapsulations are commonly used for this application. Wallboards and ceiling panels made of encapsulated PCM are used for thermal comfort in buildings for a shorter duration of time. He et al. (2019) experimentally studied the performance of a novel radiative cooling PCM wall to reduce the cooling load and improve the thermal comfort of the occupant. Three rooms are used for the experiment out of which the first is made of brick, the second is made of PCM wall, and the third is made of radiatively cooled PCM wall, which is shown in Figure 13.7. From the results of the study, a radiatively cooled PCM wall shows good potential for load reduction and fossil fuel consumption in the summer season.

The numerical analysis of micro-dispersed PCM composite boards for different seasons was performed by El Omari et al. (2016). The potential of using PCM composite instead of pure insulation material is discussed in this work. Souayfane et al. (2019) experimentally performed the energy and economic analysis of combined

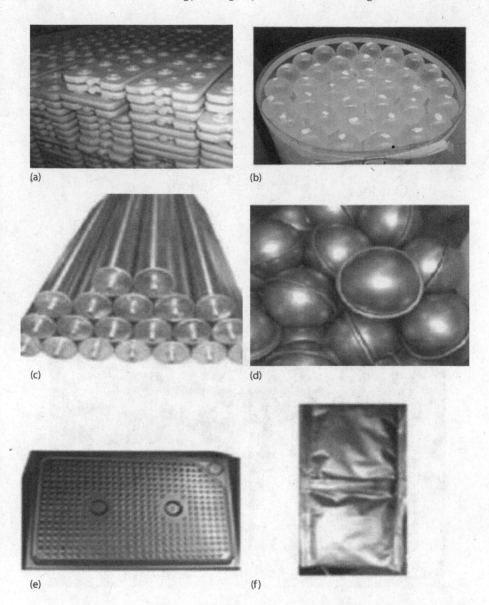

(a)

(b)

(c)

(d)

(e)

(f)

FIGURE 13.6 (a) Commercial rectangular macro-encapsulated PCM, (b) PCM encapsulated in spheres, (c) PCM tube encapsulation, (d) PCM metal ball encapsulation, (e) PCM encapsulated in aluminum panel, and (f) PCM encapsulated in aluminum pouches. (From Raj, V.A.A. and Velraj, R., *Renew. Sustain. Energy Rev.*, 14, 2819–2829, 2010.)

transparent insulation material with a PCM-incorporated wall as a building envelope under different climatic conditions. The experimental system is simulated using a validated numerical model. The result shows that the use of the TIM-PCM wall as shown in Figure 13.8 has a high potential for energy saving with a payback period of 7.8 to 10.5 years.

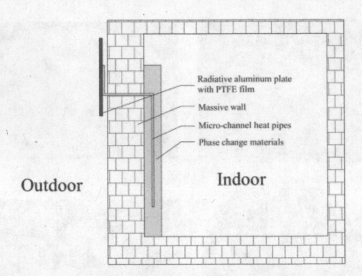

FIGURE 13.7 Structure of the RC-PCM wall. (From He, W. et al., *Energy Build.*, 2019.)

FIGURE 13.8 TIM-PCM wall from the outside. (From Souayfane, F. et al., *Energy*, 1274–1291, 2019.)

13.6 DISTRICT HEATING AND COOLING USING THERMAL ENERGY STORAGE

The district heating and cooling system, which is generally referred to as the district energy systems, is used to effectively and efficiently utilize the thermal energy in a local area. This system is classified as hourly, daily, weekly, and

FIGURE 13.9 District cooling plant at Halmstad University campus operated by Halmstad Energy and Environment. (From Cabeza, L.F., *Advances in Thermal Energy Storage Systems: Methods and Applications*, Elsevier, Amsterdam, the Netherlands, 2014.)

seasonal storage. The hourly and daily storage systems are meant for short-term storage, which supplies the energy during expected peak demand. Weekly and seasonal storage is long-term storage used to match the demand for a longer duration. In the future, the district heating and cooling systems will be combined with conventional power systems to match the energy requirement in domestic and commercial applications.

The district cooling plant owned and operated by Halmstad Energy and Environment at Halmstad University campus is shown in Figure 13.9 (Cabeza 2014). This unit consists of a 3.6 MW capacity absorption chiller operated by heat from the incineration process through the district heating system. This is a combined heating and cooling model.

13.7 CONCLUSION

Building-integrated storage systems are being used to shave the peak load and also to increase the usage of renewable energy power in building energy consumption. For space-heating applications, sensible heat storage is commonly used. In the case of indoor thermal comfort applications, latent heat thermal energy storage employs PCM as the thermal energy storage is used. The technologies that use PCM are developed, and in the future, most of the construction will be integrated with thermal energy storage. As the building sector plays a major role in overall peak energy demand, thermal energy storage helps in matching the regular and irregular energy requirements so that the energy consumption is streamlined.

REFERENCES

Cabeza, L.F., 2014. *Advances in Thermal Energy Storage Systems: Methods and Applications.* Amsterdam, the Netherlands: Elsevier.

Cui, M.Y., Riffat, S., 2011. *Review on Phase Change Materials for Building Applications 1958–1962.* https://doi.org/10.4028/www.scientific.net/AMM.71-78.1958

Dheep, G.R., Sreekumar, A., 2015. *Phase Change Materials—A Sustainable Way of Solar Thermal Energy Storage.* New Delhi, India: Springer, pp. 217–244. https://doi.org/10.1007/978-81-322-2337-5_9

El Omari, K., Le Guer, Y., Bruel, P., 2016. Analysis of micro-dispersed PCM-composite boards behavior in a building's wall for different seasons. *J. Build. Eng.* 7, 361–371. https://doi.org/10.1016/j.jobe.2016.07.013

Hahne, E., n.d. Storage of Sensible Heat. Stuttgart, Germany, University of Stuttgart.

He, W., Yu, C., Yang, J., Yu, B., Hu, Z., Shen, D., Liu, X., Qin, M., Chen, H., 2019. Experimental study on the performance of a novel RC-PCM-wall. *Energy Build.* https://doi.org/10.1016/j.enbuild.2019.07.001

International Energy Agency IEA, 2018. Global energy demand report.

International Energy Agency IEA, 2011. Worldwide Trends in Energy Use and Efficiency. Authoring Institution: International Energy Agency (IEA).

Khadiran, T., Hussein, M.Z., Zainal, Z., Rusli, R., 2016. Advanced energy storage materials for building applications and their thermal performance characterization: A review. *Renew. Sustain. Energy Rev.* 57, 916–928. https://doi.org/10.1016/j.rser.2015.12.081

Li, G., Hwang, Y., Radermacher, R., 2012. Review of cold storage materials for air conditioning application. *Int. J. Refrig.* 35, 2053–2077. https://doi.org/10.1016/j.ijrefrig.2012.06.003

Lizana, J., Chacartegui, R., Barrios-Padura, A., Ortiz, C., 2018. Advanced low-carbon energy measures based on thermal energy storage in buildings: A review. *Renew. Sustain. Energy Rev.* 82, 3705–3749. https://doi.org/10.1016/j.rser.2017.10.093

Osterman, E., Butala, V., Stritih, U., 2015. PCM thermal storage system for "free" heating and cooling of buildings. *Energy Build.* https://doi.org/10.1016/j.enbuild.2015.04.012

Osterman, E., Tyagi, V.V., Butala, V., Rahim, N.A., Stritih, U., 2012. Review of PCM based cooling technologies for buildings. *Energy Build.* 49, 37–49. https://doi.org/10.1016/J.ENBUILD.2012.03.022

Raj, V.A.A., Velraj, R., 2010. Review on free cooling of buildings using phase change materials. *Renew. Sustain. Energy Rev.* 14, 2819–2829. https://doi.org/10.1016/j.rser.2010.07.004

Sharif, M.K.A., Al-Abidi, A.A., Mat, S., Sopian, K., Ruslan, M.H., Sulaiman, M.Y., Rosli, M.A.M., 2015. Review of the application of phase change material for heating and domestic hot water systems. *Renew. Sustain. Energy Rev.* 42, 557–568. https://doi.org/10.1016/j.rser.2014.09.034

Souayfane, F., Biwole, P.H., Fardoun, F., Achard, P., 2019. Energy performance and economic analysis of a TIM-PCM wall under different climates. *Energy* 1274–1291. https://doi.org/10.1016/j.energy.2018.12.116

Uddin, M., Romlie, M.F., Abdullah, M.F., Abd Halim, S., Abu Bakar, A.H., Chia Kwang, T., 2018. A review on peak load shaving strategies. *Renew. Sustain. Energy Rev.* 82, 3323–3332. https://doi.org/10.1016/J.RSER.2017.10.056

Zalba, B., Marín, J.M., Cabeza, L.F., Mehling, H., 2004. Free-cooling of buildings with phase change materials. *Int. J. Refrig.* 27, 839–849. https://doi.org/10.1016/J.IJREFRIG.2004.03.015

14 Comparative Assessment on the Use of Energy Storage in the Building Envelopes
A Review

Maitiniyazi Bake, Ashish Shukla, Shuli Liu,
Avlokita Agrawal, and Shiva Gorijan

CONTENTS

14.1   INTRODUCTION

With the development of the global economy and urbanization, human beings are now facing a huge challenge for energy conservation, environmental pollutions, and climate change [1,2]. The building sector covers 44% of total energy usage as one of the highest energy-consuming sectors in the UK, making it more important if greenhouse gas emissions needed to be reduced [3]. A similar trend of the highest percentage share of energy consumption from the building sector is also observed across the European Union (EU) and worldwide [4]. Various strategies have been proposed by the researchers, for example, energy efficiency including improvement in the building design [5], use of the renewable energy [6], introduction of the benchmarking/building certificates [7], use of the novel materials [8], building operation strategies [9], efficient HVAC design [10], efficient glazing [11], building topology [12], end-user engagement [13], building retrofit [14], and so on.

In general, the operational energy of the buildings mainly includes lighting, ventilation, and heating where heating is responsible for approximately 39% of the total energy usage [15]. The previous studies show that the biggest proportion of energy usages in the buildings comes from the HVAC services and hot water usages [16]. In order to minimize the energy consumption of the buildings, several possible solutions have been investigated, for example, passive energy-saving technologies such as heat pumps [17], natural ventilation for the passive cooling [18], mixed-mode ventilation for passive heating with ventilation [19], energy storage systems [20], and the application of renewable or sustainable energy sources, including the solar energy or wind energy into the building energy systems [21].

In order to achieve zero- or low-carbon buildings, two main strategies are adapted (1) use of passive technologies using energy from the natural environment [22] and (2) the renewable energy integration to the buildings [22,23]. However, on several occasions, solar energy availability and its utilization are not possible simultaneously, leading to the necessity for energy storage. Several studies have been reported on energy storage as electrical [24] and thermal energy [25].

Building envelopes are the critical element that determine the quality and control of the indoor environment condition [26]. In general, the building envelops include walls, windows, roofs, foundation, thermal insulation, thermal mass, and external or internal shading devices such as blinds and curtains [27]. Building envelopes are responsible for the biggest percentage of heating and cooling energy consumption because of heat loss throughout the building envelope [28]. Hence, high-performance building envelopes have been popular for design with the aim of providing a good thermally comfortable indoor environment. Currently, several technologies have been developed, such as using passive solar [29], solar wall (transpired solar collector applied to build south-face external wall) [30], and new materials for the building envelope including double-skin façade [31], ventilated façade [32], and high insulated building envelope components [33–36]. However, there is no, or very limited, study that particularly focuses on the use of energy storage in building envelope design and its impact on building energy performance.

The proposed review paper investigates the pieces of literature related to the performance of energy storage in building envelopes and discusses case studies in the context. It has been found that the building envelope provides a greater opportunity to integrate and exploit several methods for energy storage, which can play a vital role in the energy demand reduction of a building.

14.1.1 CLASSIFICATION OF BUILDING ENVELOPES

The functions of the building envelopes generally can be classified into three different categories: support, control, and finish (to meet the desire of people on the inside and outside of the buildings) [37]. The building envelope determines how much energy is consumed for heating and cooling due to the heat loss and indoor heat gain [38]. However, the energy loss through the building fabric is variable and affected by different components, for example, age and type of the buildings, orientation, location, construction techniques, operation, maintenance, and occupant behaviors [39]. Buildings lose heat to the surroundings throughout the building envelope via a combination of air leakage (air infiltration) and thermal conduction [40].

The heat loss through the walls and roofs as shown in Figure 14.1 is mostly responsible for a higher percentage of the total heating loss accounting 35% and 25%, respectively [41], followed by the heat loss (15%) through ventilation or draught.

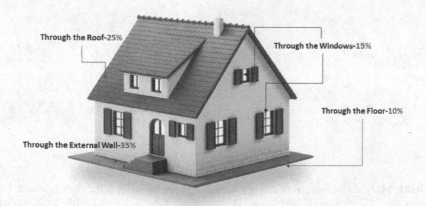

FIGURE 14.1   Possible heat loss from a house through the building envelope [41].

Thermal insulation—construction material with low thermal conductivity less than 0.1 W/mk—reduces building energy usage by preventing heat gain/loss through the building envelope [42,43]. In walls, the position of any thermal insulation will determine the thermal properties of the wall. It is very important for measuring the heat loss through the building fabric (for instance, walls), which is crucially important for determining how much energy is needed for the building operations such as heating and cooling of the buildings [44]. What is more, heat losses through external walls and roofs present more than 70% of the total for the existing buildings [50].

14.2 RATIONALE FOR ENERGY STORAGE IN THE BUILDING ENVELOPE DESIGN

Energy storage technologies for building envelopes provide several benefits [45,46]:

- Capturing the energy for later use, reducing the energy costs for consumers and businesses;
- Decreasing the usage of fossil fuels by reducing energy demand and enabling a greener energy supply mix where it can enable the interaction of more renewables (especially solar thermal); and
- Improving the envelope efficiency.

The energy storage technologies are mainly classified into three different categories: mechanical energy storage, electrochemical energy storage, and thermal storage (Figure 14.2) [45]. In the first place, the mechanical energy can be stored as the kinetic energy of linear or rotational motion, and there are three main mechanical energy storage types: hydrostorage (pumped storage), compressed-air energy storage, and flywheels [47]. Second, electrochemical energy storage often uses conventional

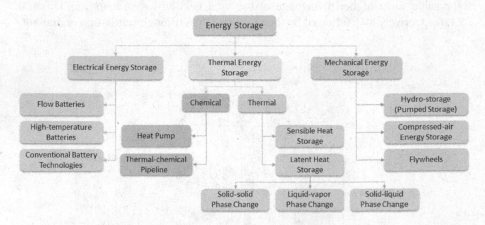

FIGURE 14.2 Classification of energy storage. (From Sadineni S.B., et al., *Renew. Sust. Energy Rev.*, 15, 3617–3631, 2011; Zhou, D. et al., *Appl. Energy*, 92, 593–605, 2012; Akeiber, H. et al., *Renew. Sust. Energy Rev.*, 60, 1470–1497, 2016; Xu, J. et al., *Sol. Energy*, 103, 610–638, 2014; Mahlia, T.M.I. et al., *Renew. Sust. Energy Rev.*, 33, 532–545, 2014.)

battery technologies, high-temperature batteries, and flow batteries. Additionally, thermal energy storage (TES) is divided into sensible thermal heat storage and latent heat storage [48–50].

TES can capture thermal energy as stored energy, which can be used at a later time for heating and cooling applications and the generation of power [50,55]. In the first place, latent heat storage has a higher energy density with a smaller storage volume than the sensible heat storage [25]. The latent heat storage materials, phase change materials (PCMs), become more popular for their application in the building envelopes since they have a high-energy storage density and the capacity to store energy at a constant temperature or over a limited range of temperature variations [50].

14.2.1 PHASE CHANGE MATERIALS

PCMs can be classified into four states: solid-solid, solid-liquid, gas-solid, and gas-liquid. There is a wide range of PCMs with different melting point ranges. According to their chemical composition, PCMs can be categorized as organic compounds (e.g., paraffin), inorganic (salt and salt hydrates) and eutectic mixtures [50].

14.2.1.1 Potential Benefits of PCMs

In the first place, the paraffin is one of the most popular organic PCMs used because it is non-corrosive and non-subcooling, and paraffin materials also are seen to be safe, reliable, cheap, and have high latent heat [52,54]. In particular, the thermal properties of greatest interest in PCMs are the melt temperature, the latent heat, specific heat, and thermal conductivity [53]. More importantly, the improvement in specific heat is desirable in the next-generation PCMs because it increases the amount of energy stored during the sensible heating period of operations. In addition, materials with high latent heats of fusion also tend to have relatively high specific heats that refer to the amount of energy, which would be key benefits for thermal heat storage [25]. However, one of the disadvantages of this organic PCM type is its high flammability, which limits it from being exposed to high temperatures [25,53].

14.2.1.2 Challenges and Solutions for PCM Implementation

On the contrary, the PCMs also have their own low thermal conductivity, which has a great impact on the reduction of effectiveness in both energy storage and thermal management application. Low thermal conductivity creates high thermal resistance and prevents heat flow from being effective and penetrating into PCM and initiating the melting process, which leads to the isolation of the melting process [25,50]. Further, there are some challenges including developing exact analytical solutions to the melting and solidification process and the response time. Popular methods have been found to increase the effective thermal conductivity for PCMs as in following [53]:

- The use of macroscale metallic inclusions, such as fins, meshes, or foams;
- The use of macroscale carbon insulations; and
- The use of nanoscale materials to create a colloidal PCM suspension with improved thermal properties.

14.2.2 Current Implementation and Challenges of the Thermal Energy Storage System

In terms of its implementations, TES can also be used with a solar thermal system intended to heat the air, and it is also effective when used with domestic hot water systems, solar-assisted heat pumps, and ground source heat pumps [53]. PCMs are popularly using in different areas, such as military, consumer products, and especially in buildings such as domestic hot water, space heating, and air-conditioning systems [53,56]. Especially, PCMs in buildings can meet the demands in energy and thermally comfortable environment as free cooling methods when being applied to the building envelope [57,58]. However, there are some challenges for applying TES system in buildings:

- Thermal stability or leakage [53]; and
- Value of power density that can be delivered to or withdrawn from a storage unit [45].

Challenges still exist in terms of achieving the theoretical performance in terms of charge/discharge rates for many PCM systems and the overall economic evaluation of a TES system needed to be developed [49].

14.3 ROLE OF ENERGY STORAGE IN THE BUILDING ENVELOPES

Energy storage types have been playing an important role in energy conservation, in particular, TES in buildings [52,53,59]. Also, PCM-enhanced building envelopes offer higher per unit heat storage capacity than conventional building materials and provide increased thermal mass for lightweight building structures [52,53,60]. Building walls with PCMs can make walls with high thermal capacity and thus contributes to decreasing indoor temperature fluctuation, reducing the heating and cooling loads, and lowering the energy consumption of buildings. For instance, the thermal storage can be part of the building structure even for lightweight building with gypsum board, plaster, concrete, or other wall covering materials including PCMs [25,52,53,59,61].

14.3.1 Application of PCMs into Building Components

TES with PCM on buildings is mainly classified into two different groups: active energy storage systems and passive energy storage system. Each of them are also divided into different categories. Figure 14.3 displays the classification of PCMs on various building envelopes including building materials. PCMs have been placed into different building envelopes, such as the solar wall, wallboard, floor, ceiling or roofs, concrete, and insulation materials due to more effective heat transfer [52,58,60,62,63].

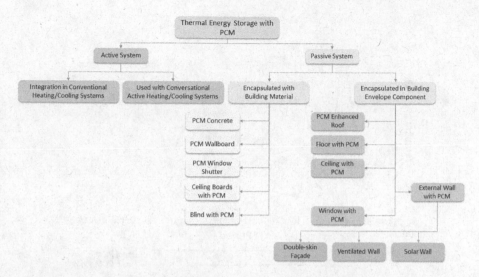

FIGURE 14.3 Classification of thermal energy storage with PCM and its application on various building envelopes, including building materials. (From Sadineni S.B., et al., *Renew. Sust. Energy Rev.*, 15, 3617–3631, 2011; Akeiber, H. et al., *Renew. Sust. Energy Rev.*, 60, 1470–1497, 2016; Omrany, H. et al., *Renew. Sust. Energy Rev.*, 62, 1252–1269, 2016; Lee, K.O. et al., *Appl. Energy*, 137, 699–706, 2015; Mirrahimi, S. et al., *Renew. Sust. Energy Rev.*, 53, 1508–1519, 2016; Mohammad, S. and Shea, A., *Buildings*, 3, 674–688, 2013; Saadatian, O. et al., *Renew. Sust. Energy Rev.*, 16, 6340–6351, 2012.)

Studies show that the building construction materials and building components are normally filled with PCMs, such as walls with PCM, PCM integrated into an external wall, ceilings or roofs, windows, and blinds as well, which would be demonstrated in details in the following sections.

14.3.2 PCM-ENHANCED BUILDING MATERIALS

There is a detailed implementation of PCMs in different building materials, including concrete, plaster, wallboard, and lightweight envelope. The wallboards or boards are cheap and widely used in a variety of applications, making them very suitable for PCM encapsulation because of their larger heat exchange area [25,57]. The PCM-enhanced wallboards and concrete have the ability to reduce the energy cost, the scale of air-conditioning, peak indoor air temperature, and the fluctuation of indoor temperatures [67–70]. Also, PCM for the lightweight building envelope is the most suitable solution for implementing PCM into buildings. In addition, they can be very effective for transferring the heat and cooling loads away from the peak demand times. Figure 14.4 summarizes the impacts of PCMs in various building envelope materials including concrete, lightweight wall, and wallboard. Therefore, Figure 14.4a shows that the PCM-enhanced concrete cubicles present much higher fluctuations (up to 18°C).

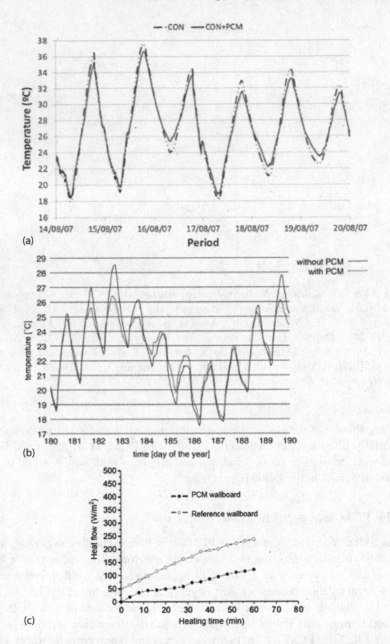

FIGURE 14.4 (a) Internal air temperature evolution for the concrete cubicles (From Castell, A. and Farid, M.M., *Energ. Buildings*, 81, 59–71, 2014.); (b) room temperature profile for a lightweight construction office with and without PCM plaster (From Schossiga, P. et al., *Sol. Energy Mater. Sol. Cells*, 89, 10, 2005.); and (c) heat flow of PCM wallboard and reference wallboard. (From Liu, S. et al., *Renew. Sust. Energy Rev.*, 73, 14, 2017.)

In addition to the energy saved by the reduced cooling load, the lower surface temperatures of the walls result in greater comfort (Figure 14.4b). Also, the heat flow of the
wallboard can be reduced from 8.5% to 77.9% using PCMs (Figure 14.4c).

14.3.3   DESIGN OF VARIOUS EXTERNAL WALLS WITH PCM

14.3.3.1   PCM in External Walls

External walls generally present a majority portion of the building envelope and
allow larger solar radiation to pass through due to a large surface [74]. PCMs in
the external wall reduce the temperature fluctuations in terms of solar radiation
loads [53,60,74]. Additionally, Figure 14.5a shows that the cubicle walls with PCM

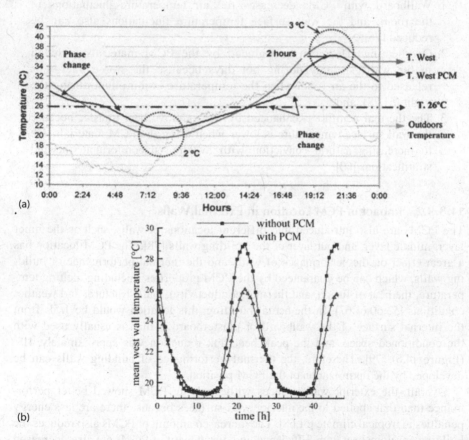

(a)

(b)

FIGURE 14.5   (a) Outside temperature and temperature of the west wall with and without
PCM with closed windows tested in July 2005 (From Ling, T.-C. and Poon, C.-S., *Constr.
Build. Mater.*, 46, 55–62, 2013.); and (b) experimental mean temperatures of the interior sur-
faces of the west wall. (From Kuznik, F. and Virgone, J., *Appl. Energy*, 86, 2038–2046, 2009.)

always obtained a smaller temperature value (2°C–3°C) than the cubicle walls without PCM, and the cubicles with PCM reached the same temperature of the cubicles without PCM about 2 h later. Moreover, the results show that the PCM tested maintains the room air temperature within the comfort zone by decreasing the maximum air temperature of the room to a maximum value of 2.9°C (Figure 14.5b).

Also, the room in the west-facing case integrated with PCM has a better performance in terms of decreasing the interior surface temperature up to a maximum of 41.4%. An energy savings of 2.9% in the air-conditioning system for a year is achieved for the west-facing case (Figure 14.5b) [77]. Moreover, there are key benefits for PCM composite walls as follows:

1. Wallboard with PCMs decreases the air temperature fluctuations in the room, and the wall surface temperature fluctuations also can be reduced [76];
2. Overheating effect can be reduced by the PCM materials included in the walls strongly during hot days because the stored energy is released to the air room when the temperature is minimum during the night [51,78]; and
3. The thermal comfort is enhanced by the radiative heat transfer because the wall surface temperature is lower when using the PCM materials due to increasing natural conviction with avoiding uncomfortable thermal stratifications [60].

14.3.3.2 Impact of PCM Location in External Walls

The PCMs are also introduced into different locations of walls, such as the inner layer, middle layer, and out layer of the building walls [78]. The PCM location has a great effect on the performance of walls and the thermal performance of building walls, which can be guaranteed by the PCM properties (including melting temperature, the heat of fusion, and thermal conductivity), wall structure, and weather conditions [52,60,61,67]. In the actual building, this location would be 1/5L from the internal surface of the wallboard or plasterboard, which is usually used with the conditioned space, and the peak heat flux reduction was approximately 41% (Figure 14.6) [79]. Therefore, the thermal performance of building walls can be developed by the optimization of the PCM position.

Overall, the exterior wall surfaces with applying PCM showed better performance than that applied to the interior wall surfaces because there are less energy penalties in tropical climate [61,68]. The increased amount of PCMs also reduces the building envelope heat gain. Efficiency and cost benefit of PCMs are also decreased due to the increased thickness of the PCM layer [69].

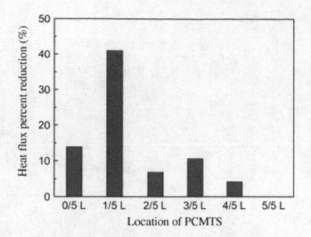

FIGURE 14.6   Heat flux percent reduction as a function of PCMTS location.

14.3.3.3   Solar Wall Enhanced with PCM

Solar wall (Trombe wall) with PCMs is designed to trap and transmit solar energy efficiently into a building. It was first described by Edward S. Morse in 1881. Trombe walls have been seen as a wall system capable of significantly reducing the energy consumption of buildings via improving the performance of the walls [66]. A solar wall with latent heat storage is more efficient than the conventional concrete walls [76]. Particularly, the PCM enhanced the Trombe wall to store the solar thermal gained from incoming solar radiation during the day and release it into the building space overnight [34]. Simply, the heat is stored in the Trombe wall with PCMs in the daytime; at night, the stored heat is discarded by ventilation to the outside of the buildings, which can prevent overheating problems in the daytime during the summer period for building side environment [54,57,61]. Moreover, careful consideration of Trombe wall design parameters improves the overall performance of the wall including Trombe wall ration, glazing emittance, wall thickness, and wall thermal conductivity [34,56,66,80].

Figure 14.7 displays the configuration of the tested Trombe wall model with PCM, and the space between the wall and glazing is regarded as the important element for preventing the heat loss through the glass. The results show that the Trombe wall with PCM of smaller thickness was more desirable in comparison to an ordinary masonry wall for providing efficient TES. Also, the Trombe wall for building ventilation absorbs solar energy into PCMs, and the stored heat was used for heating the air for the ventilation of the house. The efficiency of the absorption was found to be 79%.

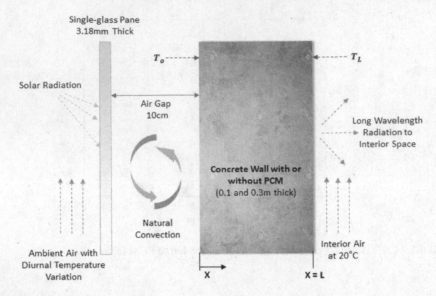

FIGURE 14.7   Configuration of the tested Trombe wall model with PCM. (From Sharma, A. et al., *Renew. Sust. Energy Rev.*, 13, 318–345, 2009.)

14.3.4   Ceilings with PCM

Ceiling with PCMs can absorb, store, and releases excess internal heat gain to provide a lightweight thermal mass solution as well as thermal comfort through passive free cooling [53]. The novel system stores the coolness in the PCMs in the off-peak time and releases the energy stored during peak time. In this system, the melting point for the PCM used was in the range of 20°C–30°C, and it is almost equal to the expected suitable room temperature [59]. The operation modes of the ventilated cooling ceiling are as follows [81]:

- During the daytime, the ventilation is in purely circulating operation guiding the warm room air onto the PCM; and
- During the nighttime, cool outside air is used to regenerate the PCM.

Peak cooling loads can be reduced because of cool storage within the ceiling and adjoining structural elements [67,81]. PCM-enhanced ceiling cooling systems can improve their energy performance and also decrease the risk of moisture condensation [82]. The increased thermal mass only has an influence on the PCM operation temperature range [70]. The regeneration of the PCM in the cooling ceiling during the night was a problem since no direct infiltration of the suspended ceiling with cool outer air could be realized in the test rooms [59,81].

14.3.5 PCM-ENHANCED ROOF

The roof absorbs more solar radiation compared to the other areas of building as a critical area for the building envelope to improve energy consumption [83]. Also, the roof filled with PCM could be the best option for dealing with direct sunlight and solar radiation. PCM embedded with the vertical building envelopes could be more effective by mutual shading and mutual reflection of the surrounding buildings [84–87]. The thermal performance of different PCM roofs in terms of different PCM layer thickness was investigated in Figure 14.8. From the figure, the peak time of average temperature for four kinds of PCM roofs delays with the increase of the PCM layer thickness, which occurs at 4:00 PM, 4:30 PM, 5:00 PM, and 6:00 PM when the PCM thickness is 40, 60, 80, and 100 mm, respectively. The results show that the effect of PCM layer thickness on the average temperature, heat flux of the upper surface, and the utilization rate of the PCM is huge.

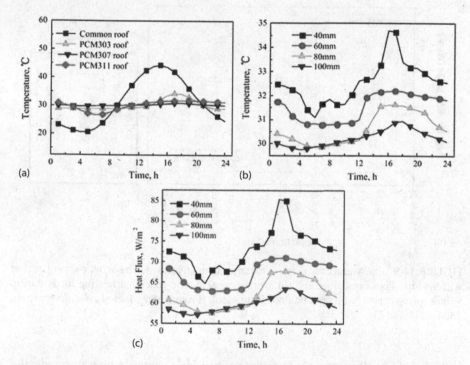

FIGURE 14.8 The mean temperature of (a) the upper surface in the base layer, the upper surface in the base layer; (b) the average temperature; and (c) the heat flux of the upper surface in the base layer under varying thicknesses of a PCM layer. (From Li, D. et al., *Energy Convers. Manag.*, 100, 147–156, 2015.)

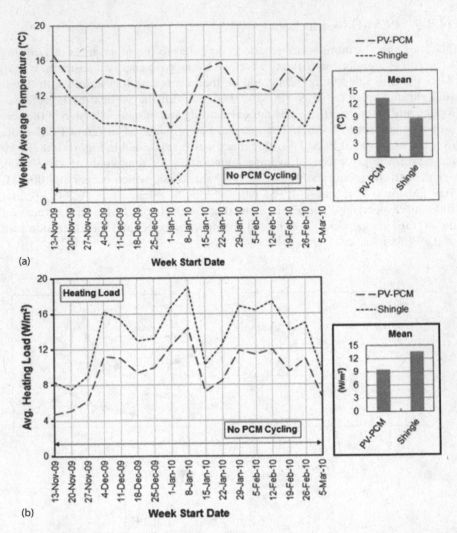

FIGURE 14.9 (a) Average weekly attic center temperatures during winter-spring period with no attic floor insulation; and (b) average weekly attic-generated heating loads during winter-spring period (with no attic floor insulation). (From Kośny, J. et al., *Sol. Energy*, 86, 2504–2514, 2012.)

Figure 14.9a illustrates the weekly average attic center temperatures during the period defined as winter-spring, between November 13, 2009 and March 11, 2010. The results show the weekly minimum PV-PCM attic air temperatures were higher on average by 8.9°C when compared to the shingle attic during that time (Figure 14.9a). Moreover, Figure 14.9b displays the average weekly heat flow into the attic from the conditioned space below, between November 13, 2009 and March 11, 2010. The results explain that the photovoltaic (PV)-PCM attic performed significantly better than the shingle attic by reducing the heating load by about 30%. The PV-PCM and shingle attics generated average heating loads of 9.55 and 13.54 W/m², respectively (Figure 14.9b).

Moreover, the heat gain is decreased when the PCM is incorporated in the roof, and the heat flux at the indoor space can be reduced by up to 39% [87]. Also, a double layer PCM incorporated in the roof also minimizes the transmission of dynamic thermal excitations by shaving and shifting dynamic loads [89].

14.3.6 WINDOWS WITH PCM

PCM can also be popular for windows because they produce a significant part of building thermal loads. Generally, window glazing with PCM is to decrease the thermal loads produced via absorbing the heat gain (especially solar heat gain) before it comes to the indoor space [90]. Simply, PCM-based windows change their phase from the solid-state to the liquid state when heated by absorbing the available solar energy, and then the stored heat is released during the night when the ambient temperature drops [69,91,92]. Additionally, the heat flux and the air temperature are the main components to evaluate the thermal performance of the PCM-enhanced windows, and the U-values of the windows are also increased with the addition of PCM [2]. The efficient work of windows with PCM depends on different factors such as climate conditions, desired comfort temperatures, and the ability to absorb and release an amount of heat energy [61,93]. Figure 14.10 displays the various impacts of PCM-enhanced window (PCMW).

Figure 14.10 shows that the peak heat flux on the interior surface of the PCMW is 11.4 W/m² lower than the hollow window (HW). The heat transferred through the PCMW increases by 4.9% compared with the HW. Also, the use of the PCMW can mitigate the temperature fluctuation of indoor air and shift the peak load of the air

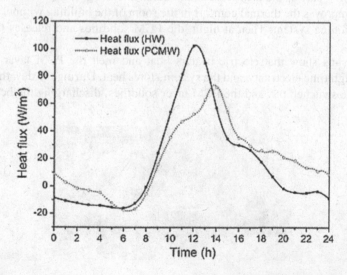

FIGURE 14.10 Heat flux on the interior surfaces of the PCMW and HW on a sunny summer day. (From Zhong, K. et al., *Energ. Buildings*, 106, 87–95, 2015.)

conditioning system. However, it slightly increases the energy consumption of the air conditioning system in a rainy summer day [91].

14.3.7 BLINDS INTEGRATED WITH PCM

Another possible implementation of PCM into building components is blinds [59]. The external and internal blinds have advantage and drawbacks during the process of reducing the direct heat gain form glass facades [94]. The external blinds are required to cost more construction money for installation, and high-rise buildings demand special maintenance for keeping external blinds clean. Additionally, the advantage of internal blinds is releasing the heat after their temperature is raised due to absorbing transmitted solar thermal through the windows [57,95–98]. For example, PCM enhanced in internal blinds can keep the indoor air temperature in the comfortable range to improve the building's thermal performance [82,92]. In addition, the PCM-integrated blinds can also delay the temperature rise and consequently delay the heat release into the building inside. It is good for preventing overheating problems [92,99]. Thus, internal blinds with PCM can lower the building energy consumption by decreasing the cooling loads via limited overheating problems [92].

14.3.8 FLOORS WITH PCM

The floor is also one of the important parts of building envelope components. Thermal mass incorporated into a floor heating system can be used for off-peak thermal energy to increase thermal comfort level [100]. The purpose of a hollow concrete floor panel filled with PCM (Figure 14.11) is that during the daytime the PCM absorbs part of the heat from solar radiation through the process of melting, and thus improves the thermal comfort in the room of the building without using the air-conditioning system. Then at night, the PCM solidifies and releases the stored heat [70].

The results show that electric heaters heat and melt the PCM layer by using cheaper nighttime electricity, and the system stores heat. During the day, the electric heaters are switched off, and the PCM layer solidifies, discharging the heat stored,

FIGURE 14.11 Diagram of a concrete hollow floor slab with PCM. (From Zhou, D. et al., *Appl. Energy*, 92, 593–605, 2012; Karim, L. et al., *Energ. Buildings*, 68, 703–706, 2014.)

during the nights also, with reducing the costs of electric energy consumption. PCM-enhanced floors have great benefits on reducing the building energy consumption, especially heating in the winter due to limiting the heat loss through building envelope [59,69].

14.3.9 Ventilated Façade or Double-Skin Façades with PCM

The ventilated double-skin façade (VDSF) offers the possibility of improving the energy efficiency of the building and providing better energy performance rather than the conventional facades [32,102–105]. The double-skin façade (DSF) can be defined as a special type of building envelope, where a second skin, usually transparent glazing is placed in front of a regular building façade [106]. PCMs in a ventilated façade capture solar radiation and use it for heating purposes, as a cold storage system during the summer season to reduce the energy consumption of the HVAC system as well [32,104]. Moreover, VDSF presents three potential benefits: free cooling, cold storage, and prevention of solar radiation incidence in different climate conditions [103].

The use of PCM layers with the DSF integrated with the PV modules proved to be effectively reducing the cooling load of an indoor space as well as increasing the solar to the electrical conversion efficiency of PV modules [107]. Moreover, there are various advantages for the application of DSFs in building envelopes, including permitting radiation to pass through the building due to its transparent characteristics, as well as providing a visual connection with surroundings [56].

14.4   SUMMARY AND DISCUSSION

The pivotal comparison of the impact of different types of building envelopes or building materials with and without PCM in building performance is presented in Table 14.1. Mainly most of the previous studies are based on the organic PCM for building passive cooling and even applied into different building sections including walls, ceilings, windows, and floors. The fundamental principle or benefits of PCM are good to provide free cooling at the night mostly. The reason for this focus might be explained by its heat of fusion, considerable compatibility with most building envelopes, and chemical stability, which may lead to overcooling, fluctuation, and thermal leakage problems or challenges.

Table 14.1 shows the development of building envelopes utilizing PCMs. In addition, the table also illustrates the application of PCMs on different building envelopes or building walls, such as in roofs, external walls, wallboards, plasters and concrete, glazing, various types of roofs, and DSFs, due to different research studies in different countries as well. More importantly, the benefits of PCM enhancement within various types of building envelopes dramatically reduce around 20%–40% on the total energy consumption annually. For instance, the PCM drywalls in light steel-framed buildings can improve the energy efficiency of buildings by 10%–60%, depending on the climate zone. Moreover, the solar wall with PCM as an external wall of a building is good for reducing the heating energy demand by approximately 16%. However, it is difficult to decide the energy

TABLE 14.1

Summary of Previous Researches Done on the Impact of Various Building Envelopes with PCM and Without on Building Performances

Author	Year	Building Envelope Materials	Method of Study	Building Performance	Climate/ Location
Athienitis et al. [108]	1997	PCM-based wall lining	Experimental and numerical simulation study	Lowered the maximum room temperature by 4°C and reduced the heating demand during nights.	Canada
Balaras et al. [109]	2000	Thermal insulation	EPIQR methodology and software	Reduced energy consumption by 20%–40%.	Greece
Bojic et al. [110]	2002	External shadings, light-colored roof and external walls		Reduced the space cooling load by 30% and 2%–4%, respectively.	
Bojic et al. [110]	2002	External wall thermal insulation	Numerical simulation	Reduced cooling energy and peak cooling load up to 20% and 29%, respectively.	Hong Kong
Ciampi et al. [32]	2003	Fully designed ventilated façade	Mathematical modeling	Achieved 40% summer cooling energy savings typically.	Italy
Cheung et al. [111]	2005	Lighter color building envelope	Experimental study	30% reduction in solar absorption can achieve a 12.6% savings in the annual required cooling energy.	Hong Kong
Liu & Minor [112]	2005	The green roofs	Experimental study	Reduced the total annual heat gain through the roof by 95% but the heat loss by only 23%.	Toronto, CA
Cabeza et al. [113]	2007	PCM in concrete walls	Experimentally investigated	The indoor temperature of the PCM-enhanced concrete building was 1°C lower than the reference building without PCM inclusion.	Spain

(Continued)

TABLE 14.1 (Continued)
Summary of Previous Researches Done on the Impact of Various Building Envelopes with PCM and Without on Building Performances

Author	Year	Building Envelope Materials	Method of Study	Building Performance	Climate/ Location
Zhang et al. [114]	2008	PCM wallboard	Mathematical and numerical analysis	The most energy-efficient approach of applying PCM in a solar house is to apply it in its internal wall.	China
Zhou et al. [115]	2008	Shape-stabilized phase change material plates	Enthalpy model and numerical modeling	Create a heavyweight response to lightweight constructions with an increase of the minimum room temperature at night by up to 3°C.	China
Han et al. [85]	2009	Lightweight roof	Mathematical and numerical analysis	Saved 53.8% of the peak cooling load compared to a dark painted roof with glass wool insulation. Space cooling load also can be reduced up to 20%.	Hong Kong
Kuznik and Virgone [76]	2009	PCM-composite wall boards	Experimental assessment	A decrease in maximum room temperature by 4.2°C.	France
Phelan et al. [116]	2010	Organic-based phase change materials	Experimental study	Maximum energy savings of about 30%.	USA
Xu et al. [117]	2011	Cool roof with coatings	Innovative field-based analytical method	Achieve up to 20% in cooling energy use savings for commercial buildings.	India
Alawadhi and Alqallaf [87]	2011	Building roof with conical holes containing PCM	Numerical study	Heat flux at the indoor surface of the roof can be reduced up to 39% on a certain type of PCM and geometry of PCM cone frustum holes.	Kuwait
He et al. [118]	2011	Double-glazed PV glass window	Experimental and numerical study	54% reduction in energy consumption.	East China

(Continued)

TABLE 14.1 (Continued)

Summary of Previous Researches Done on the Impact of Various Building Envelopes with PCM and Without on Building Performances

Author	Year	Building Envelope Materials	Method of Study	Building Performance	Climate/ Location
Sun et al. [119]	2011	Photovoltaic Trombe wall (PVTW)	Experimental and numerical studies	Total efficiency of solar usage is reduced by 5%. Also, the electric conversion efficiency of the PVTW achieves 11.6% while the glazing is fully filled with PV cells.	China
Košny et al. [88]	2012	Naturally ventilated solar roof with PCM	Experimental study	About 30% heating and 50% cooling load reductions are possible.	USA
Peng et al. [120]	2013	Double-skin façade	Experimental study	A reduction of annual cooling energy consumption up to 26%.	Hong Kong
		Single-glazed windows		26% reduction of energy consumption.	
Shi et al. [121]	2014	PCM in a concrete wall	Experimental study	Reduce the maximum indoor temperature up to 4°C and relative humidity by 16%.	Hong Kong
Soares et al. [122]	2014	PCM drywalls in lightweight steel-framed buildings	A multi-dimensional optimization study	Improve the energy efficiency of buildings by 10%–60% depending on the climate zone.	Portugal
Weinläder et al. [81]	2014	Ventilated cooling ceiling with PCM	Monitoring	Reduced the maximum operative room temperature in the office rooms by up to 2 K compared to a reference room without the cooling system.	Germany
Belmonte et al. [123]	2015	Chilled ceiling coupled with a floor containing PCM	Numerical analysis with TRNSYS	Reduce the cooling energy demand more than 50% compared to the cooling energy demand of the same building without PCM. Also, the predicted dissatisfied (PPD) increases by 2%–5%.	Spain

(Continued)

TABLE 14.1 (Continued)
Summary of Previous Researches Done on the Impact of Various Building Envelopes with PCM and Without on Building Performances

Author	Year	Building Envelope Materials	Method of Study	Building Performance	Climate/Location
Lee et al. [60]	2015	PCM layer into the residential building walls	Experimental study	30%–50% of peak heat flux reductions, 2–6 h delay in peak heat flux, and maximum daily heat transfer reductions were estimated as 3%–27%.	USA
Li et al. [124]	2015	PCM roof	Numerically investigated	PCM roofs effect on the temperature delay in the room beyond 3 hours than a common roof.	China
Lei et al. [61]	2016	PCMs in building envelope	Numerical method	Reduce the heat gain effectively in a range of 21%–32% throughout the whole year.	Singapore
Li et al. [124]	2016	PCM-filled window	Experimental research	The heat transferred into room through the triple-pane window plus phase change material (TW + PCM) is reduced by 16.6% and 28% compared with the double-pane window filled with phase change material (DW + PCM) and (triple-pane window) TW, respectively.	China
Liu et al. [86]	2016	PCM-filled double glazing roof	Mathematical and numerical modeling	Decrease building energy consumption and improve thermal comfort by enhancing its thermal energy storage capacity.	China
Marin et al. [125]	2016	PCM in gypsum board	Numerical modeling and validation	The potential of energy consumption reduction due both for heating and cooling periods in arid and warm temperate main climate areas.	Spain

(Continued)

TABLE 14.1 (Continued)

Summary of Previous Researches Done on the Impact of Various Building Envelopes with PCM and Without on Building Performances

Author	Year	Building Envelope Materials	Method of Study	Building Performance	Climate/Location
Mi et al. [77]	2016	PCM in building walls	Numerical simulation with Energy-Plus	Reduce the heating requirements within the office building and more energy savings during summer and winter period, especially in Shenyang.	China
Nghana and Tariku [126]	2016	PCMs in building envelope	Experimental study	Reducing indoor air and wall temperature fluctuations by 1.4°C and 2.7°C, respectively. Also, lowering the heating energy demand by up to 57% during the winter.	Canada
Park and Krarti [127]	2016	Reflectivity coatings on roofs and wall	Numerical simulation	Reduce annual energy use in commercial buildings up to 11% for office buildings	Chicago
Silva et al. [92]	2016	PCM window shutter	Numerical validation and experimental testing	(1) Decreased the maximum indoor temperature up to 8.7% for the warming period; (2) increased 16.7% the minimum indoor temperature for the night period and; (3) increased the time delay 1 h for the maximum temperature peak and 30 min for the minimum temperature peak.	Portugal
Weinläder et al. [128]	2016	PCM cooling ceilings	Monitoring	Passive cooling powers of 10–15 W per m² PCM cooling ceiling area was measured for a globe temperature of 26°C.	Germany
Yu et al. [129]	2017	Solar wall	Experimental testing and verification	In winter, heat loss through the solar wall is heavily reduced by about 80%–88% than the separate wall.	China

performance in the building envelope due to different aspects, including the limitation of PCMs and challenges of PCMs in building envelopes as follows:

- Heat loss is the main challenge for PCM-enhanced building envelopes because of the temperature difference and the thermal leakage [40];
- The fluctuation, instability based on the thermal's performance with uncertain time for the phase change period, and its low thermal conductivity [25,53];
- Thermal stability or leakage [53]; and
- The value of power density that can be delivered to or withdrawn from a storage unit [45].

In addition, achieving the theoretical performance in terms of charge/discharge rates for many PCM systems and the overall economic evaluation of a TES system needs to be developed [49]. Because the energy storage and building performance are so location dependent, general saving assumptions are not valid. Therefore, there is further scope to see if there is research with the same technology for the various regions and changes in performance.

14.5 CASE STUDIES OF THE PERFORMANCE OF PCM IN BUILDING ENVELOPES

PCM storage is likely to become a viable technology in the next few years. For instance, it can be concluded that PCM-based thermal storage in conjunction with an electric air-source heat pump, offered as part of a Green Deal, could be technically possible in retrofit buildings. Also, the introduction of a thermal store as part of a heating system offers a potential economic impact, through the requirements for initial installation and ongoing maintenance. Moreover, the use of latent heat energy storage is finding applications in the built environment with PCMs used in building cooling systems to displace peak cooling loads and by using microencapsulated PCMs in the building fabric. This paper also displays some case studies on the performance of PCMs in building envelope as follows.

14.5.1 COOL-PHASE® NOTRE DAME SCHOOL BUILDING

Two COOL-PHASE systems with PCM were installed in an IT classroom (approximately 70 m^2) where the system was required to overcome high internal heat gains through 30 PCs and glazing in April at Notre Dame School Building (London) [130]. Also, two control rooms (one room with similar internal heat gain and another one with much lower internal and external heating load) were chosen to provide a comparison to the performance of the COOL-PHASE systems. During the experiment, temperature and CO_2 levels were monitored every minute and collected by data logger during the spring term. Therefore, the results show that the average temperatures increased in the control room slightly between spring and summer term due to warmer weather (Figure 14.12).

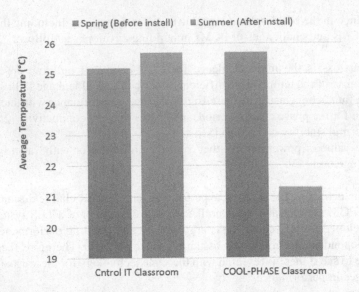

FIGURE 14.12 Comparison of the temperature before and after the install. (From COOL-PHASE system in IT classroom, Available: http://monodraught.vertouk.com/documents/downloads/download_34.pdf, May 17, 2011.)

However, the room with the COOL-PHASE system has seen a significant reduction in the average temperatures before and after the install. Therefore, it is clear that the COOL-PHASE system has had a significant impact on average temperatures as PCM with high-energy storage density stores energy over a limited range of temperature variations. Also, it would deliver financial savings of more than 26% (approximately £10,000) and CO_2 savings of 12.8 tons over the anticipated system lifespan of 20 years.

14.5.2 COOL-PHASE OWEN BUILDING AT SHEFFIELD HALLAM UNIVERSITY

COOL-PHASE was also installed in a particular "problem" room in the Owen Building at Sheffield Hallam University [131]. It is a 90-person teaching room on the tenth floor that consistently exhibited poor air quality and high temperatures. The room had no mechanical ventilation and was fitted with opening windows to one side; but due to the height of the building, these were restricted to a 100 mm maximum opening, limiting the effectiveness of natural ventilation. In November 2009, an initiated environmental check was taken by the university's Estates Department to record the CO_2 levels change and temperature change. COOL-PHASE was installed in March 2012, and the performance was monitored throughout the late March warm spell and recorded a room temperature peak of 23°C when outside temperatures were hovering around 21°C. The results displayed that the temperatures were consistently recorded at 25°C at a time when November's outside temperatures were reaching only 5°C. Moreover, the COOL-PHASE system was working harder to bring the CO_2 levels down during these peaks and, once the levels had been controlled, COOL-PHASE maintained CO_2 levels at around 600 ppm for the rest of the time. Therefore, the system has dramatically reduced CO_2 levels and is controlling previously excessive room temperatures well within normal comfort zones. All this while reducing energy consumption by up to 90% compared to conventional mechanical cooling systems.

14.5.3 New Ford Retail Car Showroom

A new Ford retail car showroom and used car sales office completed in January 2013 is equipped with COOL-PHASE, a low-energy cooling, ventilation, and heat recovery system [132]. The result can be estimated that the system uses intelligently controlled PCM to actively ventilate and cool buildings, maintaining temperatures within the comfort zone, while radically reducing energy consumption by up to 90% compared to conventional cooling systems. Also, COOL-PHASE reduces the running costs of buildings while creating a fresh and healthy indoor environment without the use of compressors or hazardous coolants.

14.5.4 Sustainable Building Envelope Center

A composite flooring system that incorporates under-floor heating and PCM in the Sustainable Building Envelope Centre (SBEC) (Figure 14.13). A prefinished steel composite floor has been installed on the first-floor level of the building with PCM added to the concrete mixture just above the steel deck. In this application, PCM is being used to store, buffer, and release heat via the high emissivity prefinished

FIGURE 14.13 PCM in composite floor decking at SBEC, Tata Steel, Shotton works, UK. (From PCM in composite floor decking at SBEC, May 22, 2011.)

steel floor deck that acts as a ceiling to the meeting room below. Excess heat load from usage is absorbed by the PCM and released when the heat load is reduced, providing a constant temperature. Therefore, the room temperature of the building can be controlled without heavy reliance on carbon-intensive fuels. The system allows cool or warm water to circulate through a network of water pipes embedded into the concrete to thermally activate the floor, providing highly effective cooling and heating method for the structure. It also can provide a constant and comfortable room temperature to be maintained during use, thereby reducing CO_2 emission.

14.5.5   Sir John Liang Building Coventry University

PCM technology with 1590 PCM TubeICE in total was installed within the Architecture Studio and two offices within the John Laing Building at Coventry University (Figure 14.14). The PCM Tubes are installed and respond to the surrounding temperature of the room. At the beginning of the day, the TubeICE are frozen, and as the room heats up due to body heat and heat from the sun, the PCM Tubes passively cool the room by absorbing the heat until completely melted. Over the night,

FIGURE 14.14 Schematic of PCM TubeICE installation in architecture studio in John Laing Building at Coventry University, UK. (From PCM TubeICE installation in Architecture studio in John Laing Building at Coventry University, UK Available: www.architect-bim.com/resseepe-project-sustainable-innovations-installed-coventry-university/#.WSROqOsrKUk, May 22, 2016.)

TABLE 14.2
Comparison of Energy Consumption of Building with and Without PCM TubeICE

	Boiler Energy (MWh)	Total System Energy (MWh)	Total Nat. Gas (MWh)	Total Carbon Emissions (KgCO$_2$)	Total Energy (MWh)	Total Energy (MWh / m^2)
Previous energy consumption before PCM TubeICE installation	418.80	448.80	418.8	106064.00	448.80	0.12
Present energy consumption after PCM TubeICE installation	399.30	428.90	399.3	101614.00	428.90	0.11
Energy savings (%)	11.35	10.58	11.35	9.67	10.58	10.58

Source: PCM TubeICE Passive Cooling at Coventry University, UK, Available: www.pcmproducts.net/files/John%20Laing%20Coventry-UK-%20Passive%20Cooling.pdf, May 23, 2016.

as the temperature cools, so does the PCM. The PCM effectively loses energy to the immediate surroundings, charging for the next day.

In this project, the PCM technology provides free cooling with 230 kWh energy storage. The spaces and tubes will be energy monitored over the next year to gather full performance data, and the result shows that the new PCM system could save 10% total energy compared to the previous system (Table 14.2). Also, CO$_2$ emissions will be 48.15 kg/m^2 per year, corresponding to more than 60% reduction. Associated investment costs to building renovation are expected to represent a maximum of 19% on average of the total costs of building an equivalent new building in the same location.

It can be concluded from the preceding case studies that PCM technology with passive cooling is increasingly being used in building envelopes because it has various benefits:

- Very low running cost;
- No external units are required;
- Highly energy-efficient system;
- Long life and a warranty for several years;
- Creates a healthy and productive environment for occupancy;
- High-performance ventilation and cooling system; and
- Environmentally friendly and sustainable solution that uses no refrigerants.

14.6 CONCLUSIONS

This article presents a detailed review of the literature that deals with the introduction to various building envelope components and their impact on building energy performance. The performance of PCMs on various building envelope components has also been critically analyzed, and the following conclusions have been made:

1. Building envelopes play a vital role in the overall energy performance of the buildings. Various building envelope components have a different impact on building energy performance Also, the heat loss through the walls and roofs has a higher percentage of the total heating loss form buildings with 35% and 25%, respectively. Also, the advanced glazing leads to cooling energy savings of the building, ranging from 3.4% to 6.4%.
2. The shape factor of the buildings has a significant impact on energy demands; it can account for up to 10%–20% variation for their final energy demand.
3. PCMs have been applied to different building envelope components, such as roofs, ceilings, and different positions on the external walls as passive energy technology to reduce building energy consumption. Their application varies from free heating/cooling to an active role in the novel HVAC designs.
4. Case studies have provided significant information on PCM-enhanced building envelope components, proving their potential to reduce the total energy demand in the range of 20%–40%.
5. Novel PCM-embedded building components, for example, solar wall with the energy storage can improve the energy efficiency of the buildings. It has been reported that improvements can be made in the range of 10%–60%, which varies significantly because of the size of the building fabric area, location of the building, type of the building, number of occupants, age of the buildings, and so on.

ACKNOWLEDGMENTS

This research was fully supported by the Engineering and Physical Sciences Research Council—EPSRC, EP/N007557/1, for the project Active-LIVing Envelopes (ALIVE). Underlying research material for this project can be accessed by contacting the corresponding author and principal investigator of this project, Dr. Ashish Shukla.

REFERENCES

1. E. Commission, "Air pollution and climate change," *Science for Environment Policy,* no. 24, 2010.
2. V. Ramanathan and Y. Feng, "Air pollution, greenhouse gases and climate change: Global and regional perspectives," *Atmospheric Environment,* vol. 43, pp. 37–50, 2009.
3. L. Pérez-Lombard, J. Ortiz, and C. Pout, "A review on buildings energy consumption information," *Energy and Buildings,* vol. 40, pp. 394–398, 2008.

4. X. Cao, X. Dai, and J. Liu, "Building energy-consumption status worldwide and the state-of-the-art technologies for zero-energy buildings during the past decade," *Energy and Buildings,* vol. 128, pp. 198–213, 2016.

5. C. Llatas and M. Osmani, "Development and validation of a building design waste reduction model," *Waste Management,* vol. 56, pp. 318–36, 2016.

6. P. Shen and N. Lior, "Vulnerability to climate change impacts of present renewable energy systems designed for achieving net-zero energy buildings," *Energy,* vol. 114, pp. 1288–1305, 2016.

7. A. Capozzoli, M. S. Piscitelli, F. Neri, D. Grassi, and G. Serale, "A novel methodology for energy performance benchmarking of buildings by means of Linear Mixed Effect Model: The case of space and DHW heating of out-patient Healthcare Centres," *Applied Energy,* vol. 171, pp. 592–607, 2016.

8. A. Mija, J. C. van der Waal, J.-M. Pin, N. Guigo, and E. de Jong, "Humins as promising material for producing sustainable carbohydrate-derived building materials," *Construction and Building Materials,* vol. 139, pp. 594–601, 2017.

9. U. Y. A. Tettey, A. Dodoo, and L. Gustavsson, "Primary energy implications of different design strategies for an apartment building," *Energy,* vol. 104, pp. 132–148, 2016.

10. R. Lathia and J. Mistry, "Process of designing efficient, emission free HVAC systems with its components for 1000 seats auditorium," *Pacific Science Review A: Natural Science and Engineering,* vol. 18, no. 2, pp. 109–122, 2016.

11. T. Gao, B. P. Jelle, and A. Gustavsen, "Building integration of aerogel glazings," *Procedia Engineering,* vol. 145, pp. 723–728, 2016.

12. M. Donofrio, "Topology optimization and advanced manufacturing as a means for the design of sustainable building components," *Procedia Engineering,* vol. 145, pp. 638–645, 2016.

13. S. Niu, W. Pan, and Y. Zhao, "A virtual reality supported approach to occupancy engagement in building energy design for closing the energy performance gap," *Procedia Engineering,* vol. 118, pp. 573–580, 2015.

14. A. Giovanardi, A. Passera, F. Zottele, and R. Lollini, "Integrated solar thermal façade system for building retrofit," *Solar Energy,* vol. 122, pp. 1100–1116, 2015.

15. U. N. E. Programme, "Buildings and climate change: Status, challenges and opportunities," pp. 1–65, 2007.

16. Eskom, "Heating, Ventilation and Air Conditioning (HVAC) systems: Energy-efficient usage and technologies," 2016.

17. T. Sivasakthivel, K. Murugesan, and P. K. Sahoo, "A study on energy and CO_2 saving potential of ground source heat pump system in India," *Renewable and Sustainable Energy Reviews,* vol. 32, pp. 278–293, 2014.

18. A. Tejero-Gonzalez, M. Andres-Chicote, P. Garcia-Ibanez, E. Velasco-Gomez, and F. J. Rey-Martinez, "Assessing the applicability of passive cooling and heating techniques through climate factors: An overview," *Renewable and Sustainable Energy Reviews,* vol. 65, pp. 727–742, 2016.

19. S. Ezzeldin and S. J. Rees, "The potential for office buildings with mixed-mode ventilation and low energy cooling systems in arid climates," *Energy and Buildings,* vol. 65, pp. 368–381, 2013.

20. G. Comodi, F. Carducci, J. Y. Sze, N. Balamurugan, and A. Romagnoli, "Storing energy for cooling demand management in tropical climates: A techno-economic comparison between different energy storage technologies," *Energy,* vol. 121, pp. 676–694, 2017.

21. A. Chel and G. Kaushik, "Renewable energy technologies for sustainable development of energy efficient building," *Alexandria Engineering Journal,* vol. 57, no. 2, pp. 655–669, 2018.

22. H. Ma, W. Zhou, X. Lu, Z. Ding, and Y. Cao, "Application of low cost active and passive energy saving technologies in an ultra-low energy consumption building," *Energy Procedia,* vol. 88, pp. 807–813, 2016.
23. D. Ma and Y. Xue, "Solar energy and residential building integration technology and application," *International Journal of Clean Coal and Energy,* vol. 2, no. 2, pp. 8–12, 2013.
24. N. Soares, "Thermal energy storage with phase change materials (PCMs) for the improvement of the energy performance of buildings, PHD Thesis, Sustainable Energy Systems. Department of Mechanical Engineering-University of Coimbra, 2015," 2015.
25. M. M. Farid, A. M. Khudhair, S. A. K. Razack, and S. Al-Hallaj, "A review on phase change energy storage: Materials and applications," *Energy Conversion and Management,* vol. 45, pp. 1597–1615, 2004.
26. G. K. Oral, A. K. Yener, and N. T. Bayazit, "Building envelope design with the objective to ensure thermal, visual and acoustic comfort conditions," *Building and Environment,* vol. 39, pp. 281–287, 2004.
27. F. Asdrubali and U. Desideri, "Building envelope." In *Handbook of Energy Efficiency in Buildings: A Life Cycle Approach,* K. McCombs, (ed.), Butterwoth-Heinemann, Oxford, pp. 295–441, 2018.
28. M. H. Sherman and I. S. Walker, "Heat recovery in building envelopes," *ASHRAE/DOE/BTEC Thermal Performance of Exterior Envelopes of Buildings VIII,* pp. 1–20, 2001.
29. M. Olenets, J. Z. Piotrowski, and A. Stroy, "Heat transfer and air movement in the ventilated air gap of passive solar heating systems with regulation of the heat supply," *Energy and Buildings,* vol. 103, pp. 198–205, 2015.
30. R. Ogden, R. Hall, L. Elghali, and X. Wang, "Transpired solar collectors for ventilation air heating," *Proceedings of the ICE-Energy,* vol. 164, pp. 101–110, 2011.
31. M. M. S. Ahmed, A. K. Abel-Rahman, A. H. H. Ali, M. Suzuki, and E. Resources, "Double skin façade: The state of art on building energy efficiency," *Journal of Clean Energy Technologies,* vol. 4, no. 1, pp. 84–89, 2016.
32. M. Ciampi, F. Leccese, and G. Tuoni, "Ventilated facades energy performance in summer cooling of buildings," *Solar Energy,* vol. 75, pp. 491–502, 2003.
33. H.-Y. Chan, S. B. Riffat, and J. Zhu, "Review of passive solar heating and cooling technologies," *Renewable and Sustainable Energy Reviews,* vol. 14, pp. 781–789, 2010.
34. S. Jaber and S. Ajib, "Optimum design of Trombe wall system in Mediterranean region," *Solar Energy,* vol. 85, pp. 1891–1898, 2011.
35. A. Ghaffarianhoseini, A. Ghaffarianhoseini, U. Berardi, J. Tookey, D. H. W. Li, and S. Kariminia, "Exploring the advantages and challenges of double-skin façades (DSFs)," *Renewable and Sustainable Energy Reviews,* vol. 60, pp. 1052–1065, 2016.
36. E. Iribar-Solaberrieta, C. Escudero-Revilla, M. Odriozola-Maritorena, A. Campos-Celador, and C. Garcia-Gifaro, "Energy performance of the opaque ventilated facade," *Energy Procedia,* vol. 78, pp. 55–60, 2015.
37. S. A. Olaniyan, Ayinla, A.K. and Odetoye, A.S., "Building envelope vis-a-vis indoor thermal discomfort in tropical design: How vulnerable are the constituent elements?" *Science, Environment,* vol. 2, no. 5, p. 10, 2013.
38. F. Goia, B. Time, and A. Gustavsen, "Impact of opaque building envelope configuration on the heating and cooling energy need of a single family house in cold climates," *Energy Procedia,* vol. 78, pp. 2626–2631, 2015.
39. C. Pearson, "Thermal imaging of building fabric," *A BSRIA Guide,* 2011.
40. C. Younes, C. A. Shdid, and G. Bitsuamlak, "Air infiltration through building envelopes: A review," *Journal of Building Physics,* vol. 35, no. 3, pp. 267–302, 2011.
41. R. Armani, P. Zangheri, M. Pietrobon, and L. Pagliano, "Heating and cooling energy demand and loads for building types in different countries of the EU," p. 86, March 2014.
42. A. M. Papadopoulos, "State of the art in thermal insulation materials and aims for future developments," *Energy and Buildings,* vol. 37, pp. 77–86, 2005.

43. F. Baccilieri, R. Bornino, A. Fotia, C. Marino, A. Nucara, and M. Pietrafesa, "Experimental measurements of the thermal conductivity of insulant elements made of natural materials: Preliminary results," *10th AIGE 2016 and 1st AIGE/IIETA International Conference,* vol. 34, pp. 1–7, 2016.

44. R. Ji, Z. Zhang, Y. He, J. Liu, and S. Qu, "Simulating the effects of anchors on the thermal performance of building insulation systems," *Energy and Buildings,* vol. 140, pp. 501–507, 2016.

45. R. E. Association, "Energy storage in the UK an overview," p. 32, 2015.

46. P. Komor and J. Glassmaire, "Electricity storage and renewables for Island power," *Irena,* p. 48, 2012.

47. R. Amirante, E. Cassone, E. Distaso, and P. Tamburrano, "Overview on recent developments in energy storage: Mechanical, electrochemical and hydrogen technologies," *Energy Conversion and Management,* vol. 132, pp. 372–387, 2017.

48. H. Chen, T. N. Cong, W. Yang, C. Tan, Y. Li, and Y. Ding, "Progress in electrical energy storage system: A critical review," *Progress in Natural Science,* vol. 19, pp. 291–312, 2009.

49. P. Eames, D. Loveday, V. Haines, and P. Romanos, "The Future Role of Thermal Energy Storage in the UK Energy System: An Assessment of the Technical Feasibility and Factors Influencing Adoption Research Report The Future Role of Thermal Energy Storage in the UK Energy System: An assessment of the Techni," *Ukerc,* 2014.

50. L. Navarro, A. De Gracia, S. Colclough, M. Browne, S. J. McCormack, P. Griffiths, L. F. Cabeza, "Thermal energy storage in building integrated thermal systems: A review. Part 1. active storage systems," *Renewable Energy,* vol. 88, pp. 526–547, 2016.

51. S. B. Sadineni, S. Madala, and R. F. Boehm, "Passive building energy savings: A review of building envelope components," *Renewable and Sustainable Energy Reviews,* vol. 15, pp. 3617–3631, 2011.

52. D. Zhou, C. Y. Zhao, and Y. Tian, "Review on thermal energy storage with phase change materials (PCMs) in building applications," *Applied Energy,* vol. 92, pp. 593– 605, 2012.

53. H. Akeiber, P. Nejat, M. Z. Majid, M. A. Wahid, F. Jomehzadeh, I. Z. Famileh, J. K. Calautit, B. R. Hughes, and S. A. Zaki, "A review on phase change material (PCM) for sustainable passive cooling in building envelopes," *Renewable and Sustainable Energy Reviews,* vol. 60, pp. 1470–1497, 2016.

54. J. Xu, R. Z. Wang, and Y. Li, "A review of available technologies for seasonal thermal energy storage," *Solar Energy,* vol. 103, pp. 610–638, 2014.

55. T. M. I. Mahlia, T. J. Saktisahdan, A. Jannifar, M. H. Hasan, and H. S. C. Matseelar, "A review of available methods and development on energy storage; technology update," *Renewable and Sustainable Energy Reviews,* vol. 33, pp. 532–545, 2014.

56. H. Omrany, A. Ghaffarianhoseini, A. Ghaffarianhoseini, K. Raahemifar, and J. Tookey, "Application of passive wall systems for improving the energy efficiency in buildings: A comprehensive review," *Renewable and Sustainable Energy Reviews,* vol. 62, pp. 1252–1269, 2016.

57. E. Osterman, V. Butala, and U. Stritih, "PCM thermal storage system for 'free' heating and cooling of buildings," *Energy and Buildings,* vol. 106, pp. 125–133, 2015.

58. M. Thambidurai, K. Panchabikesan, N. Krishna Mohan, and V. Ramalingam, "Review on phase change material based free cooling of buildings-The way toward sustainability," *Journal of Energy Storage,* vol. 4, pp. 74–88, 2015.

59. A. Sharma, V. V. Tyagi, C. R. Chen, and D. Buddhi, "Review on thermal energy storage with phase change materials and applications," *Renewable and Sustainable Energy Reviews,* vol. 13, pp. 318–345, 2009.

60. K. O. Lee, M. A. Medina, E. Raith, and X. Sun, "Assessing the integration of a thin phase change material (PCM) layer in a residential building wall for heat transfer reduction and management," *Applied Energy,* vol. 137, pp. 699–706, 2015.

61. J. Lei, J. Yang, and E.-H. Yang, "Energy performance of building envelopes integrated with phase change materials for cooling load reduction in tropical Singapore," *Applied Energy,* vol. 162, pp. 207–217, 2016.
62. D. Densley Tingley, A. Hathway, and B. Davison, "An environmental impact comparison of external wall insulation types," *Building and Environment,* vol. 85, pp. 182–189, 2015.
63. F. Abbassi, N. Dimassi, and L. Dehmani, "Energetic study of a Trombe wall system under different Tunisian building configurations," *Energy and Buildings,* vol. 80, pp. 302–308, 2014.
64. S. Mirrahimi, M. F. Mohamed, L. C. Haw, N. L. N. Ibrahim, W. F. M. Yusoff, and A. Aflaki, "The effect of building envelope on the thermal comfort and energy saving for high-rise buildings in hot-humid climate," *Renewable and Sustainable Energy Reviews,* vol. 53, pp. 1508–1519, 2016.
65. S. Mohammad and A. Shea, "Performance evaluation of modern building thermal envelope designs in the semi-arid continental climate of Tehran," *Buildings,* vol. 3, pp. 674–688, 2013.
66. O. Saadatian, K. Sopian, C. H. Lim, N. Asim, and M. Y. Sulaiman, "Trombe walls: A review of opportunities and challenges in research and development," *Renewable and Sustainable Energy Reviews,* vol. 16, pp. 6340–6351, 2012.
67. A. De Gracia and L. F. Cabeza, "Phase change materials and thermal energy storage for buildings," *Energy and Buildings,* vol. 103, pp. 414–419, 2015.
68. X. Jin, S. Zhang, X. Xu, and X. Zhang, "Effects of PCM state on its phase change performance and the thermal performance of building walls," *Building and Environment,* vol. 81, pp. 334–339, 2014.
69. S. E. Kalnæs and B. P. Jelle, "Phase change materials and products for building applications: A state-of-the-art review and future research opportunities," *Energy and Buildings,* vol. 94, pp. 150–176, 2015.
70. A. L. S. Chan, "Energy and environmental performance of building façades integrated with phase change material in subtropical Hong Kong," *Energy and Buildings,* vol. 43, pp. 2947–2955, 2011.
71. A. Castell and M. M. Farid, "Experimental validation of a methodology to assess PCM effectiveness in cooling building envelopes passively," *Energy and Buildings,* vol. 81, pp. 59–71, 2014.
72. P. Schossiga, H.-M. Henninga, S. Gschwandera, T. Haussmannb, "Micro-encapsulated phase-change materials integrated into construction materials," *Solar Energy Materials and Solar Cells,* vol. 89, no. 2–3, p. 10, 2005.
73. S. Liu, C. Zeng, and A. Shukla, "Adaptability research on phase change materials based technologies in China," *Renewable and Sustainable Energy Reviews,* vol. 73, p. 14, 2017.
74. D. A. Chwieduk, "Dynamics of external wall structures with a PCM (phase change materials) in high latitude countries," *Energy,* vol. 59, pp. 301–313, 2013.
75. T.-C. Ling and C.-S. Poon, "Use of phase change materials for thermal energy storage in concrete: An overview," *Construction and Building Materials,* vol. 46, pp. 55–62, 2013.
76. F. Kuznik and J. Virgone, "Experimental assessment of a phase change material for wall building use," *Applied Energy,* vol. 86, pp. 2038–2046, 2009.
77. X. Mi, R. Liu, H. Cui, S. A. Memon, F. Xing, and Y. Lo, "Energy and economic analysis of building integrated with PCM in different cities of China," *Applied Energy,* vol. 175, pp. 324–336, 2016.
78. M. A. Izquierdo-Barrientos, J. F. Belmonte, D. Rodríguez-Sánchez, A. E. Molina, and J. A. Almendros-Ibáñez, "A numerical study of external building walls containing phase change materials (PCM)," *Applied Thermal Engineering,* vol. 47, pp. 73–85, 2012.

79. X. Jin, M. A. Medina, and X. Zhang, "On the importance of the location of PCMs in building walls for enhanced thermal performance," *Applied Energy,* vol. 106, pp. 72–78, 2013.

80. P. Sharma and S. Gupta, "Passive solar technique using Trombe wall—A sustainable approach," *IOSR Journal of Mechanical and Civil Engineering (IOSR-JMCE),* vol. AETM'16, pp. 77–82, 2016.

81. H. Weinläder, W. Körner, and B. Strieder, "A ventilated cooling ceiling with integrated latent heat storage—Monitoring results," *Energy and Buildings,* vol. 82, pp. 65–72, 2014.

82. F. Souayfane, F. Fardoun, and P. H. Biwole, "Phase change materials (PCM) for cooling applications in buildings: A review," *Energy and Buildings,* vol. 129, pp. 396–431, 2016.

83. T. Xu, J. Sathaye, H. Akbari, V. Garg, and S. Tetali, "Quantifying the direct benefits of cool roofs in an urban setting: Reduced cooling energy use and lowered greenhouse gas emissions," *Building and Environment,* vol. 48, pp. 1–6, 2012.

84. D. Li, Y. Zheng, C. Liu, and G. Wu, "Numerical analysis on thermal performance of roof contained PCM of a single residential building," *Energy Conversion and Management,* vol. 100, pp. 147–156, 2015.

85. J. Han, L. Lu, and H. Yang, "Investigation on the thermal performance of different lightweight roofing structures and its effect on space cooling load," *Applied Thermal Engineering,* vol. 29, pp. 2491–2499, 2009.

86. C. Liu, Y. Zhou, D. Li, F. Meng, Y. Zheng, and X. Liu, "Numerical analysis on thermal performance of a PCM-filled double glazing roof," *Energy and Buildings,* vol. 125, pp. 267–275, 2016.

87. E. M. Alawadhi and H. J. Alqallaf, "Building roof with conical holes containing PCM to reduce the cooling load: Numerical study," *Energy Conversion and Management,* vol. 52, pp. 2958–2964, 2011.

88. J. Kośny, K. Biswas, W. Miller, and S. Kriner, "Field thermal performance of naturally ventilated solar roof with PCM heat sink," *Solar Energy,* vol. 86, pp. 2504–2514, 2012.

89. A. Pasupathy and R. Velraj, "Effect of double layer phase change material in building roof for year round thermal management," *Energy and Buildings,* vol. 40, pp. 193–203, 2008.

90. W. J. Hee, M. A. Alghoul, B. Bakhtyar, O. Elayeb, M. A. Shameri, M. S. Alrubaih, K. Sopian, "The role of window glazing on daylighting and energy saving in buildings," *Renewable and Sustainable Energy Reviews,* vol. 42, pp. 323–343, 2015.

91. K. Zhong, S. Li, G. Sun, S. Li, and X. Zhang, "Simulation study on dynamic heat transfer performance of PCM-filled glass window with different thermophysical parameters of phase change material," *Energy and Buildings,* vol. 106, pp. 87–95, 2015.

92. T. Silva, R. Vicente, C. Amaral, and A. Figueiredo, "Thermal performance of a window shutter containing PCM: Numerical validation and experimental analysis," *Applied Energy,* vol. 179, pp. 64–84, 2016.

93. V. V. Tyagi, S. C. Kaushik, S. K. Tyagi, and T. Akiyama, "Development of phase change materials based microencapsulated technology for buildings: A review," *Renewable and Sustainable Energy Reviews,* vol. 15, pp. 1373–1391, 2011.

94. M. A. Kamal, "A study on shading of buildings as a preventive measure for passive cooling and energy conservation," *International Journal of Civil & Environmental Engineering,* vol. 10, no. 6, pp. 19–22, 2010.

95. S. A. Olaniyan, A. K. Ayinla, and A. S. Odetoye, "Building envelope vis-a-vis indoor thermal discomfort in tropical design: How vulnerable are the constituent elements?" *International Journal of Science, Environment,* vol. 2, pp. 1370–1379, 2013.

96. E. Gratia and A. De Herde, "The most efficient position of shading devices in a double-skin facade," *Energy and Buildings,* vol. 39, pp. 364–373, 2007.

97. F. Ascione, R. F. De Masi, F. de Rossi, S. Ruggiero, and G. P. Vanoli, "Optimization of building envelope design for nZEBs in Mediterranean climate: Performance analysis of residential case study," *Applied Energy,* vol. 183, pp. 938–957, 2016.

98. H. B. Madessa, "A review of the performance of buildings integrated with phase change material: Opportunities for application in cold climate," *Energy Procedia,* vol. 62, pp. 318–328, 2014.

99. T. Silva, R. Vicente, F. Rodrigues, A. Samagaio, and C. Cardoso, "Development of a window shutter with phase change materials: Full scale outdoor experimental approach," *Energy and Buildings,* vol. 88, pp. 110–121, 2015.

100. P. Ogley, D. Pickles, C. Wood, and I. Brocklebank, "Energy Efficiency in Historic Buildings-Insulating solid ground floors," p. 19, 2010.

101. L. Karim, F. Barbeon, P. Gegout, A. Bontemps, and L. Royon, "New phase-change material components for thermal management of the light weight envelope of buildings," *Energy and Buildings,* vol. 68, pp. 703–706, 2014.

102. G. Diarce, A. Campos-Celador, K. Martin, A. Urresti, A. Garcia-Romero, and J. M. Sala, "A comparative study of the CFD modeling of a ventilated active facade including phase change materials," *Applied Energy,* vol. 126, pp. 307–317, 2014.

103. A. de Gracia, L. Navarro, A. Castell, and L. F. Cabeza, "Energy performance of a ventilated double skin facade with PCM under different climates," *Energy and Buildings,* vol. 91, pp. 37–42, 2015.

104. A. Gagliano, F. Nocera, and S. Aneli, "Thermodynamic analysis of ventilated façades under different wind conditions in summer period," *Energy and Buildings,* vol. 122, pp. 131–139, 2016.

105. A. De Gracia, L. Navarro, A. Castell, and L. F. Cabeza, "Numerical study on the thermal performance of a ventilated facade with PCM," *Applied Thermal Engineering,* vol. 61, pp. 372–380, 2013.

106. H. Poirazis, "Double skin façades for office buildings literature review," *Technical Report,* p. 196, 2004.

107. H. Elarga, F. Goia, A. Zarrella, A. Dal Monte, and E. Benini, "Thermal and electrical performance of an integrated PV-PCM system in double skin facades: A numerical study," *Solar Energy,* vol. 136, pp. 112–124, 2016.

108. A. K. Athienitis, C. Liu, D. Hawes, D. Banu, and D. Feldman, "Investigation of the thermal performance of a passive solar test-room with wall latent heat storage," *Building and Environment,* vol. 32, pp. 405–410, 1997.

109. C. A. Balaras, K. Droutsa, A. A. Argiriou, and D. N. Asimakopoulos, "Potential for energy conservation in apartment buildings," *Energy and Buildings,* vol. 31, pp. 143–154, 2000.

110. M. Bojic, F. Yik, and P. Sat, "Energy performance of windows in high-rise residential buildings in Hong Kong," *Energy and Buildings,* vol. 34, pp. 71–82, 2002.

111. C. K. Cheung, R. J. Fuller, and M. B. Luther, "Energy-efficient envelope design for high-rise apartments," *Energy and Buildings,* vol. 37, pp. 37–48, 2005.

112. K. K. Y. Liu and J. Minor, "Performance evaluation of an extensive green roof," *Greening Rooftops for Sustainable Communities,* pp. 1–11, 2005.

113. L. F. Cabeza, C. Castéllón, M. Nogués, M. Medrano, R. Leppers, and O. Zubillaga, "Use of microencapsulated PCM in concrete walls for energy savings," *Energy and Buildings,* vol. 39, pp. 113–119, 2007.

114. Y. Zhang, K. Lin, Y. Jiang, and G. Zhou, "Thermal storage and nonlinear heat-transfer characteristics of PCM wallboard," *Energy and Buildings,* vol. 40, pp. 1771–1779, 2008.

115. G. Zhou, Y. Zhang, K. Lin, and W. Xiao, "Thermal analysis of a direct-gain room with shape-stabilized PCM plates," *Renewable Energy,* vol. 33, pp. 1228–1236, 2008.

116. P. Phelan, K. Muruganantham, P. Horwath, D. Ludlam and T. McDonald, "Experimental investigation of a bio-based phase change material to improve building energy performance," *ASME 2010 4th International Conference on Energy Sustainability,* 2010, pp. 979–984.

117. T. Xu, H. Akbari, H. Taha, C. Wray, and J. Sathaye, V. Garg, S. Tetali, M. H. Babu and K. N. Reddy, "Using cool roofs to reduce energy use, greenhouse gas emissions, and urban heat-island effects: Findings from an India experiment," vol. 51, 2011.

118. W. He, Y. X. Zhang, W. Sun, J. X. Hou, Q. Y. Jiang, and J. Ji, "Experimental and numerical investigation on the performance of amorphous silicon photovoltaics window in East China," *Building and Environment,* vol. 46, pp. 363–369, 2011.

119. W. Sun, J. Ji, C. Luo, and W. He, "Performance of PV-Trombe wall in winter correlated with south facade design," *Applied Energy,* vol. 88, pp. 224–231, 2011.

120. J. Peng, L. Lu, and H. Yang, "An experimental study of the thermal performance of a novel photovoltaic double-skin facade in Hong Kong," *Solar Energy,* vol. 97, pp. 293–304, 2013.

121. X. Shi, S. A. Memon, W. Tang, H. Cui, and F. Xing, "Experimental assessment of position of macro encapsulated phase change material in concrete walls on indoor temperatures and humidity levels," *Energy and Buildings,* vol. 71, pp. 80–87, 2014.

122. N. Soares, A. R. Gaspar, P. Santos, and J. J. Costa, "Multi-dimensional optimization of the incorporation of PCM-drywalls in lightweight steel-framed residential buildings in different climates," *Energy and Buildings,* vol. 70, pp. 411–421, 2014.

123. J. F. Belmonte, P. Eguía, A. E. Molina, and J. A. Almendros-Ibáñez, "Thermal simulation and system optimization of a chilled ceiling coupled with a floor containing a phase change material (PCM)," *Sustainable Cities and Society,* vol. 14, pp. 154–170, 2015.

124. S. Li, G. Sun, K. Zou, and X. Zhang, "Experimental research on the dynamic thermal performance of a novel triple-pane building window filled with PCM," *Sustainable Cities and Society,* vol. 27, pp. 15–22, 2016.

125. P. Marin, M. Saffari, A. de Gracia, X. Zhu, M. M. Farid, L. F. Cabeza, and S. Ushak, "Energy savings due to the use of PCM for relocatable lightweight buildings passive heating and cooling in different weather conditions," *Energy and Buildings,* vol. 129, pp. 274–283, 2016.

126. B. Nghana and F. Tariku, "Phase change material's (PCM) impacts on the energy performance and thermal comfort of buildings in a mild climate," *Building and Environment,* vol. 99, pp. 221–238, 2016.

127. B. Park and M. Krarti, "Energy performance analysis of variable reflectivity envelope systems for commercial buildings," *Energy and Buildings,* vol. 124, pp. 88–98, 2016.

128. H. Weinläder, F. Klinker, and M. Yasin, "PCM cooling ceilings in the Energy Efficiency Center—passive cooling potential of two different system designs," *Energy and Buildings,* vol. 119, pp. 93–100, 2016.

129. G. Yu, P. Zhao, D. Chen, and Y. Jin, "Experimental verification of state space model and thermal performance analysis for active solar walls," *Solar Energy,* vol. 142, pp. 109–122, 2017.

130. COOL-PHASE system in IT classroom. (2011, May 17). Available: http://monodraught. vertouk.com/documents/downloads/download_34.pdf

131. COOL-PHASE® system is specified for 'Problem Room' at Sheffield Hallam University. (2012, May 18). Available: www.monodraught.com/blog/latest-posts/monodraughts-cool-phase-system-is-specified-for-problem-room-at-sheffield-hallam-university

132. Cool-phase® helps Ford Retail car showroom. (2013, May 17). Available: www.build-ingservicesindex.co.uk/entry/123546/Monodraught/Coolphase-helps-Ford-Retail-car-showroom-go-green/#description-block

133. PCM in composite floor decking at Sustainable Building Envelope Centre (SBEC). (2011, May 22).

134. PCM TubeICE installation in Architecture studio in John Laing Building at Coventry University, UK. (2016, May 22). Available: www.architect-bim.com/resseepe-project-sustainable-innovations-installed-coventry-university/#.WSROqOsrKUk

135. PCM TubeICE Passive Cooling at Coventry University, UK. (2016, May 23). Available: www.pcmproducts.net/files/John%20Laing%20Coventry-UK-%20Passive%20 Cooling.pdf

15 Passive and Free Cooling of Buildings

Sašo Medved, Suzana Domjan, and Ciril Arkar

CONTENTS

15.1   INTRODUCTION

Residential sectors in EU are responsible for 40% use of final energy and almost 50% of total final energy used for space heating and cooling. Although the energy demand for space cooling is not well known, it is estimated (Persson and Werner, 2015; HRE4 project, 2017) that currently 47 to 77 TWh per year of cooling energy is supplied, which is only 2 to 2.55% of total final energy demand in the residential sector. Nevertheless, in some states (e.g., Cyprus and Malta), space cooling contribute to almost half of the final energy demand in buildings, and in Greece and Spain the energy demand for space cooling is much above the EU average (HRE4 project, 2017). Meanwhile, the energy demand for heating will decrease due to more stringent energy efficiency of buildings requirements and global warming. It is expected that the energy demand for cooling will significantly rise in the future. There are several reasons for that, such as global warming, increased urban heat islands, demographic trends, and substantial increase of cooled buildings is expected. Although 80% of residential buildings are mechanical cooled in Cyprus and 65% in Malta, less than 5% of residential buildings are mechanical cooled in the EU member states in average.

Local climate conditions related to the needs for space cooling are expressed by cooling degree-hours (CDH) or cooling degree-days (CDD) expressed in K · hour or K · day per year. CDH and CDD are determined by the expression:

$$CDH = \sum_{i=1}^{8760} \left(\theta_{e,i} - \theta_{e,base}\right)^{+}; CDD = \sum_{j=1}^{365} \left(\overline{\theta}_{e,avg,j} - \theta_{e,base}\right)^{+} \text{ where } \left(X\right)^{+} \text{ is max}\left(X,0\right)$$

(15.1)

where $\theta_{e,i}$ is the outdoor air temperature of the i-th hour over the year, $\overline{\theta}_{e,avg,j}$ is the average daily outdoor air temperature of the j-th month over the year, and $\theta_{e,base}$ is the outdoor temperature at which the building doesn't need cooling. The value of the so-called "base" temperature of $\theta_{e,base}$ of 18°C is commonly used (European Environment Agency, 2019). For the territory of the Europe, the CDD map is shown in Figure 15.1. Significant trends were noticed in the last three decades showing that the heating degree day (HDD) decreases, meanwhile CDD increases (Figure 15.2).

Electricity is the primary energy carrier that is used for compressor-driven mechanical cooling nowadays. In spite of the technological development that led to the improved energy efficiency of cooling aggregates (with coefficient of performance [COP] up to 8), the use of electricity has a large impact on the environment.

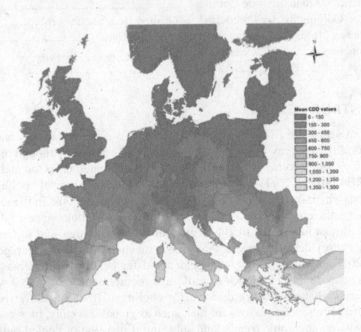

FIGURE 15.1 Cooling degree days for the territory of the Europe. (From Jakubcionis, M. and Carlsson, J., *Energy Policy*, 113, 223–231, 2018.)

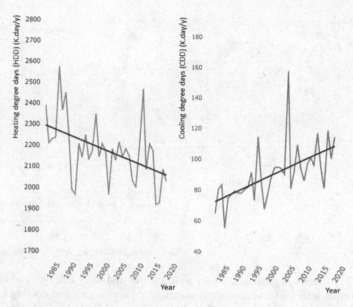

FIGURE 15.2 Changes of HDD and CDD in Europe (European Environment Agency, Data and maps, Heating and cooling degree days, Prod-ID: IND-348_en, 2019); values are weighted by the population exposed to the climate changes.

More environmentally friendly cooling technologies include PV and thermal solar-driven space cooling. Nevertheless, the electricity demand can be decreased applying passive and free cooling techniques.

Meanwhile, the CDD gives appropriate insight in the energy demand for mechanical cooling of the building. They are not suitable for assessment of the potential of the passive and the free cooling, at least not for natural cooling by ventilation. In this case, it is more appropriate that meteorological data for the site are expressed by average daily outdoor air temperature $\overline{\theta}_{e,avg}$ and daily amplitude of the outdoor air temperature $\hat{\theta}_e$. An example for selected sites is presented in Figure 15.3 together with the conditions favored for the efficient passive and free cooling of the buildings by ventilation. Values for the period between June and August are shown.

It is expected that climate changes will have an impact on the efficiency of the passive and the free cooling of the buildings due to the decreased amount of environment cold and increased overheating of the buildings. With the following example, we want to show which of these changes, which otherwise has the opposite effect, will have a greater impact on the efficiency of natural cooling. Four estimated climate change scenarios (A1T, A1B, A1F1C, and B1) were investigated by dynamic thermal response modeling of a reference building (Figure 15.30) using meteorological data in form of Test Reference Year (TRY) for The City of Ljubljana (Cfb Köppen-Geiger classification). TRY data was corrected (CTRY) using Fienkelstein-Schafer statistics (adapted from Vidrih and Medved, 2008): (1) moderate temperature increase (CTRY A ~ A1T)—average temperature rise of 1°C; (2) moderate temperature and solar-radiation energy increase (CTRY B ~ A1B)—average temperature rise of 1°C and average solar radiation energy increase of 3%; (3) significant

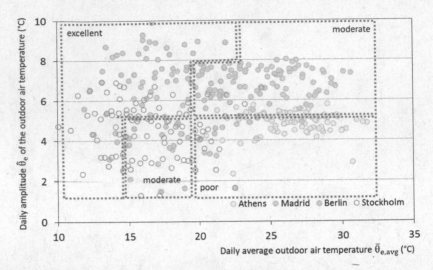

FIGURE 15.3 Daily average outdoor air temperatures and daily amplitude of outdoor air temperature for selected sites; conditions suitable for passive and free cooling by ventilation are presented. (From Meteonorm, Global Meteorological Database for Solar Energy and Applied Climatology; Version 5.1: Edition 2005; Software and Data on CD-ROM; Meteotest AG, Bern, Switzerland, 2005.)

temperature increase (CTRY C ~ A1F1C)—average temperature rise of 3°C; and (4) significant temperature and solar-radiation energy increase (CTRY D ~ B1): average temperature rise of 3°C and average solar-radiation energy increase of 6%. Beside reference climate data (TRY), the measured data at the site from the year 2003 was analyzed because the outdoor air temperature was close to the most severe climate change scenario. Results of dynamic thermal response of the free running reference office (without mechanical cooling) are presented in Figure 15.4. The occupancy

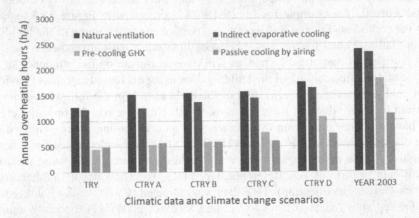

FIGURE 15.4 Annual overheating hours of a reference free running office at different climate conditions. (Adapted from Vidrih, B. and Medved, S., *Int. J. Energy Res.*, 32, 1016–1029, 2008.)

schedules and internal gains were assumed according to Vidrih and Medved (2008) and ISO 18523-1 (2016). It can be seen that the technique of the natural cooling will continue to be effective in spite of expected climate change.

15.2 PASSIVE AND FREE COOLING OF BUILDINGS

The common feature of both techniques of natural cooling is that they exploit the natural processes of cold generation, such as radiative or evaporative cooling and environment cold. The cold of the environment is a phenomenon that results from the transfer of heat from the surface of the earth through an "atmospheric window" to the space, whose temperature is near to the absolute zero (0 K). This process occurs throughout the day but is more noticeable at night, since the surfaces then do not receive solar irradiation. The "atmospheric window" is radiative property of Earth's atmosphere that transmits almost all irradiation with wavelengths between 8 and 12 μm, which are exactly the wavelengths at which the environment surface areas and buildings envelope structures emit the largest heat flux. In nature, the effect of environment radiate cooling can be noticed in the cold stored in ambient air as well as in the shallow layer of the soil. Another natural phenomenon that can be utilized by passive as well as by free cooling is the evaporation of water droplets into the outdoor or indoor air. To evaporate, energy needed is transferred from the surrounding air. This has happened on the leaves of greenery and trees and on water surfaces. As a result, air is moisturized but cooled as well.

. The passive and free cooling of the buildings will be more efficient if some measures are taken into account when planning buildings. The efficiency of the passive and free cooling will be increased as well, if the outdoor environment will be designed in the urban planning process to mitigate the urban heat island by greened structures and city parks. Figure 15.5 shows the non-mechanical cooling techniques as well as processes that must be implemented to increase the efficiency of passive and free cooling techniques.

Free cooling is sometimes related to the use of ground, sea, or lake water and thermally activated building structures. Free cooling is also one of the operation modes of compressor-driven cold generators where the refrigerant is cooled in by-pass mode in an air heat exchanger at low ambient temperature. Those technologies are not discussed in this chapter, since they are used in combination with mechanical space heating and space cooling systems. Details can be found in Medved et al. (2019).

15.2.1 ADAPTATION OF URBAN ENVIRONMENT FOR EFFICIENT PASSIVE AND FREE COOLING OF BUILDINGS

Replacements of natural ecosystems with the building blocks of the urban environment have significant influence on the thermal and hydrological balance of the urban environment. The phenomenon is known as urban heat island (UHI) and is related to higher air temperatures in cities compared to the suburban or rural areas.

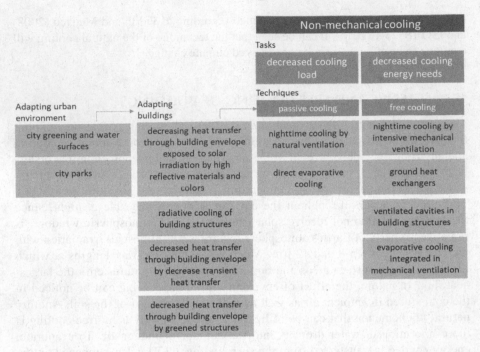

	Non-mechanical cooling	
	Tasks	
	decreased cooling load	decreased cooling energy needs
	Techniques	
Adapting urban environment → Adapting buildings	passive cooling	free cooling
city greening and water surfaces	decreasing heat transfer through building envelope exposed to solar irradiation by high reflective materials and colors	nighttime cooling by natural ventilation	nighttime cooling by intensive mechanical ventilation
city parks		direct evaporative cooling	ground heat exchangers
	radiative cooling of building structures		ventilated cavities in building structures
	decreased heat transfer through building envelope by decrease transient heat transfer		evaporative cooling integrated in mechanical ventilation
	decreased heat transfer through building envelope by greened structures		

FIGURE 15.5 Non-mechanical cooling techniques for cooling of the building as well as measures that increase the efficiency of passive and free cooling techniques.

Recent research on UHI carried out in Europe indicated different values of the UHI from slight around 0.1°C up to 10°C (Santamouris, 2007) in the cities with 1 million inhabitants. The most effective strategies to mitigate UHI are reducing of solar radiation absorptivity of urban surfaces by using the materials with high reflectance (albedo) of solar radiation, green roofs, and urban vegetation. Santamouris reports that the increase of the albedo of the built environment, which is close to the value 0.35 in a typical city by 0.1, results in decreasing of the average outdoor air temperature by 0.3°C (Santamouris, 2007). An overview study made by Yang et al. (2015) shows that an increase of albedo in cities up to 45% decreased the intensity of UHI up to 3.1°C.

On the smaller scale, the local urban heat island (LUHI) that is in street canyons is defined by the thermal response and the velocity patterns in the settlements. LUHIs could be even more intense than UHIs, and therefore, they have a major impact on both the energy demand for space cooling and the effectiveness of the natural cooling techniques. The most effective way to mitigate the LUHI with thoughtful urban planning of greening the built-up areas and incorporation meadows, water areas, and parks into the settlements. Figure 15.6 shows an example of a LUHI that occurs in the street canyons of row building settlements. The results were obtained by numerical modeling using the CFD technique and are presented for a selected typical summer sunny day in the city with a central European climate (Medved and Vidrih, 2015).

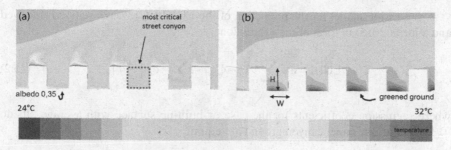

FIGURE 15.6 An example of CFD modeled temperatures in (a) street canyons of the row-building settlement in case of typical albedo-built surfaces and (b) grassy street ground. (From Medved, S. and Vidrih, B., Evaluating the potential of local urban heat island mitigation by the whitening and greening of the settlement surfaces, World Renewable Energy Congress XIV-Clean Energy for a Sustainable Development, 2015.)

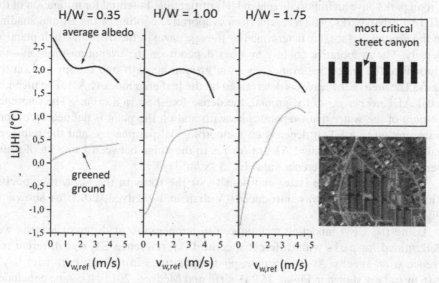

FIGURE 15.7 LUHI in the most critical street canyon in the row-house settlement with texture of the street canyon ground surface.

On the basis of the CFD simulations, LUHI was determined for the most critical street canyon in a row-building settlement, depending on the street height-to-width ratio and reference wind speed out of the settlement (Figure 15.7). It can be seen that at low wind speed ($v_{w, ref}$ < 1.5 m/s, which is quite common) the LUHI is up to 3 K lower in case of grassy street canyons ground compared to a settlement without greened surfaces, and the highest daily temperatures are even lower than air temperatures outside the settlement. At higher wind speed, the mitigation of LUHI is smaller, due to the swirling stream of air in the street canyon.

A regression model for the prediction of the LUHI (°C) was developed (Medved and Vidrih, 2015) in form of the expression:

$$LUHI = a + \sum_{i=1}^{4} b_i \cdot v_{w,ref}^i + \sum_{i=1}^{4} c_i \cdot \left(\frac{H}{W}\right)^i \qquad (15.2)$$

where regression coefficients for the case of built-up surface with average albedo (0.35) and grassy street canyon ground are equal:

	a	b_1	b_2	b_3	b_4	c_1	c_2	c_3	c_4
Albedo 0.35	2.89	−1.47	0.99	−0.25	0.02	−0.02	−0.28	−0.04	0.08
Grassy street canyons	−0.82	1.22	−0.38	0.06	−0.01	−0.17	0.02	0.04	0.02

Urban parks have an important role in UHI mitigation. Potential for mitigation of the UHI by urban parks, which results from evaporative cooling of the air and shading of the ground surface, are determined by the age (size) of the trees and their planting density. The evaporative cooling by trees depends on the evapotranspiration—the amount of the water that evaporates from ground beneath tree canopy and on the leafs. The area of the leaves is determined by the leaf area index (LAI). The meadow has LAI 1 m² per m² of the ground, the dense forest up to 8 m²/m². The reference amount of the water that evaporates beneath and on the plant is defined by evapotranspiration rate ET. It depends on meteorological parameters, and the reference value (for the plant having LAI 1 m²/m²) is in the range between 1 to 8 kg of water per m² per day, with reference value ET_o 5 kg/m²day.

To combine the age (size) and density of the trees in the park, the specific dimensionless LAI_{sp} was introduced (Vidrih and Medved, 2013) as shown in Figure 15.8.

Using the CFD modeling technique, the temperature at different lengths was determined for parks with different specific LAI, reference ET_o and different reference wind speeds. As an example, air temperatures in the city pars with LAI_{sp} 3.16 m²/m² are shown in Figure 15.9 (Vidrih and Medved, 2013). It can be concluded that old, dense parks have a very big impact on the urban microclimate. Because the value of the LUHI is negative, the park cooling island (PCI) can be defined instead of LUHI. A regression model for PCI was developed in the form:

$$PCI(L) = 0,27 + LAI_{sp} \cdot \left(-0.32 + 7.1461 \cdot 10^{-3} \cdot L - 3.258 \cdot 10^{-4} \cdot L^2\right.$$

$$\left. + 1.869 \cdot 10^{-6} \cdot L^3 - 2.895 \cdot 10^{-9} \cdot L^4\right) \qquad (15.3)$$

The model is limited to the length L of the park 120 m, but it can be seen from Figure 15.9 that larger parks have a minor impact on urban climate mitigation, since the PCI increases only slightly with L > 100 m.

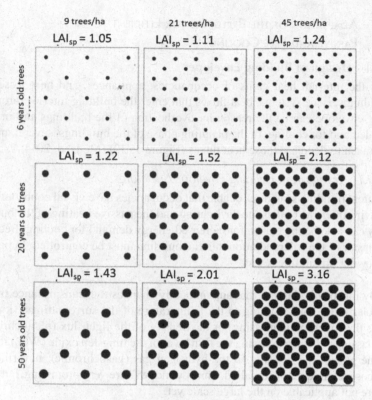

FIGURE 15.8 Specific LAI$_{sp}$ determined according to the tree age (size) and the density of the tress in the city park. (From Vidrih, B. and Medved, S., *Urban For. Urban Green.*, 12, 220–229, 2013.)

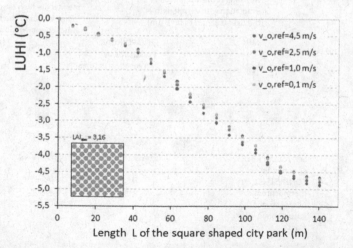

FIGURE 15.9 Air temperature in the square-shaped park at different lengths; air temperatures shown are average temperatures in the pedestrian zone (0.1 to 1.8 m) in plane at distance L from the edge of the part. (From Vidrih, B. and Medved, S., *Urban For. Urban Green.*, 12, 220–229, 2013.)

15.2.2 ADAPTATION OF THE BUILDINGS FOR EFFICIENT PASSIVE AND FREE COOLING

15.2.2.1 Design of Building Envelopes

Besides the internal heat gains of occupancies, appliances, and heat losses from the building service systems, heat fluxes that enter the building interior through the building envelope can be the reason for overheating of the buildings and increased energy demand for cooling. In the planning phase of the buildings, several measures can be taken to decrease those heat fluxes (Figure 15.10) (Medved, 2014).

15.2.2.1.1 Transparent Building Envelope Structures

Transparent building structures enable that occupancies have visual contact with outdoor environment as well as the daylighting and the passive heating of the buildings, but they can also be the reason for increased energy demand for cooling. Because of that, transfer of the solar radiation into the building must be controlled. In principle, the following possibilities exist:

- By installing glassing that automatically changes the transmittance of the solar radiation according to the temperature of the surroundings, (known as thermochromic glassing) or the density of the light flux (photochromic glassing) or when exposing the thin layer of the tungsten oxide (WO_3) inside the gap in the glassing to the hydrogen gas (gasochromic); nevertheless those techniques of so-called "smart windows" are very promising, but they are not applicable on the large scale yet.

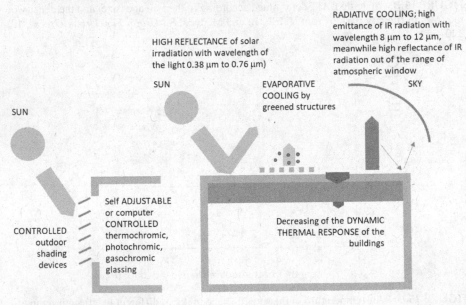

FIGURE 15.10 The effectiveness of the passive and free cooling can be significantly increased by decreasing the heat fluxes that enters the building through transparent (left) and opaque building envelope structures (right).

FIGURE 15.11 Movable, high solar reflective outdoor shades installed 30 cm in front of the facade for buoyancy-driven convective cooling. Shades are computer controlled; however, manual intervention of the users must be enabled in any case.

- By glass panes with very thin metal or oxides layer(s) that reflect a large part of the solar radiation; such glassing has low transmittance, although it cannot be adjusted to the needs of the occupancies (and the building).
- As most efficient, by shading devices installed on the outer side of the window or glassed facade, such devices must be movable, high reflective for the solar radiation, and must be installed in the way that the air gap between shades and glassing is formed to enable buoyancy-driven convective cooling of shades (Figure 15.11).

15.2.2.1.2 Opaque Building Envelope Structures

The heat flux that passes sunlit opaque building structures can be decreased by painting the outer surface of those structures with "cold color." These are paints have high reflectance of the visual part of the solar irradiation (wavelengths λ between 0.38 and 0.76 μm), which causes this color to be bright white. Figure 15.12 shows heat flux that passes the flat massive roof structure with a darker surface and cold-paint coating.

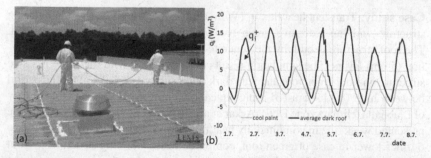

FIGURE 15.12 (a) Painting the roof with "cool paint" (From LEXIS Coatings, 2019); transient thermal response of the concrete roof with 5 cm thick outside thermal insulation layer with thermal transmittance U 0.65 W/m²K in case of dark and "cool paint" coating; and (b) in the numerical simulations meteorological data from TRY for Ljubljana (Cfb) was used. (From Arkar, C. et al., *Earth Environ. Sci.*, 323, 2019.)

Case study: Transient heat flux $\dot{q}_i$ (W/m²) on the interior surface of the flat concrete roof was determined over the period between June 1 to August 31 using typical meteorological year data for the City of Ljubljana (climate zone Cfb). The concrete layer is 20 cm thick, and the roof has a 5-cm thick thermal insulation layer on the outer side. No other layer was taken into account. Figure 15.12 shows $\dot{q}_i$ of the roof with an average dark surface (with absorptivity of solar radiation α_s equal 0.7) and a roof painted with "cool paint" (α_s 0.18) for the first week of modeled period. The total heat gains determined by q_i^+ (integrated over the time when $\dot{q}_i > 0$ W/m²) over the observed period are 10.13 kWh/m² of the roof with dark surface and 2.48 kWh/m² in case of the roof with "cool paint."

Due to the energy, the environmental and the social benefits, green roofs and facades are becoming predominant solutions in urban planning and building envelope retrofitting, especially in the form of extensive green roofs because of low additional structural load, low maintenance, and low cost in comparison to intensive solutions. Extensive green roofs consist of (from top to bottom) a vegetation layer, a thin organic soil layer, a lightweight rock mineral wool growing media 2 to 5 cm thick, a drainage system, and a root membrane. Environmental effects are the consequence of the absorption of the greenhouse CO_2, particles and heavy metals in plants and the soil, as well as urban heat island mitigation. At least as important is the role of a green roof in retention and detention of the precipitation. When completely saturated, extensive green roofs store between 25 and 55 kg of water per m² area. Social benefits can be seen from new urban areas intended for socializing of inhabitancies, urban food production in cooperatives, and even the settings up of beehives. Green roofs, however, also have a considerable influence on the thermal response of building structures due to the latent heat transfer by ET and freezing of the water in growing media (Arkar et al., 2018).

Case study: Transient heat flux $\dot{q}_i$ (W/m²) at the interior surface of the building structure with an extensive green roof (Figure 15.13, left) was determined. The roof structure is built by the concrete (15 cm) and the outer thermal insulation (5 cm) layers. The simulation results are compared with the same massive structure without green roof and with dark outer surface (α_s 0.7; Ca) for the period between June 1 and August 31 using typical meteorological year data for the City of Ljubljana (climate zone DbF) (Arkar et al., 2019). It can be seen that during the presented week, the maximum daily heat flux toward the building interior is 5 to 6 times lower in case of green roof, compared to the building structure without vegetation (Figure 15.13, right). The total heat gains q_i^+ of the structure with the green roof over the observed period are 1.67 kWh/m², which is significantly lower compared to the structure without the green roof (q_i^+ equals 10.13 kWh/m²).

(Continued)

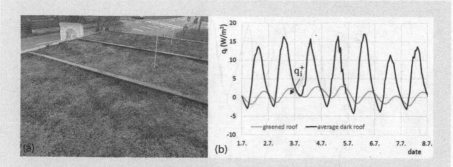

FIGURE 15.13 (a) Extensive green roofs in test stand at Laboratory for Sustainable Technologies in Buildings; and (b) transient thermal response of the massive roof structure with green roof and the same roof structure without vegetation and with dark outer surface in the first week of July.

Building structures on the envelope of the buildings exchange long-wavelength radiation (IR radiation) with their surroundings and in case they are horizontal, mostly with the sky. The temperature of the sky is strongly dependent on the cover of the sky with clouds. When the sky is cloudy, the apparent sky temperature is only few K below the outdoor air temperature; meanwhile, in the case of a clear sky, the apparent sky temperature is tens of degrees lower than the outdoor air temperature. If the surface of the building structures has selective radiative properties, designed for radiative cooling, this means that such surfaces have thermal emissivity e_{IR} in the range of the wavelengths λ of the atmospheric window (8 to 12 μm) close to 1. The reason is that the atmosphere in a clear sky is almost totally transparent for those wavelengths, meaning, that building structures exchange the radiation with the space, which has the temperature of only few K. For all other wavelengths of the thermal radiation, including solar radiation and IR radiation outside of atmospheric window wavelengths, the surface of building structure must have a reflectance close to 1, to reflect all incoming radiation. In such a case, it is possible that the surfaces of building structures cool below the outdoor air temperatures, and it is possible that heat flux is directed outside of the building as it is shown in Figure 15.14. The figure shows the heat flux at the inner surface of the massive roof building structure with ideal radiative properties for radiative cooling ($e_{IR} = 1$ for wavelengths 8 mm $< \lambda <$ 12 μm and $\rho_{IR} = 1$ for wavelengths 8 μm $> \lambda >$ 12 μm) during the first week in July (TRY Ljubljana).

The thermal response of the building structures depends not only on the boundary conditions of the outer surface of the structure but on the capability of the construction to accumulate the heat during the periodic daily transient heat transfer as well. In engineering practice (EN ISO 13786:2017), the dynamic boundary conditions at the outside surface of the building structure are defined by variation (amplitude) of the air temperature $\hat{\theta}_e$ or the heat flux $\hat{q}_e$ around average value $\overline{\theta}_e$, $\overline{q}_e$ and described by a sine function of time. The period of the process, which is in nature 24 hours, is replaced in calculation method by frequency w (rad), and because of that, the dynamic heat transfer in the building structure is determined by the heat transfer matrix Z of the

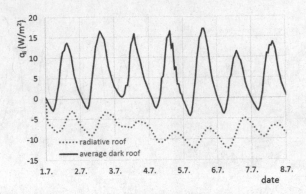

FIGURE 15.14 Heat flux at inner surface of the massive roof structure with ideal selective radiative properties for radiative cooling and at the same structure with common nonselective coating; term selective radiative properties indicate that one or more radiative properties (including absorptivity, reflectivity, transmittivity, and emissivity) are adjusted to the particular heat transfer problem.

building component. To evaluate the dynamic thermal response of the building structure, decrement factor f and time shift Δt are used. Decrement factor f is defined as the ratio of the modulus of the periodic thermal transmittance to the steady-state thermal transmittance U. In other words, f indicates how dynamic boundary conditions at the outer surface is reflected at the inner building structure. If the value of f is close to 1, the building structure will not suppress any change in the outdoor state; meanwhile, the building structures characterized by f equal 0 will act as perfect adiabatic structures. Therefore, the building built by structures with low decrement factor (<0.3) will have better predispositions for efficient passive and free cooling due to lower heat flux $\dot{q}_i$. Besides that, the maximum value of $\dot{q}_i$ over the day will enter the interior of the building with a larger time shift Δt as well (Figure 15.15).

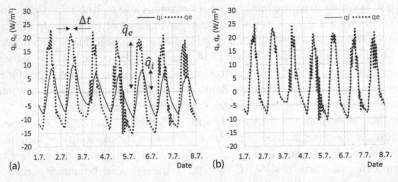

FIGURE 15.15 (a) Heat flux at the outer and the inner surface of the massive; and (b) lightweight structure; both structures have equal thermal transmittance U 0.615 W/m²K, but obviously very different dynamic thermal properties. This can be noticed by decrement factor f equal 0.315 for massive and 0.998 for lightweight building structure.

Case study: Two building structures have equal (static) thermal transmittance, U, but are built as massive (including 15-cm-thick concrete layer) and lightweight structures built by thermal insulation layer only. The dynamic thermal response during real environment conditions was modeled, and results are shown in Figure 15.15. It can be seen that (1) despite both structures having equal thermal transmittance, heat flux that enters (and heat up) the building is significantly larger in lightweight building structures (there is no difference in and $\hat{q}_e \sim \hat{q}_i$) and (2) there is almost no time delay (Δt) of heat flux in lightweight building structures. The theoretical decrement factor f for the massive structure is 0.315 and for lightweight structure it is 0.998.

15.2.2.2 Design of the Building Interiors for Efficient Passive and Free Cooling

The amount of cold that can be stored in the building structures is determined by (1) indoor air temperature fluctuation over the day determined by acceptable indoor thermal comfort and (2) the capacity of the building structure to accumulate the cold. This means that the building structure is cooled during the nighttime by one of the passive or free cooling techniques and stores the heat at lower indoor air temperatures during the daytime to prevent overheating and decreasing the energy needs for mechanical cooling.

15.2.2.2.1 With Respect to the Indoor Thermal Comfort

Acceptable range of indoor air temperatures is prescribed by indoor thermal comfort requirements. Standard EN 15251 (2017) defines the quality classes I, II, III, and IV of indoor air temperature θ_i for building different categories (use). For example, in office buildings, the summertime indoor air temperature should be in the time of occupancy in the range between 22°C and 27°C (Category III of thermal comfort); meanwhile, Category I of thermal comfort indoor air temperature should be in the range between 23.5°C and 25.5°C. Most often, the indoor thermal comfort Class III is required. It can be concluded that passive and free cooling can be effective without undesirable effect on the indoor thermal comfort. Taken into account indoor thermal comfort requirements, office buildings, kindergartens, schools, and department stores are among most suitable for efficient passive and free cooling.

15.2.2.2.2 With Respect to the Accumulation of the Cold in the Building Structures

In passive or free-cooled buildings, building structures are exposed to the periodic changes of indoor air temperature, which enables a periodic process of storing and releasing the heat in the interior of buildings. The non-stationary indoor air temperature is modeled as a periodic function $\theta_i(t)$:

$$\theta_i(t) = \overline{\theta_i} + \widehat{\theta_i} \cdot \sin\left(\frac{2 \cdot \pi}{t_p} - \varphi\right) \tag{15.4}$$

where $\overline{\theta_i}$ (°C) and $\widehat{\theta_i}$ (°C) are the daily average and amplitude of indoor air temperature, t_p is the period (24 h or 86,400 s), and φ is the time lag of amplitude according to twelve o'clock. It is assumed that there is no thermal resistance to the surface heat transfer between the indoor air; therefore, $\theta_i(t)$ is equal to the $\theta_{si}(0,t)$. Assuming that transient heat transfer is one-dimensional and a homogeneous single-layer building construction, the solution of the differential equation of transient heat transfer in building structure is given by (EN 15251:2017):

$$\theta(x,t) = \overline{\theta_{si}} + \widehat{\theta_{si}} \cdot e^{-x\sqrt{\frac{\pi}{t_p \cdot a}}} \cdot \sin\left(\frac{2 \cdot \pi}{t_p} \cdot t - x\sqrt{\frac{\pi}{t_p \cdot a}}\right) \tag{15.5}$$

where $\theta(x, t)$ is the temperature (°C) of building structure at a depth x of the building structure at a time t, $\overline{\theta_{si}}$ (°C) and $\widehat{\theta_{si}}$ (°C) are the average surface temperature and the amplitude of the building structure in the period t_p (24 h or 86,400 s), a is the thermal diffusivity (m²/s), and t is the time (s). The temperature at a depth x will be the highest when the part defining the time shift of the amplitude is 0. From that fact the effective thickness d_p of the building structure can be developed. It is defined as the depth x of the structure where the amplitude of the temperature decreases by a factor of 1/e to $0.367 \cdot \widehat{\theta_i}$. It is defined with the expression:

$$d_p = \sqrt{\frac{t_p \cdot a}{\pi}} = 165.8 \cdot \sqrt{a} \tag{15.6}$$

where a is the thermal diffusivity of materials (m²/s).

The double $(2d_p)$ and triple $(3d_p)$ effective thickness are defined similarly, whereby the amplitude is reduced by a factor $1/(2 \cdot e)$ to $0.184 \cdot \widehat{\theta_i}$, or by a factor of $1/(3 \cdot e)$ to $0.122 \cdot \widehat{\theta_i}$. In the design of capacity of the heat storage of the building structure, the double thermal response depth $2d_p$ is usually selected. The amount of the specific heat that is accumulated in m² of building structure area during half of period t_p (an "realized" during the second half period) is equal to (Medved et al., 2019):

$$q_{accu} = 2 \cdot \sqrt{\frac{t_p}{2 \cdot \pi} \cdot \lambda \cdot c_p \cdot \rho} \cdot \overset{\widehat{\theta_i}}{\overbrace{\widehat{\theta_{si}}}} = 234.53 \cdot b \cdot \widehat{\theta_i} \ \left[kJ/m^2\right] \tag{15.7}$$

where λ is the thermal conductivity of single-layer building structure (W/mK), c_p is the specific heat capacity (J/kgK), ρ is the density (kg/m³), and b is the thermal effusivity (kJ/m²Ks⁰·⁵).

Case study: Determine the effective thickness and required area of the building structure made of solid concrete (a = 68×10^{-8} m²/s, b = 2.18 kJ/(m²Ks⁰·⁵)) and by solid wood (a = 14×10^{-8} m²/s, b = 0.38 kJ/(m²Ks⁰·⁵)) that will in the theory accumulate 1 kWh of daily solar gains. Assume that $\widehat{\theta}_i$ is equal to 3K. The effective thickness $2 \times d_p$ of the concrete building structure is ~250 mm, and of the wooden one is ~120 mm. To store 1 kWh of solar gain without overheating of the room ~2.5 m² of concreate and ~14 m² of wooden construction is needed.

In the case explaining the procedure for determining the effective thickness of the building structure for periodic accumulation of the cold, we assumed that indoor air temperature θ_i is equal to the surface temperature θ_{si} of the building structure. This means that convective heat transfer resistance $R_{c,si}$ was assumed to be 0 ($h_{c,si} = \infty$). In reality, this is not the case, and the capacity of cold accumulation in building structures is lower.

The convective surface's heat transfer coefficient used to determine heat transfer in building structures is listed in ISO 6846:2017. The standard suggests $h_{c,si}$ 2.5 W/m²K for vertical walls, 0.7 W/m²K for floors, and 5.0 W/m²K for ceiling. Nevertheless, the temperature difference between the indoor air and surface temperatures of building structures in the passive cooled room by ventilation is much higher (6 to 8 K) than in the temperature-controlled buildings. Meanwhile, the indoor air velocity is much higher compared to the thermal buoyancy-driven convection (v_i is in the range between 0.15 and 0.25 m/s) in the closed room. The maximum air exchange rates can be 10 h⁻¹ or above, which significantly increase convective heat transfer at the surface of building structures. This is especially the case in free cooled buildings, where air exchange rates could be even larger.

Some examples available in the literature of the approximation models (ASHRAE, 2001; Khalifa and Marshall, 1989) for determination of the average surface convective heat transfer coefficient $h_{c,\,avg,\,wall}$ for vertical walls dependent on the temperature difference $\Delta\theta$ between indoor air θ_i and surface temperature θ_{si} of the building structure are shown in Figure 15.16. In can be seen that $h_{c,\,avg,\,wall}$ is in the range between 1.8 and 3.3 W/m²K taking into account the temperature difference $\Delta\theta$ 4K and 8K.

In the case of free cooling, room air exchange rates (ACH) are even larger, and surface convection becomes mixed, and the surface convective heat transfer coefficient combines

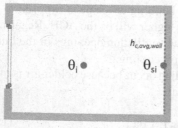

$$h_{c.avg.wall,[ASHRAE]} = 1.31 \cdot \Delta\theta^{0.33} \,(W/m^2K)$$

$$h_{c.avg.wall,[Khalifa]} = 1.983 \cdot \Delta\theta^{0.25} \,(W/m^2K)$$

$$h_{c.avg.wall,[Khalifa]} = 0{,}609 \cdot ln(\Delta\theta) + 1.182 \,(W/m^2K)$$

FIGURE 15.16 Approximation models for convective surface heat transfer coefficient dependant on temperature difference $\Delta\theta_i$ between indoor air temperature and surface temperature of wall θ_{si}.

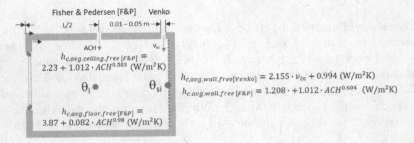

FIGURE 15.17 Approximation models for convective surface heat transfer coefficient in case of free cooling of the buildings dependent on temperature difference of the supply air velocity vi_n and ACH.

natural and forced convection phenomena. Among several researches, an approximation model for local surface heat transfer coefficient for mix convection caused by supply air jet blown near a vertical wall was developed by Venko et al. (2015). An approximation model of the average convective surface heat transfer coefficient that includes inlet velocity of adiabatic supply air temperature and a model that includes ACH as an independent variable were presented by Fisher and Pedersen (1997). The approximation models are shown in Figure 15.17. Taking into account the range ACH between ACH 10 h^{-1} and 20 h^{-1}, the $h_{c, avg, wall, free}$ is in the range between 6.24 and 7.40 W/m^2K.

It can be concluded that in numerical modeling of thermal response of non-mechanical cooled buildings, the standardized convective surface heat transfer coefficient should be replaced with one that includes bought temperature difference and ACH as influence variables.

15.3 MODELING OF PASSIVE AND FREE COOLING TECHNIQUES

15.3.1 BUOYANCY AND WIND-DRIVEN PASSIVE COOLING BY VENTILATION

Passive cooling is a technique of cooling that can be achieved by high air exchange rates ACH (h^{-1}) during the daily period when outdoor temperatures are at least several K below the indoor temperature by natural means. Therefore, no auxiliary energy is needed. Air flow between environment and building's interior is a consequence of the air pressure difference, which results from the difference in the outdoor and indoor air temperature or because the building envelope is exposed to the wind. The higher the air pressure difference, the higher will be the ACH. Regardless of the reason of the pressure difference, adequate ventilation openings in the building envelope must be provided for efficient passive cooling.

Temperature or buoyancy-driven passive cooling occur because colder air is more dense than hotter air. Pressure difference is equal:

$$\Delta p_{i-o} = H \cdot g \cdot (\rho_{\theta i} - \rho_{\theta e}) \quad [Pa] \tag{15.8}$$

where H (m) is the vertical deference in the envelope opening, g (m/s^2) is the acceleration due to gravity, and $\rho_{\theta i}$ and $\rho_{\theta e}$ are densities of the air at indoor θ_i and outdoor

H=15 m

FIGURE 15.18 Building Research Establishment office building is passively cooled by stack ventilation. (From bre.co.uk)

θ_e temperature. Pressure difference that occurs is normally less than 2 Pa in case of 2-m-high ventilation opening but can be significantly higher in stack buoyancy ventilation (Figure 15.18).

When wind at speed v_{win} (m/s) is stopped at the envelope of the building, kinetic energy of the control volume of the air is transformed into the pressure energy, and pressure difference is equal:

$$\Delta p_{i-o} = \frac{1}{2} \cdot \rho_e \cdot v_{win}^2 \quad [\text{Pa}] \tag{15.9}$$

At a wind speed up to 1.5 m/s, the pressure difference is similar to that of buoyancy-driven ventilation, but at a wind speed (in front of a building) of 5 m/s it rises to 16 Pa, and at a wind speed of 10 m/s to a difference of 65 Pa, compared to the steady indoor air.

Ventilation openings can be large or small, and buildings can be passive cooled by a single side (through ventilation opening(s) installed in the same wall) or cross (through ventilation opening in opposite wall(s)) ventilation. In passive cooling through large ventilation openings, air enters and leaves the building through same opening and pressure-neutral level established somewhere in the area of the opening. Meanwhile, in passive cooling through small openings, air enters from surroundings through openings that are below the neutral pressure level and exit the building through openings above the neutral pressure level (Figure 15.19).

For the purpose of the engineering practice, the volume air flow rate $\dot{V}(m^3/s)$ due to the buoyancy and wind-driven single side airing can be determined by the empirical expressions (EN 15242:2007; Allard, 2012).

For buoyancy (temperature)-driven airing through large opening, the volume air flow rate $\dot{V}_b$ (m³/s) is:

$$\dot{V}_b = C_d \cdot C_\alpha \cdot \frac{A}{3} \cdot \sqrt{\frac{|\theta_i - \theta_e| \cdot g \cdot H}{\frac{T_i + T_e}{2}}} \quad \left[\frac{m^3}{s}\right] \tag{15.10}$$

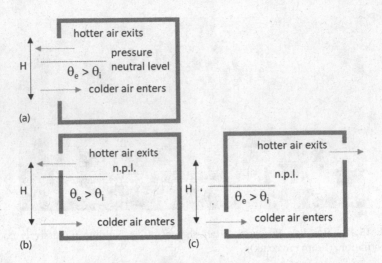

FIGURE 15.19 (a) Passive cooling by ventilation through single-side large ventilation opening, (b) single side, and (c) cross-ventilated building through small opening.

where C_d is the discharge coefficient of opening (~0.6); A is the window opening area (m²), θ_i, T_i (°C, K) and θ_e, T_e (°C, K) are the indoor and outdoor air temperatures; g is the acceleration due to the gravitation (m/s²); H is the free area height of the opening (m); and C_α is the coefficient of effective open area of the window determined for the bottom hung window with expression:

$$c_\alpha = 2.6 \cdot 10^{-7} \cdot \alpha^3 - 1.0 \cdot 10^{-4} \cdot \alpha^2 + 1.86 \cdot 10^{-2} \cdot \alpha \quad (15.11)$$

where α is the tilt angle of bottom hung window wing with value 0.9 at tilt angle 90° and 1° at tilt angle of 180°.

Case study: At indoor air temperature θ_i 20°C and outdoor air temperature θ_e 0°C, the volume air flow rate $\dot{V}_b$ buoyancy-driven airing through a single side window with an area A of 0.35 m² and height H 1.4 m large with bottom hung wing at the tilt angle 10° (Figure 15.20) will be ~44 m³/h. Meanwhile, at outdoor air temperature θ_e 15°C, the $\dot{V}_b$ will decrease to the half (~21.5 m³/h).

(Continued)

FIGURE 15.20 Window with bottom hung window wing from case study. In the presented case, tilt coefficient C_α is equal 0.176.

For multiple single-side ventilation openings:

$$\dot{V}_b = C_d \cdot A \cdot \left[\frac{\epsilon \cdot \sqrt{2}}{(1+\epsilon) \cdot (1+\epsilon^2)^{0.5}} \right] \cdot \sqrt{\frac{|\theta_i - \theta_e| \cdot g \cdot H}{\frac{T_i + T_e}{2}}} \quad \left[\frac{m^3}{s} \right] \qquad (15.12)$$

where C_d is the discharge coefficient of opening (~0.6), A is the total are of ventilation openings $(A_i + A_o)$ (m²), and ϵ is the ration between outlet and inlet ventilation openings (Figure 15.19).

For wind-driven airing through a large opening, the volume air flow rate $\dot{V}_w$ (m³/s) is

$$\dot{V}_w = 0.025 \cdot A \cdot v_w \quad \left[\frac{m^3}{s} \right] \qquad (15.13)$$

where A (m²) is the window opening (corrected with tilt coefficient C_α in case of bottom hung window), and v_w is the wind speed (m/s). At wind speed 5 m/s in a direction perpendicular to the window volume air flow rate $\dot{V}_w$ through window shown in Figure 15.20, it will be ~28 m³/h.

It is common that a larger value max $(\dot{V}_b, \dot{V}_w)$ is in the calculation of passive cooling effectivity.

15.3.2 Free Cooling by Ventilation

In the free cooling systems, in which cold is transported into the building by cold environment air, a fan must be installed. The electrical power of the fan depends on airflow rate and total pressure difference that must be provided by the fan. If total pressure drop in distribution channels and other elements (e.g., heat exchangers or air filters) is not known, approximate fan power can be determined by the expression:

$$P = \frac{\dot{V}_{fan} \cdot \Delta p_{tot}}{\eta_{fan}} \tag{15.14}$$

where P is the electrical power of the fan, $\dot{V}_{fan}$ is the volume air flow for free cooling (m_3/s), and η_{fan} is the efficiency of the fan, including typical values 0.4 to 0.5. Typical total pressure drop in the free cooling system with distribution channel is 0.5 to 1 Pa per 1 m of the channel, and the total pressure drop is in the range between 200 and 500 Pa. In the case that hydraulic characteristics of the distribution system are not known, the electrical power of the fan can be predicted from energy efficiency guidelines stated in the EN 13779 (2007) standard, taking into account the fan efficiency category (SFP). SFP expresses the specific fan power in W's installed to provide the air flow rate of 1 m³/s. A category not higher than SFP 3 should be predicted for free cooling by ventilation. If free cooling is provided without the distribution channels by a propeller fan installed in the external wall of the building, the total pressure drop Δp_{tot} is significantly lower. This is very important because the all-year ratio between delivered cold and the use of electricity for the operation of the free cooling system $COP_{free, an}$ expressed by:

$$COP_{free,an} = \rho \cdot c_p \cdot \frac{V}{3600} \cdot \frac{\sum_{j=1}^{k} (ACH_{free,j}) \cdot (\theta_{i,j} - \theta_{e,j})^+}{\sum_{j=1}^{k} P_{fan,j}} \tag{15.15}$$

should be at least 10 to justify the cost of the free cooling by ventilation (Medved et al., 2014). In the expression, j is the numerator of operation hours, k is the total hours of operation over the year, ρ is the density of the air (kg/m³), c_p is the specific heat capacity (J/kgK) of the air, V is the net volume of the building, $ACH_{free,j}$ is the air exchange rate (h^{-1}) provided by free cooling ventilation system in the j-th hour, $\theta_{i,j}$ and $\theta_{e,j}$ are the indoor and outdoor air temperatures (°C), and $P_{fan,j}$ (W) is the electrical power of the fan (and actuators and control equipment) at j-th hour, assuming that ventilation system can operate with variable air volume (VAV). "+" indicates that free cooling system operates only when $\theta_{i,j} > \theta_{e,j}$.

A big advantage of free cooling over the passive cooling is that operation of the free cooling system by ventilation can be controlled and adapted to the requirements of indoor thermal comfort.

An appropriate control algorithm must take into account the optimal switch on/off temperature difference ($\Delta\theta_{on}$ 6–10 K, $\Delta\theta_{off}$ 0K) frequency-controlled fan motor for efficient adjusted air flow rate and weather forecasting (Vidrih et al., 2016).

Case study: All-year $COP_{free, \, an}$ was determined for free cooling by ventilation system for a reference office. The occupancy schedule and internal gains are taken from ISO 18523-1 (2016). The fan with energy efficiency category SFP 3 was assumed (Vidrih et al., 2016) and $\Delta\theta_{on}$ 6 K. From results shown in Figure 15.21, it can be seen that efficiency of the free cooling is significantly higher in massive built buildings (bottom) than in lightweight built ones, and at average daily outdoor air temperatures $\overline{\theta}_e$ below 20°C if daily amplitude of the outdoor air temperatures are above the 5K. In the case of lightweight buildings, the average outdoor air temperatures should be below 16°C to reach the same efficiency of free cooling by ventilation.

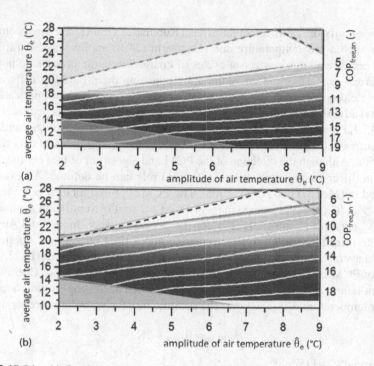

FIGURE 15.21 (a) Overall coefficient of efficiency $COP_{free, \, an}$ of free cooling by ventilation system for the reference office build by lightweight, (b) or massive building structures; gray areas represent environmental conditions that probably will not appear in the EU climate.

As a rule, free cooling by ventilation is only possible during the night. A cold storage can be used to eliminate this disadvantage. Due to the small difference in the temperature of the cold source θ_e and the indoor air temperature θ_i, storages with phase change material (PCM) are especially suitable. Many variations of PCM storage integration into ventilation systems were developed; nevertheless, the system shown in Figure 15.22 is common.

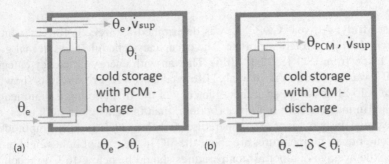

FIGURE 15.22 (a) Free cooling system by ventilation with integrated PCM cold storage, nighttime operation, and (b) daytime operation.

Pure or a mixture of different paraffins (Rubitherm GmbH) are most commonly used for such a low temperature and short-term (24 h) applications. Paraffins are suitable because a large number of cycles of charge/discharge process do not affect their thermal properties and, in addition, by adjusting the mixture of different paraffins, a melting point can be adapted that is optimal for particular application. An additional advantage is that melting/solidification without hysteresis occurs at a wide (up to 15 K) temperature range, which is important because average outdoor air temperatures $\overline{\theta}_e$ (°C) and their amplitudes $\hat{\theta}_e$ (°C) change significantly over the summer period. Although the selection of the PCM and algorithm of free cooling system operation differ among applications, a general role can be defined. An example is presented in (Medved and Arkar, 2008). The PCMs optimal melting point temperature $\theta_{PCM, m}$ (°C), specific storage mass of PCM per unit of ventilation air mass flow rate m_{PCM} (kg/(m³/h), and CDH (Kh) for the period between June 1 and August 31 were determined according to the average summer time outdoor air temperatures $\overline{\theta}_{e,avg}$ and average daily amplitudes $\hat{\theta}_{e,avg}$. Results are shown in Figure 15.23. For the design of PCM cold storage, two regression models were developed. Optimal melting point temperature $\theta_{PCM, m}$ (°C) can be determined according to the average outdoor air temperature $\overline{\theta}_{e,sum}$ (°C) over the summer months with the expression:

$$\theta_{PCM,m} = \overline{\theta}_{e,avg} + 2K \qquad (15.16)$$

and seasonal CDH (Kh) is:

$$CDH = 1200 \cdot \hat{\theta}_{e,avg} \qquad (15.17)$$

Free cooling of the building can be provided by mechanical ventilation of the cavity embedded in the building envelope structure (Figure 15.24). In this case, free cooling is achieved by decreased heat flux that passes the building envelope toward the interior during the daytime and by supplying air with volume flow rate $\dot{V}_{sup}$ (m³/s) through the cavity during the nighttime. Due to radiative cooling, not only is the supply air temperature lower than the ambient air temperature, but the heat flux through the building envelope can be positive (directed toward the outdoors).

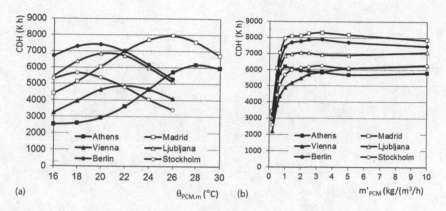

(a)   (b)

θ_PCM,m (°C)   m'_PCM (kg/(m³/h))

FIGURE 15.23 (a) Summertime (June 1–August 31) CDH provided by PCM storage integrated into the free cooling by ventilation system at different climate conditions according to the PCM's melting point temperature (temperature at which maximum specific heat occurs) and (b) specific mass of PCM expressed per 1 m³/h volume flow rate of ventilation air; the range of the melting temperature 12K was assumed in simulations; climate data $\bar{\theta}_{e,avg}$ and $\hat{\theta}_{e,avg}$: Ath 26.4°C; 4.87K; Mad 23.9°C; 6.74K; Vie 19.9°C; 4.00K; Lju 19.0°C; 5.50K; Ber 18.0°C; 6.23K; Sto 16.5°C; 4.55K.

Case study: 2D transient heat flux at the inner surface $\dot{q}_{i,L,t}$ (W/m²) of the roof structure with mechanical ventilated cavity (2 cm wide) with thermal transmittance U equal 0.467 W/m²K (Figure 15.24) was modeled in a 10-minute timestep by discrete Fourier series (Černe and Medved, 2007). Results for two different air flow rates $\dot{V}'_v$ expressed per m² of the roof area are shown in Figure 15.25 for a clear summer day with daily solar radiation 5200 Wh/m², the maximum solar irradiation 950 W/m², the average outdoor temperature $\bar{\theta}_e$ 27.5°C, and the outdoor air temperature amplitude $\hat{\theta}_e$ 6K.

Note: Roof was coated by non-selective paint with uniform emissivity ε in the range of wavelengths between 0.3 and 20⁺ μm. In the case of selective coating with advance (selective) radiative properties for radiative cooling, maximum heat flow toward outdoors during nighttime operation 12 W/m² has been measured.

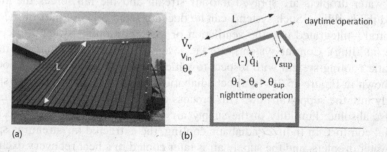

(a)   (b)

FIGURE 15.24 (a) Free cooling can be established by mechanical ventilated cavity embedded in the building envelope structure; pilot roof structure with length L equal 9 m (From Černe, B. and Medved, S., *Build. Environ.*, 42, 2279–2288, 2007.); and (b) operational principle of free cooling.

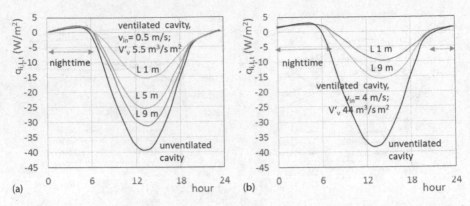

FIGURE 15.25 2D transient heat flux at inner surface of the roof structure with mechanical ventilated cavity. (a) with specific air flow rate 5.5 m³/s per m² of the roof structure and (b) with specific air flow rate 44 m³/s per m² of the roof structure.

15.3.3 ADIABATIC FREE COOLING

In nature, air always contains water in different aggregate states of matter, mostly as water vapor. The amount of the water vapor in the air depends on its temperature. When the maximum air content is possible, the amount of the water vapor is saturated and relative humidity of the air is 100%. Internal energy of the air consists of sensible heat, which can be expressed by air temperature and latent heat of water vapor, which depends on the amount of the water vapor in the air and is measured as enthalpy in Joules, most often as specific enthalpy content in 1 kg of air. When water droplets are introduced into the air, (e.g., by the fountain), they evaporate using enthalpy of the surrounding air, which results in decreased temperature of the air. While there is no energy supplied to drive evaporation, the process is adiabatic cooling. During adiabatic cooling, the enthalpy of the air remains constant as the temperature decreases; meanwhile, air humidity increases. This technique can be used for passive as well as free cooling of the buildings. The most efficient passive adiabatic cooling takes place on the leaves of plants; meanwhile, free cooling systems consist of the humidification chamber where small water droplets are sprayed into air stream and the fan forces the air flow toward the building. Such systems can be decentralized (to cool part of the room) or central—integrated into the ventilation or air-conditioning system (to cool the whole building). Central systems can be designed as the direct and the indirect adiabatic cooling system with respect to which airflow is adiabatically cooled, as it is shown in Figure 15.26. In direct adiabatic cooling water, droplets are sprayed directly into the supply airflow. This process can be efficient in a hot, dry climate because absolute humidity of the supply air increases and can affect the indoor thermal comfort. In indirect adiabatic cooling, the extracted air stream is sprayed with water droplets, and the supply air is later cooled in a heat recovery exchanger. Such a process has no effect on the indoor air humidity, because absolute humidity of the supply air remains unchanged, compared to the outdoor air. Besides that, larger cooling flux can be transferred into the buildings in this case.

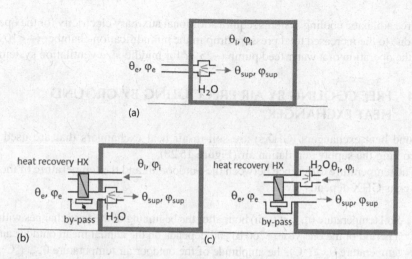

FIGURE 15.26 (a) Passive adiabatic cooling, (b) direct free adiabatic cooling, and (c) indirect free adiabatic cooling.

Case study: Building is ventilated with supply air flow rate 280 m³/h. Outdoor air temperature θ_e is 32°C, and relative humidity φ_e is 35% (absolute humidity x_e 10.39 g/kg). Determine the supply air temperature θ_{sup} in the case of passive, direct free, and indirect free cooling. In all cases, supply air relative humidity φ_{sup} should not exceed indoor theme comfort limit 65%. Efficiency of heat recovery exchanger in direct and indirect free cooling is 82%. Results are shown in Mollier's h-x diagram presented in Figure 15.27. In passive adiabatic cooling, the supply air temperature θ_{sup} is equal to 25.0°C (taking into account limited φ_{sup}), and in direct adiabatic cooling θ_{sup} is 23.5°C. Meanwhile, in the case of indirect adiabatic cooling, the supply air temperature is 21.5°C. There is a significant difference in the enthalpy of cooling flux transferred into the buildings: −480 W in direct and −1010 W in the indirect adiabatic cooling.

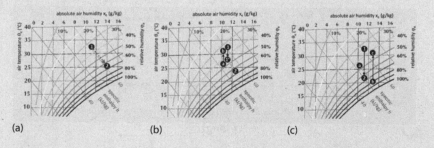

FIGURE 15.27 (a) Process of the passive cooling, (b) the direct free cooling, and (c) the indirect free adiabatic cooling shown in the Mollier's h-x diagram.

Free adiabatic cooling systems require additional auxiliary electricity for the operation due to the increased total pressure drop in the humidification chamber (+<10 Pa) and the operation of a water feed pump (~25 W for middle-size ventilation systems).

15.4   FREE COOLING BY AIR PRECOOLING BY GROUND HEAT EXCHANGER

Ground heat exchangers (GHXs) are soil-to-air heat exchangers that are used for precooling the supply ventilation air (Figure 15.28).

The temperature difference between the outdoor air and the temperature of the air that exits GHX depends on:

- Soil temperature $\theta_{grn,t}$ at (t-th) hour after the beginning of the year changes with a period of the one year (8760 h) and depends on the annual mean outdoor air temperature $\overline{\theta}_{e,an}$ (°C), the amplitude of the outdoor air temperature $\hat{\theta}_{e,an}$ (°C), which is determined by taking into account the average outdoor air temperature of the hottest month in the year $\theta_{e,max,m}$, the depth Z(m) of the GHX below the surface and thermal properties of the soil. Soil temperature at the depth Z in the hour t is defined according to EN 16798-5-1 (2017) standard by expression:

$$\theta_{grn,t} = \overline{\theta}_{e,an} - \overbrace{\left(\theta_{e,max,m} - \overline{\theta}_{e,an}\right)}^{\hat{\theta}_e} \cdot e^{-Z \cdot C_{gm}} \cdot \cos\overbrace{\left(\frac{2 \cdot \pi}{8760} \cdot t - Z \cdot C_{grn} - f_t\right)}^{RAD} \quad (15.18)$$

where soil thermal properties are combined in soil thermal properties factor C_{grn} with value 0.443 for the dry sand, 0.360 for moist soil, and 0.314 for wet, t is the annual hour (with t = 0 at the beginning of the year) and f_t is time shift factor that indicate annual hour at which the soil temperature is the lowest over the year – for central EU climate equal to $1.08 \cdot \pi$.

- The heat transfer coefficient U_{GHX} (W/m²K); assuming common diameter, thickness, and material of the GHX in form of pipe, $U_{GHX,t}$ at t-th annual hour can be approximated by expression:

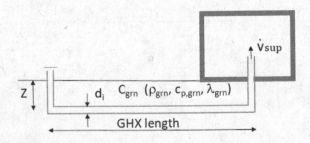

FIGURE 15.28 Design parameters of the ground heat exchanger.

$$U_{GHX,t} = \left(4.13 + 0.23\frac{\theta_{e,t}}{100} - 0.0077\left(\frac{\theta_{e,t}}{100}\right)^2\right) \cdot \frac{v^{0.75}}{d_i^{0.25}} \qquad (15.19)$$

where $\theta_{e,t}$ (°C) is the outdoor air temperature at t annual hour after the beginning of the year, v (m/s) is the velocity of air in GHX, and d_i is its inner diameter (m). The temperature difference of the air between inlet (outdoor) and outlet air temperature is equal to:

$$\Delta\theta_{sup,t} = \left(\theta_{grn,t} - \theta_{e,t}\right) \cdot \left[1 - e^{-\left(\frac{U_{GHX,t} \cdot A_{GHX}}{\dot{V}_{sup} \cdot \rho_a \cdot c_{p,a}}\right)}\right] \qquad (15.20)$$

$$\theta_{sup,t} = \theta_{e,t} - \Delta\theta_{sup,t} \qquad (15.21)$$

where A_{GHX} (m²) is the heat transfer area of GHX, $\dot{V}_{sup}$ is the air flow rate (m³/s) of supply air in GHX, and ρ_a and $c_{p,a}$ are the density (kg/m³) and the specific heat capacity (J/kgK) of the air, respectively.

The GHX increases the pressure drop in the ventilation system and therefore of electricity for the operation of the fan. It can be assumed that the total pressure drop and is up to 1 Pa per 1 m length of the pipe; nevertheless, filters and additional heat exchangers (if installed) must be taken into account (e.g., +100 Pa if F5-F7 filter is installed or +40 Pa for coil heat exchanger) (EN 13779, 2007).

Case study: What will be the temperature of supply air at the outlet of GHX at t = 4392 and 5136 h annual hour in the site with average annual outdoor air temperature $\bar{\theta}_{e,an}$ 9.5°C, and average outdoor air temperature in July $\theta_{e,max,m}$ 26°C (hottest month of the year)? Combined soil thermal properties C_{grn} is 0.360. The length of the GHX is 18 m, GHX is built 1.8 m below the ground surface, and the inner diameter d_i of the GHX is 180 mm (A_{GHX} is 10.1 m²). The supply airflow rate in GHX $\dot{V}_{sup}$ is 280 m³/h; therefore, the air velocity v is 3 m/s. Outdoor air temperatures are $\theta_{e,4393}$ 30.5°C and $\theta_{e,4392}$ 28.2°C. The soil temperature at depth 1.8 m $\theta_{grn,4392}$ is 14.9°C, and $\theta_{grn,5136}$ is 17.6°C. $U_{GHX,4392}$ is equal to $U_{GHX,5136}$ which is equal 14.7 W/m²K. Finally, the temperature of the supply air $\theta_{sup,4392}$ is 18.3°C, and $\theta_{grn,5136}$ is 19.9°C. Figure 15.29 shows the process of air cooling in GHX in the Mollier's h-x diagram and the cooling heat flux provided by GHX. It can be also seen that relative humidity at 4392 annual hour exceeds 80%, which may accelerate the growth of microorganisms. To avoid this, the fan can be switched off or airflow rate can be increased.

(Continued)

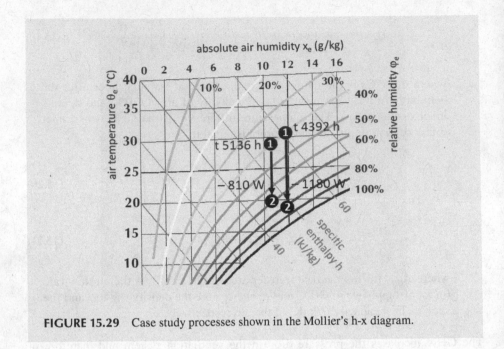

FIGURE 15.29   Case study processes shown in the Mollier's h-x diagram.

15.5   ENERGY NEEDS FOR COOLING OF REFERENCE BUILDING

The effect of passive and free cooling techniques on the energy needs for cooling $Q_{c,nd}$ was analyzed for a reference office shown in Figure 15.30. Transient thermal response was modeled by TRNSYS computer code. The office has useful area of 14 m² and operation patterns (occupancy schedules, use of equipment, thermal comfort conditions) was taken from ISO 18523-1 (2016) (Category Of-1). Meteorological data from TRY for Ljubljana was used in computer simulations.

Thermal transmittance of the outer wall (south orientated) U_{wall} 0.5 W/m²K and thermal transmittance of windows U_w 1.3 W/m²K were assumed. All other walls are adiabatic. Two options were analyzed for those walls: (1) massive build

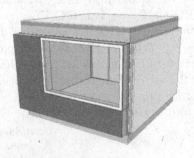

FIGURE 15.30   Reference office.

structures (by 15 cm thick concreate), and (2) as lightweight frame structures. A constant operative temperature was used as the set-point temperature ($\theta_{op, i}$ 26°C). According to the density of occupancies (0.1 occ/m²), the ACH rate during the occupancy period was 1.6 h⁻¹, and no infiltration was taken into account. Internal heat gains of occupancies (11.9 W/m²) and appliances (12 W/m²) were assumed. The office is mechanically ventilated, with heat recovery temperature efficiency 85% with controlled by-pass operation. The shading factor of the shading device S_f 0.7 was assumed. Passive cooling is provided by bottom-hung windows. The tilt angle is set to 10°, and the window is closed at an outdoor temperature θ_e below 12°C and indoor air temperature θ_i below 15°C or if $\theta_e > \theta_i$. Direct adiabatic cooling was controlled by maximum relative humidity of supply air ($\varphi_{sup} < 80\%$ and indoor air humidity ($\varphi_i \leq 50\%$). Results are presented as hourly energy needs for cooling $Q_{c,nd,h}$ (kWh/h) in Figure 15.31. Specific energy needs $Q'_{c,nd}$ (kWh/m²) for cooling per m² of useful office area are shown for the following case studies: (a) the office with massive structures and windows

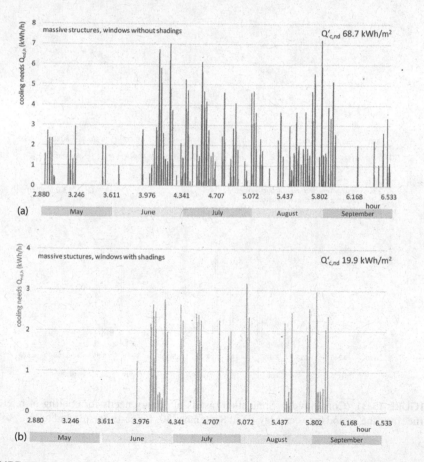

FIGURE 15.31 Simulation results of energy needs for cooling of reference office over the period between May 1 and September 30; for cases (a) and (b). *(Continued)*

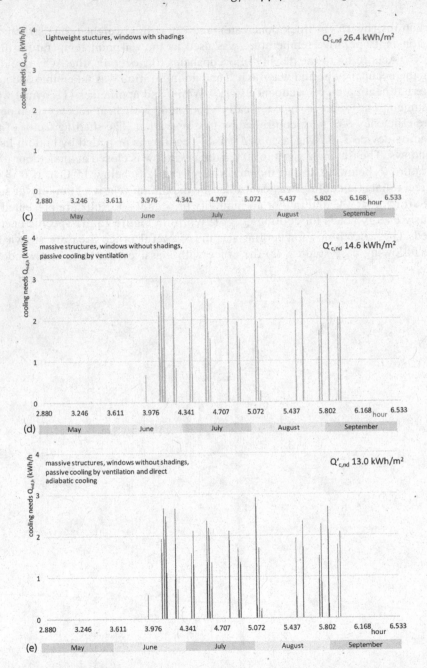

FIGURE 15.31 (Continued) Simulation results of energy needs for cooling of reference office over the period between May 1 and September 30; for cases (c), (d) and (e).

without shading, (b) the same office with shaded windows, (c) the office with lightweight indoor structures and shaded windows, (d) the office with passive cooling by ventilation, and (e) the office with passive cooling by ventilation (during nighttime) and direct adiabatic cooling (during the daytime).

15.6 CONCLUSION

In this chapter, techniques for natural cooling of the buildings by passive and free cooling exploiting environmental cold or natural processes are presented. It is shown that the effect of natural cooling depends on several areas of influence, such as urban microclimate mitigation and adaptation of the buildings to the natural cooling. Besides that, the impact of natural cooling techniques on energy demands of the building depends on, in large extent, local climate conditions and transient natural processes such as storing heat in building structures. This can only be modeled as a transient thermal response with advanced computer simulation tools. The rapidly developing building information modelling (BIM) approach will be an effective way for widely spreading the use of natural cooling in engineering practice. Passive techniques and free cooling systems require more advanced controlling, comparing to the mechanical cooling, such as adaptive comfort models and weather forecast integration, to maintain a high quality of indoor environment while improving the energy efficiency of the building. Nevertheless, a successful design of passive and free cooling techniques is only possible with understanding the operation principles and designer's engineering sense. The chapter was designed with this in mind.

REFERENCES

Allard F. Natural ventilation in buildings. IDES-EDU IEE/09/631/SI2.558225, 2012.

Arkar C., Domjan S., Medved S. Heat transfer in a lightweight extensive green roof under water-freezing conditions. *Energy and Buildings* 167 (2018), pp. 187–199.

Arkar C., Domjan S., Majkovič D., Šumi J., Medved S. Hydrological and thermal response of green roofs in different climatic conditions. *IOP Conference Series: Earth and Environmental Science* 323:012063 (2019).

ASHRAE. *2001 ASHRAE Handbook Fundamentals.* ASHRAE, Atlanta, GA, 2001.

Černe B., Medved S. Determination of transient two-dimensional heat transfer in ventilated lightweight low sloped roof using Fourier series. *Building and Environment* 42 (2007), pp. 2279–2288.

EN 13779: 2007 Ventilation for non-residential buildings – Performance requirements for ventilation and room-conditioning systems.

EN 15242:2007 Ventilation for buildings – Calculation methods for the determination of air flow rates in building including infiltration.

EN 15251:2017 Indoor environmental input parameters for design and assessment of energy performance of buildings – addressing indoor air quality, thermal environment, lighting and acoustic.

EN 16798-5-1:2017. Energy performance of buildings – Ventilation for buildings – Part 5-1: Calculation methods for energy requirements of ventilation and air conditioning systems – Method 1: Distribution and generation.

EN ISO 13786:2017 Thermal performance of building components – Dynamic thermal characteristics – Calculation method.

European Environment Agency, Data and maps, Heating and cooling degree days, Prod-ID: IND-348_en, 2019.

Fisher D.E., Pedersen C.O. Convective heat transfer in building energy and thermal load calculations, *ASHRAE Transactions* 103 (1997), Part 2.

Heat Roadmap Europe 4 (HRE4); A low-carbon heating and cooling strategy; H2020-EE-2015-3-MarketUptake; Fraunhofer Institute for Systems and Innovation Research, Karlsruhe, Germany, 2017.

ISO 18523-1:2016 Energy performance of the buildings – Schedule and condition of building, zone and spaces usage for energy calculation – Part 1: Non-residential buildings.

ISO 6846:2017 Building components and building structures – Thermal resistance and thermal transmittance – Calculation methods.

Jakubcionis M., Carlsson J. Estimation of European Union service sector space cooling potential. *Energy Policy* 113 (2018), pp. 223–231.

Khalifa A.J.N., Marshall R.H. Natural and forced convection on interior building surfaces: preliminary results. *Applied Research Conference* (1989), pp. 249–257.

LEXIS Coatings, Acworth, GA, 2019.

Medved S. Building Physics. University of Ljubljana, Faculty of Architecture (Second Edition), Ljubljana, Slovenia, 2014.

Medved S., Arkar C. Correlation between the local climate and the free-cooling potential of latent heat storage. *Energy and Buildings* 40 (2008), pp. 429–437.

Medved S., Babnik M., Vidrih B., Arkar C. Parametric study on the advantages of weather-predicted control algorithm of free cooling ventilation system. *Energy* 73 (2014), pp. 80–87.

Medved S., Domjan S., Arkar C. *Sustainable Technologies for Nearly Zero Energy Buildings: Design and Evaluation Methods.* Springer Nature, Cham, Switzerland, 2019.

Medved S., Vidrih B. Evaluating the potential of local urban heat island mitigation by the whitening and greening of the settlement surfaces, World Renewable Energy Congress XIV - Clean Energy for a Sustainable Development, 2015.

Meteonorm. Global Meteorological Database for Solar Energy and Applied Climatology; Version 5.1: Edition 2005; Software and Data on CD-ROM; Meteotest AG, Bern, Switzerland, 2005.

Persson U., Werner S. Quantifying the Heating and Cooling Demand in Europe; STRATEGO project, 2015.

Santamouris M. Heat island research in Europe: The state of the art. *Advance in Build Energy Research* 1 (2007), pp. 123.

Venko S., Pavlovič, E., Vidrih, B., Medved, S. An experimental study of mixed convection over various thermal activation lengths of vertical TABS. *Energy and Buildings* 98 (2015), pp. 151–160.

Vidrih B., Arkar C., Medved S. Generalized model-based predictive weather control for the control of free cooling by enhanced night-time ventilation. *Applied Energy*, 168 (2016), pp. 482–492.

Vidrih B., Medved S. The effect of changes in the climate on the energy demands of buildings. *International Journal of Energy Research* 32 (2008), pp. 1016–1029.

Vidrih B., Medved S. Multiparametric model of urban park cooling island. *Urban Forestry & Urban Greening* 12 (2013), pp. 220–229.

Yang J., Wang Z. and Kaloush K.E. Environmental impacts of reflective materials: Is high albedo a 'silver bullet' for mitigating urban heat island? *Renewable and Sustainable Energy Reviews* 47 (2015), pp. 830–843.

16 Sustainable Timber-Based Building Systems in the Context of Reducing Energy Performance in the Building Use Phase

Jozef Švajlenka, Mária Kozlovská, and Terézia Pošiváková

CONTENTS

16.1 INTRODUCTION

The concept of the product life cycle is now increasingly applied in the construction industry (Estanqueiro et al. 2018; Zhang et al. 2018). In construction, it is understood as the life cycle of a project. A construction's life cycle is defined as the period of time from the idea to build the construction and its transformation into an intention, through design, completion and use, to its demolition and disposal (Cabeza et al. 2014; Chau et al. 2015).

According to Beránková (2019), the creation of a building is determined by inputs, both material and energy-related. Energy and materials for operation and maintenance are consumed in the course of a construction's life cycle (Ding 2008; Gašparík et al. 2017; Mohammadhosseini and Tahir 2018). Each period of a construction's life cycle consumes vast amounts of energy and produces substantial waste and emissions (Almusaed and Almssad 2015; Tang 2018). From the economic perspective, it is the use phase that is the most intensive, representing approximately three-quarters of the total costs for the entire lifetime of the construction, of which one-third is made up of operation and maintenance costs (Beránková 2019).

Matters related to saving fuel, energy, and financial resources in particular, with respect to sustainability, are a much-debated topic (Tsai and Chang 2012; Siva et al. 2017; Le Truong et al. 2018; Zhao et al. 2018). This means that one of the main considerations in the design and construction of new buildings, as well as in the case of existing buildings, is their energy consumption during use (Vinodh et al. 2014; Antošová et al. 2017; Xie et al. 2017). Many newly constructed buildings are already designed as low-energy or passive buildings (Wagner 2014). Construction systems based on the wood are often considered for these energy standards, as they have very positive thermal-technical properties (Drozd and Leśniak 2018; García et al. 2018). Construction time for wood constructions is very short with a low—sometimes even negative—ecological footprint (Ali and Nsairat 2009; Ritchie and Stephan 2018). The use of wood and structural elements based on wood save non-renewable resources, which is also highly desirable from the perspective of a sustainability philosophy (Hafner and Schäfer 2018). Wood constructions are defined as construction projects whose main supporting structure is made from wood (De Araujo et al. 2016a). The most commonly used construction systems based on the wood are panel systems, column systems and, in some regions, log systems or modern log systems.

Modern methods of construction based on the wood are not just economically beneficial variants, but their construction is considerably less time-consuming and requires much less effort (De Araujo et al. 2016b; Watts 2016; Park et al. 2018) compared to traditional construction systems (traditional masonry constructions). Wood houses, whose supporting structure uses a column system or prefabricated panels, have properties that are comparable to those of silicate-based buildings. Wood is a renewable source with a wide range of benefits. It is an energy carrier, producing energy even during disposal (Maratovich et al. 2016). It belongs in a natural chain without a negative environmental impact, provided, of course, that land biodiversity is not neglected. It is also worth noting that in the region of Central Europe, wood has great cultural and historical significance. In the construction industry, wood is a highly versatile material. It is used in wood constructions' supporting and filling structures and also in construction carpentry products, such as wooden windows and doors, winter gardens, wooden stairs, linings, and in furniture making. Price is not the only advantage of modern construction systems based on wood. There is a whole host of other positive aspects (Toppinen et al. 2018). These include time-saving benefits compared to a traditional construction (as it dispenses with any wet processes), highly favorable thermal resistance properties of walls, whose thickness is only half that of traditional walls, thereby increasing useful floor area, in addition to benefits such as energy saving, eco-friendliness, and benefits related to installations (Motuziené et al. 2016). Because electrical, water, and sewage installations are already built into individual components, there is no need to hack into walls, as is the case with masonry walls. Equally positive is the absence of thermal bridges and the fact that such a construction settles only negligibly after completion, and if it does, it settles evenly as a unit. The biggest disadvantage of constructions based on wood is prejudice and lack of awareness (Gosselin et al. 2016), and Slovakia significantly lags behind other countries in the field of assembled houses. The lack of confidence stems from the misconception that the walls of an assembled house are not sufficiently firm, which in fact ceased to be a concern a long time ago. Another reason may be a missing accumulation layer for collecting heat. A combination with masonry is usually used in such cases (walls, floor). If this solution is necessary, the advantage of dispensing with any wet processes previously mentioned no longer applies.

16.2 SELECTED ASPECTS OF TIMBER-BASED BUILDINGS FROM THE PERSPECTIVE OF SUSTAINABILITY

Wood is originally a natural and organic material. It is clear from the composition and structure of wood that it is a non-homogeneous and anisotropic biopolymer in terms of its chemistry, anatomy, morphology, and macroscopic structure. Its structural composition reflects a tree's growth in a coded system of photosynthetic and biochemical reactions. This imparts it with an entire array of special properties.

Physical and Mechanical Properties of Wood—Wood, according to Štefko and Balent (2014), features exceptional mechanical properties and a very good ratio of construction elements' load-bearing capacities to their weight. In the case of a suitably selected structure and technology, which eliminates some of its adverse properties, it enables the construction of efficient supporting wall and frame structures, as well as large-format roofs.

Rheological Properties of Wood—Deformations worsen, wood leaks, and its rigidity reduces with longer periods under load. The most significant change in deformations and rigidity shows during extreme periods—during very short periods under load or during long-term loads, for example, during the life of a structure weighed down by roofing, snow, etc. Wood under permanent strain can be characterized as vicious and flexible, linearly responding material. The structure of wood determines its deformation properties under permanent load.

Wood Flammability/Resistance to Fire—Flammability is often used as an incorrect argument against the use of wood. Although wood does burn, in contrast with other materials it has a great advantage in that the amount of time it burns can be calculated with precision (it contains 8%–15% of water). During a large energy input, every ton of wood must, therefore, evaporate 80–150 kg of water before burning completely.

Impact of Wood on Health—Solid wood has a favorable impact on creating an internal environment for human health on the grounds that:

- Wood has an ability to "breathe" (hygroscopic qualities of wood).
- Air is filtered by passing through the wood (detoxicating effect).
- The filtration process does not filter out active oxygen ions.
- It has a natural color, mostly brown-yellow shades, causing a feeling of warmth.
- Wood is a porous material and has a low heat conductivity (values for wood between 0.11–0.21 W/(m·K)—linings and floors remain warm when touched.
- Wood is a flexible material (construction systems of floors), etc.

Aesthetic Properties of Wood—Natural wood is perceived positively by the user in terms of its color and texture. The use of wood in an environment with artificial materials achieves a surface appearance similar to the positive feedback received from the external natural environment. It has a demonstrably favorable psychological impact. Its technological properties allow shaping it into interesting structural details (admitted carpentry joints, sculptures, reliefs, polychromies, or contrasting combinations with other materials). It remains an important instrument of constructivism in architecture. Many structures and wooden structures by prominent architects such as Frank Lloyd Wright, Alvar Aalto, Dušan Samuel Jurkovič, or contemporary architects Imré Makovecz and Lubor Trubka present unique wooden tectonics based on a harmony of its unique mechanical and aesthetic properties (Štefko et al. 2010).

Technological Properties of Wood—Wood is easy to process and join, its structure allows further processing in the form of semi-products, and also serves as a raw material for large-format materials. Good technological properties together with excellent mechanical properties allow the building of economical and modern prefabricated structures.

Štefko (2014) summarized the arguments in favor of using wood in the construction industry as follows:

- Wood is a historical material in the Slovak context.
- It is a strategic domestic renewable source.
- It has no equivalent in terms of this set of properties:
  - Thermal-technical
  - Aesthetic
  - Environmental impact

- Utility
- Technological
- Mechanical
- Increased use of wood has a favorable impact on the environment:
  - Reduction of energy intensity
  - Reduction of CO_2 emissions
- The favorable economic impact in the Slovak context:
  - Added value
  - Labor market
  - Solutions to the current crisis

16.3 BASIC TYPES OF TIMBER-BASED CONSTRUCTION SYSTEMS

Current wood-based construction methods are very diverse and can be individually tailored and combined. In principle, the current wooden buildings for housing can be divided according to the nature of the vertical load-bearing structures into massive skeletal and elemental structures (Vaverka 2008). The individual groups differ considerably from each other by the used construction method, appearance, and possibilities of production of their structural elements. The foundations of the massive buildings are log buildings, which are still being built today, but nowadays they have also come up with modern massive buildings. Skeletal and elemental building groups have evolved from timber-framed buildings and represent wickerwork. The classic wooden construction methods can be supplemented by the so-called hybrid methods developed in recent years that combine wood as a building material with other building materials.

16.3.1 Massive Log Buildings

The term "massive structures" is used to refer to constructions whose load-bearing structure is made of solid wood, either solid or cross-sectioned, or bonded to one another according to the product system. Massive structures are characterized by the separation of supporting elements and insulating parts. The carrier is not reduced to the individual supports, as is the case with light wood timber systems. Nowadays, because of the increased demands on thermal protection, solid wood-bearing load-bearing walls are supplemented with thermal insulation layers. The traditional representative is log buildings (Vaverka 2008).

Log buildings belong to the original methods of realization of houses, which are basically all massive wood buildings. The construction of log buildings is based on massive logs (logs), beams, or horizontal stacked (stacked) beams connected by carpentry joints in corners. However, there are also logs with vertical beams or a combination of vertical and horizontal beams (Bílek 2005; Vaverka 2008) (Figure 16.1).

16.3.2 Modern Massive Buildings

Modern solid timber wooden buildings are becoming a current trend that tries to get as close to nature as possible while maintaining a functional and modern

FIGURE 16.1 Massive log buildings. (From Meadowlark Log Homes, (cit.03-07-2019) online: https://meadowlarkloghomes.com/.)

design. According to Kobl (2011), they have been created by the introduction of new construction systems, also thanks to the industrial manufacturing capabilities of large-scale elements. The structural systems are predominantly composed of solid wood construction elements, or rarely of wood-based panels (e.g., OSB and particle boards), solid or composite cross sections. The main part of these elements is formed by closed, in particular, massive plate cross sections, or so-called box components assembled in planar structural members. These elements always form the main bearing of the system—the so-called supporting core. A characteristic feature of these systems is the carrier system exclusively operating the flat, which uses a reinforcement plate to transfer the load. A common feature is the construction of an additional insulation system on the outside of the structure (Kobl 2011) (Figure 16.2).

FIGURE 16.2 Modern massive buildings. (From Finnish Log Homes Ltd. (cit.03-07-2019) online: www.finnishloghomes.co.uk/.)

16.3.3 SKELETAL STRUCTURES

The supporting structure of skeletal structures is, according to Vaverka (2008), assembled from rod members that transfer the applied load to the foundation without the interaction of walls or stiffening casing. They have more than 3,000 years of tradition, and from the constructional point of view, they can be used for the construction of frames, columns, and modern skeleton structures (Figure 16.3).

16.3.3.1 Half-Timbered Buildings

The historical structures of the timbered buildings can be included among the first buildings that have evolved from the historical skeletal system of buildings as the first skeleton system buildings. The architecture of half-timbered buildings has been extended in all regions of Europe, where it was necessary to limit the consumption of wood for construction. An alternative use of rather short deciduous wood elements was also preferred. A large number of timbered buildings, which are still preserved in many historic towns but also in rural areas, are mainly in Central and Eastern Europe but also in the Netherlands, northern Germany, Denmark, and Anglo-Saxon countries. Four- and more-story buildings of this type have been preserved from the past, and multi-story, but also commercial, half-timbered buildings have been shown to have a long tradition in Central Europe. In the territory of present-day Slovakia, such constructions were made in mountain and spa areas (Vaverka 2008; Kobl 2011) (Figure 16.4).

16.3.3.2 Modern Skeletal Structures—Heavy Skeletons

The influence of American construction methods has been reflected in the construction of the half-timbered buildings by omitting the horizontal and oblique reinforcement elements (struts and rails) with the modern skeletal structures. To preserve the massive elements of the construction, these structures are also called heavy wooden

FIGURE 16.3 Skeletal structures. (From Angies list, (cit.03-07-2019) online: www.angieslist.com/articles/house-framing-requires-skilled-contractor.htm.)

FIGURE 16.4 Half-timbered construction. (Half-timbered construction, (cit.03-07-2019) online: www.thoughtco.com/what-is-half-timbered-construction-177664.)

skeletons (HWS). In addition to struts and cross members, horizontal elements (threshold and skid) were excluded from the structure, eliminating the unfavorable planting of the structure caused by volume changes in the transverse direction of the wood. In the case of non-settling of the lower threshold, it is also necessary to provide structural protection of the columns anchored directly on the base in order to avoid degradation of wood due to possible permeable moisture (Štefko et al. 2009; Kobl 2011) (Figure 16.5).

16.3.3.3 Pillar Systems—Light Wooden Skeletons—Frame Structures

The concept of timber-frame construction is based on the use of posts in the supporting structure. As mentioned in the historical review of columnar structures, they have evolved from timber-framed timber houses in North America and have gradually expanded to Europe, where they have been modified over time to form a variety of systems under the frame structure name. The term "frame structures" does not relate to the static action of the house; rather, it is based on the construction of the individual frame walls of the rectangular shape formed by the lower and upper frame and the vertical posts. Also due to the use of small cross sections of the wicker elements, the name light wood skeleton (LWS) (Vaverka 2008)—has also been used for these buildings in some countries (Figure 16.6).

FIGURE 16.5 Modern skeletal structures—heavy skeletons. (From A Wooden Skeleton in XXL: T3 Office Building in Minneapolis, (cit.03-07-2019) online: www.detail-online.com/article/a-wooden-skeleton-in-xxl-t3-office-building-in-minneapolis-31385/.)

FIGURE 16.6 Pillar systems—light wooden skeletons. (From Think wood, (cit.03-07-2019) online: www.thinkwood.com/products-and-systems/light-frame-construction.)

16.4 INTERACTIONS AMONG THE TECHNICAL-STRUCTURAL PROPERTIES OF PANEL WOOD CONSTRUCTIONS IN THE USE PHASE

In Slovakia (Central Europe) and the neighboring countries, we have seen a certain upward trend in the use of wood constructions in the past 10 years. This trend is partly due to the undoubted benefits these constructions offer (O'Connor et al. 2004). Despite this, many investors and users are not sufficiently convinced about their actual benefits. This is why our research is aimed at raising awareness of constructions based on wood and at confronting theory with facts. Based on the facts stated above, our research objective was to examine the interactions among selected technical-structural properties of wood constructions in actual use.

16.4.1 ANALYSIS METHODOLOGY

Based on the main objective of our research focused on the interactions among selected technical-structural properties of panel wood constructions in actual use, we chose wood constructions in actual use as our research object. The examined wood constructions were completed using off-site construction technology. Wood constructions based on off-site methods were completed using a prefabricated construction system, where panels are prefabricated at a production plant, transported to the construction site, and assembled to build the resulting structure. Thirty-panel wood constructions in actual use were included in the research. The constructions had been used for an average period of 5 years. This provided a strong research sample for drawing valid conclusions because long-term use offers a more robust insight into the use of the constructions. The basic aspects of the examined wood constructions were monitored in terms of structural-technical characteristics. We monitored the following: energy standard, number of floors, number of rooms, type of heating medium (water), type of heating medium (heating), ventilation method, accumulation elements, useful floor area, period of use, average annual heating costs (EUR per m^2), other average annual operating costs (EUR per m^2), total average annual operating costs (EUR per m^2), average annual energy consumption for heating (kWh), and average annual energy consumption for heating (kWh per m^2). For the sake of a more valid interpretation of the findings, the statistical analysis used selected data that were unified by being converted into values per m^2. Summarized data were later analyzed by means of basic descriptive statistics, the statistical methods of Spearman Correlations (Croux and Dehon 2010), student t-test (Ruxton 2006), and ANOVA—Kruskal-Wallis Test (McKight and Najab 2010), using the STATISTICA 12 software. The Spearman Correlations statistical method was used for an initial quantitative analysis of the studied elements and the interactions among them. The interactions among the studied elements are presented in the form of a square matrix later in the study. For a more detailed qualitative assessment of the studied parameters, the student t-test and the ANOVA—Kruskal-Wallis statistical test were used at a significance level of ($P < 0.05$). A non-parametric student t-test was used for comparing two independent data sets in the context of the assessed aspects, namely, the group of constructions with the presence of accumulation elements on the one hand, and the group of constructions without the presence of accumulation elements on the other. The ANOVA—Kruskal-Wallis statistical test was used for comparing three independent data sets, namely, groups of constructions in terms of the type of heating medium and ventilation method.

16.4.2 Wood Constructions from Prefabricated Panels

Wood constructions assembled from prefabricated panels are currently the most widespread wood constructions in Central Europe. The shape and appearance of such wood constructions are sometimes almost indistinguishable from masonry houses. They are popular mainly for their quick assembly. Panels are carefully manufactured in a production plant (off-site), independently of the weather and are transported ready-made to the construction site. The fact that companies using this technology must hold a product quality certification offers added reassurance to customers. This construction system has excellent resistance to pests. The supporting structure of the panel consists of a wooden frame filled with insulation, lined with particle boards on the outside and by plasterboards on the inside. In the case of sandwich-glued panels, fire-retardant polystyrene is inserted between two oriented strand board (OSB) boards. The construction system from sandwich-insulated panels is mostly used in low-energy houses. It owes its name to its sandwich-like layering, where sound insulation is inserted between two boards, which serves as the lining. A facade from elements based on wood is often used for these houses (Figure 16.7).

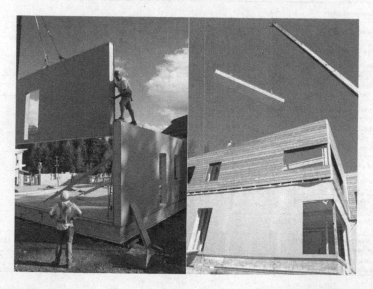

FIGURE 16.7 Wood constructions from prefabricated panels. (From Wood constructions from prefabricated panels, (cit.03-07-2019) online: https://sk.pinterest.com/pin/541769030157372253/.)

16.4.3 ANALYSIS OF RESULTS

In the following part of our evaluation of the collected data, we focus primarily on the statistically significant correlations among the examined parameters from the initial analysis in the context of the objective of our work. The initial statistical analysis of the data revealed several statistically significant interactions among the examined parameters stated in the square matrix (Figure 16.8).

Some of the statistically significant correlations among the examined parameters we found are a logical consequence, such as the increase in the number of rooms with an increase in the number of floors, etc., so it is not necessary to comment on such findings in more detail. Therefore, our further analyses focused primarily on the interactions among parameters that affect, to a statistically significant degree, the amount of financial and energy resources used during the operation of panel wood constructions.

According to Figure 16.8, the amount of financial and energy resources used during the operation was affected, to a statistically significant degree, by the following monitored parameters: type of heating medium (p = −0.5534), ventilation method

	Energy standard	Number of floors	Number of rooms	Type of heating medium (Water heating)	Type of heating medium (Heating of the building)	Ventilation method	Accumulation elements	Usable area of the building	Time of use of the building	Average annual heating costs (EUR per m²)	Others average annual operating costs (EUR per m²)	Total average annual operating costs (EUR per m²)	Average annual energy consumption for heating	Average annual energy consumption for heating (kWh per m²)
Energy standard	1.0000													
Number of floors	ns	1.0000												
Number of rooms	ns	0.4454	1.0000											
Type of heating medium (Water heating)	ns	ns	ns	1.0000										
Type of heating medium (Heating of the building)	ns	ns	ns	ns	1.0000									
Ventilation method	0.3617	ns	ns	ns	ns	1.0000								
Accumulation elements	ns	ns	ns	ns	−0.4582	ns	1.0000							
Usable area of the building	ns	0.5045	0.5891	ns	ns	ns	ns	1.0000						
Time of use of the building	ns	ns	0.3523	ns	ns	ns	ns	ns	1.0000					
Average annual heating costs (EUR per m²)	ns	ns	ns	ns	−0.5774	−0.3837	0.5459	−0.4608	ns	1.0000				
Others average annual operating costs (EUR per m²)	ns	ns	ns	ns	−0.5096	−0.4584	0.3850	−0.4020	ns	0.7568	1.0000			
Total average annual operating costs (EUR per m²)	ns	ns	ns	ns	−0.5534	−0.4687	0.5227	−0.4665	ns	0.9328	0.9290	1.0000		
Average annual energy consumption for heating (kWh)	ns	ns	ns	ns	0.7745	ns	ns	ns	ns	ns	ns	ns	1.0000	
Average annual energy consumption for heating (kWh per m²)	ns	ns	ns	ns	0.6673	ns	ns	−0.4190	ns	ns	ns	ns	0.7782	1.0000

FIGURE 16.8 Interaction among the examined parameters. Note: ns: no significant; dary grey: the correlation is significant at the level 0.0001; medium grey: the correlation is significant at the level 0.01; light grey: the correlation is significant at the level 0.05.

(p = −0.4687), presence of accumulation elements (p = 0.5227), and useful floor area (p = −0.4665). It follows from these correlations that if the panel wood constructions featured accumulation elements (masonry or concrete wall and floor structures), operating costs were lower than in constructions without accumulation elements. It was interesting to find that operating costs did not interact with the energy standard of the constructions to a statistically significant degree (Tables 16.1 and 16.2).

TABLE 16.1

Average Annual Heating Energy Consumption (kWh) Converted into Values per m² of Useful Floor Area

| | All Houses | Type of Heating Medium | | |
		Electric Energy	Gas	Biofuels
average	81.3	50.7	92.0	96.0
± std	43.3	14.5	81.4	36.4
min	36.4	36.4	38.7	49.6
max	213.3	84.9	213.3	196.4
median	76.9	48.6	57.9	88.0
		$p = 0.0010***$		

Note: ± std = standard deviation; min = minimum; max = maximum; statistical analysis of the significance level of $p < 0.05$.

TABLE 16.2

Average Annual Costs for Heating (EUR) Converted into Values per m² of Useful Floor Area

| | | Type of Heating Medium | | | | A Method of Ventilation | | The Presence of Accumulation Elements | |
	All Houses	Electric Energy	Gas	Biofuels	Mechanically, Windows	Recuperation Unit	Air-condition Unit	Yes	No
average	4.6	5.6	6.3	3.7	5.4	3.5	3.5	3.5	6.2
± std	2.5	1.6	5.6	1.4	2.9	0.8	0.5	0.7	3.2
min	1.9	4.0	2.7	1.9	1.9	2.4	3.0	1.9	2.7
max	14.7	9.3	14.7	7.7	14.7	4.6	4.3	4.6	14.7
median	4.0	5.3	4.0	3.4	4.5	3.6	3.3	3.7	5.8
		$p = 0.0051**$				ns		$p = 0.0006***$	

Note: ± std = standard deviation; min = minimum; max = maximum; statistical analysis of the significance level of $p < 0.05$; ns = no significant.

Based on a more detailed analysis, we found that operating costs during the use of panel wood constructions are divided into two main groups. The first group consisted of costs associated with heating, and the second group consisted of costs for electrical appliances, lighting, etc. The ratio of the former group to the latter was approximately 1:1. It is therefore important to note that it is necessary to monitor costs for heating panel wood constructions because they represent a significant share of the budget for operating such constructions. As for the second group of costs generated during operation, it is affected primarily by the number of users and by the way they use their constructions.

As it has already been stated, we recorded statistically significant correlations among average annual operating costs converted into values per m^2 (p = −0.5774), average annual heating energy consumption converted into values per m^2 (p = 0.6673) and the type of heating medium. This implies that if panel wood constructions were heated using gas or biofuels, average annual heating energy consumption was higher compared to wood constructions using an electrical heating medium. A reverse correlation was recorded in terms of average annual operating costs. This is affected mainly by local prices for media distribution.

Another significant finding was the fact that if panel wood constructions were equipped with air conditioning or recuperation units, their average costs for heating converted into values per m^2 were lower than those of the wood constructions without these systems. A statistically significant correlation was also recorded between the useful floor area of panel wood constructions and operating costs. This finding suggests that an increase in useful floor area proportionately reduces operating costs converted into values per m^2. Of course, this finding was also recorded with an increase in the number of floors, which only confirms the fact that constructions with two or more floors are more energy-efficient than single-story constructions.

The accumulation capacity of building materials is mainly influenced by their bulk density (Dincer 2002). Wood and wood-based materials generally have a low accumulation capacity, which need not always be a disadvantage (Hermawan et al. 2015; Němeček and Kalousek 2015). For example, in wooden constructions, the use of wooden components within a sandwich structure is effective. Thanks to the ratio of the static to the thermal-technical properties of wood, it is possible to make wood buildings energy efficient. Maintaining comparable thermal and technical characteristics of sandwich construction and traditional masonry construction, it is possible to save around 10%–20% of the floor area in the wood structure due to the thinner wall structures. On the other hand, the accumulation capacity of a sandwich structure is relatively lower than that of conventional masonry constructions (Němeček and Kalousek 2015). As it can be seen in our findings from real-world monitoring, better energy performance of wood buildings can be achieved by incorporating accumulation elements within a building. If such accumulation materials are not suitably insulated, undesirable thermal bridges with negative accompanying phenomena could occur. So-called thermal bridges are typical of conventional silica and concrete-based systems (Johra et al. 2019). Energy-efficient solutions are nowadays the subject of discussion not only in the construction sector but in almost all manufacturing sectors of the economy (Jeanjean and Olives 2013; Pomianowski et al. 2013; Johra et al. 2019). A study by Mavrigiannaki and Ampatzi (2016) points to the benefits of incorporating accumulation elements within the structural parts of buildings. These authors specified suitable materials with accumulation capacities and their use in buildings. It follows from their conclusions that it is possible to reduce costs and save energy in buildings while ensuring the user's internal comfort. We support these ideas based on our findings, in which we have seen significant interactions among the studied parameters. Therefore, it is important to pay close attention to the selection of suitable materials and their combination within energy-efficient construction systems.

16.5 CONCLUSIONS

In line with the current topic of the efficient management of resources in the construction sector and the use of constructions, our work focused on studying the aspects of the construction uses phase. Specifically, we analyzed panel wood constructions in actual use. The use of wood constructions in Central Europe is on the rise, but poor awareness of these constructions still prevents them from becoming more widespread. Therefore, our work is also a contribution in terms of expanding our understanding of these types of constructions based on wood. Based on our research of wood constructions in actual use in the context of technical-structural aspects and their effect on operating resources during use, we may state that there are a statistically significant effect and interaction among these examined parameters. Of course, operating costs are determined by several factors, so we also mentioned other statistically significant effects in the study. Other factors primarily include the thermal-technical properties of construction structures, operation factors, and the way constructions are used. Our statistical analysis revealed correlations and relations among the technical, economical, and social aspects of the operation of the constructions. The findings lead to the conclusion that combining suitable wood construction material and design solutions can lead to more efficient construction solutions and higher housing quality in the context of sustainability.

ACKNOWLEDGMENTS

The article presents a partial research result of the VEGA project—1/0557/18 "Research and development of process and product innovations of modern methods of construction in the context of the Industry 4.0 principles."

REFERENCES

A Wooden Skeleton in XXL: T3 Office Building in Minneapolis, (cit.03-07-2019) online: www.detail-online.com/article/a-wooden-skeleton-in-xxl-t3-office-building-in-minne-apolis-31385/

Ali, H. H., Al Nsairat, S. F. Developing a green building assessment tool for developing countries-case of jordan. *Building and Environment* 2009, 44, 1053–1064. doi:10.1016/j.buildenv.2008.07.015

Almusaed, A., Almssad, A. Building materials in eco-energy houses from Iraq and Iran. *Case Studies in Construction Materials* 2015, 2, 42–54. https://doi.org/10.1016/j.cscm.2015.02.001

Angies list, (cit.03-07-2019) online: www.angieslist.com/articles/house-framing-requires-skilled-contractor.htm

Antošová, N., Ďubek, M., Petro, M. Diagnostics to determine the cause of occurrence of cracks on ETICS. *Advances and Trends in Engineering Sciences and Technologies II: Proceedings of the 2nd International Conference on Engineering Sciences and Technologies*. London, UK: Taylor & Francis Group, 2017, 317–322.

Beránková, E. Životní cyklus stave. (cit.14-02-2019) online: www.tzb-info.cz/udrzba-budov/10219-zivotni-cyklus-staveb

Bílek, V. *Wooden Constructions: Design of Multi-Storey Wooden Buildings*. 1st ed. Prague, Czechia: ČVUT, 2005, 251 s.

Cabeza, L. F., Rincón, L., Vilariño, V., Pérez, G., Castell, A. Life cycle assessment (LCA) and life cycle energy analysis (LCEA) of buildings and the building sector: A review. *Renewable and Sustainable Energy Reviews* 2014, 29, 394–416. https://doi.org/10.1016/j.rser.2013.08.037

Chau, C. K., Leung, T. M., Ng, W. Y. A review on life cycle assessment, life cycle energy assessment and life cycle carbon emissions assessment on buildings. *Applied Energy* 2015, 143, 395–413. https://doi.org/10.1016/j.apenergy.2015.01.023

Croux, CH., Dehon, C. Influence functions of the Spearman and Kendall correlation measures. *Statistical Methods & Applications* 2010, 19, 497–515. doi:10.1007/s10260-010-0142-z

De Araujo, V. A., Cortez-Barbosa, J., Gava, M., Garcia, J. N., de Souza, A. J. D., Savi, A. F., Lahr, F. A. R. Classification of wooden housing building systems. *BioResources* 2016a, 11(3), 7889–7901.

De Araujo, V. A., Vasconcelos, J. S., Cortez-Barbosa, J., Morales, E. A., Gava, M., Savi, A. F., Garcia, J. N. Wooden residential buildings-a sustainable approach. *Bulletin of the Transilvania University of Brasov. Forestry, Wood Industry, Agricultural Food Engineering. Series II* 2016b, 9(2), 53.

Dincer, I. On thermal energy storage systems and applications in buildings. *Energy and Buildings* 2002, 34, 377–388. https://doi.org/10.1016/S0378-7788(01)00126-8

Ding, G. K. C. Sustainable construction—The role of environmental assessment tools. *Journal of Environmental Management* 2008, 86, 451–464. doi:10.1016/j.jenvman.2006.12.025

Drozd, W., Leśniak, A. Ecological wall systems as an element of sustainable development—cost issues. *Sustainability* 2018, 10, 2234. doi:10.3390/su10072234

Estanqueiro, B., Dinis Silvestre, J., de Brito, J., Duarte Pinheiro, M. Environmental life cycle assessment of coarse natural and recycled aggregates for concrete. *European Journal of Environmental and Civil Engineering* 2018, 22(4), 429–449. https://doi.org/10.1080/19648189.2016.1197161

Finnish Log Homes Ltd. (cit.03-07-2019) online: www.finnishloghomes.co.uk/

García, H., Zubizarreta, M., Cuadrado, J., Osa, J. L. Sustainability improvement in the design of lightweight roofs: A new prototype of hybrid steel and wood purlins. *Sustainability* 2018, 11(1), 39. doi: 10.3390/su11010039

Gašparík, J., Paulovičová, L., Szalayová, S., Gašparík, M. Method of optimal machine selection for construction processes from the point of its energy consumption minimizing with software support. In *ISARC 2017: Proceedings of the 34th International Symposium on Automation and Robotics in Construction*. Taipei, Taiwan, 2017, 730–773.

Gosselin, A., Blanchet, P., Lehoux, N., Cimon, Y. Main motivations and barriers for using wood in multi-story and non-residential construction projects. *BioResources* 2016, 12(1), 546–570.

Hafner, A., Schäfer, S. Environmental aspects of material efficiency versus carbon storage in timber buildings. *European Journal of Wood and Wood Products* 2018, 76(3), 1045–1059. https://doi.org/10.1007/s00107-017-1273-9

Half-timbered construction, (cit.03-07-2019) online: www.thoughtco.com/what-is-half-timbered-construction-177664

Hermawan, A., Prianto, E., Setyowati, E. The difference of thermal performance between houses with wooden walls and exposed brick walls in tropical coasts. *Procedia Environmental Sciences* 2015, 23, 168–174. https://doi.org/10.1016/j.proenv.2015.01.026

Jeanjean, A., Olives, R., Py, X. Selection criteria of thermal mass materials for low-energy building construction applied to conventional and alternative materials. *Energy and Buildings* 2013, 63, 36–48. http://dx.doi.org/10.1016/j.enbuild.2013.03.047

Johra, H., Heiselberg, P., Le Dréau, J. Influence of envelope, structural thermal mass and indoor content on the building heating energy flexibility. *Energy & Buildings* 2019, 183, 325–339. https://doi.org/10.1016/j.enbuild.2018.11.012

Kobl, J. *Wooden Constructions: Load-bearing Structure Systems, Cladding.* Prague, Czechia: Grada, 2011, 320s.

Le Truong, N., Dodoo, A., Gustavsson, L. Effects of energy efficiency measures in district-heated buildings on energy supply. *Energy* 2018, 142, 1114–1127. https://doi.org/10.1016/j.energy.2017.10.071

Maratovich, M. T., Anatolevich, K. V., Leonidovich, E. A., Alexandrovich, P. A. Review of the methods and the constructions for the waste wood recycling for the machine designing based on tractor Msn-10 for the pellets production. *International Journal of Applied Engineering Research* 2016, 11(22), 10945–10951.

Mavrigiannaki, A., Ampatzi, E. Latent heat storage in building elements: A systematic review on properties and contextual performance factors. *Renewable and Sustainable Energy Reviews* 2016, 60 852–866. http://dx.doi.org/10.1016/j.rser.2016.01.115

McKight, P. E., Najab, J. Kruskal-Wallis Test. *The Corsini Encyclopedia of Psychology* 2010, 1–1. https://doi.org/10.1002/9780470479216.corpsy0491

Meadowlark Log Homes, (cit.03-07-2019) online: https://meadowlarkloghomes.com/

Mohammadhosseini, H., Tahir, M. M. Production of sustainable fibre-reinforced concrete incorporating waste chopped metallic film fibres and palm oil fuel ash. *Sadhana—Academy Proceedings in Engineering Sciences* 2018, 43(10), 156. doi:10.1007/s12046-018-0924-9

Motuzienė, V., Rogoža, A., Lapinskienė, V., Vilutienė, T. Construction solutions for energy efficient single-family house based on its life cycle multi-criteria analysis: A case study. *Journal of Cleaner Production* 2016, 112, 532–541. https://doi.org/10.1016/j.jclepro.2015.08.103

Němeček, M., Kalousek, M. Influence of thermal storage mass on summer thermal stability in a passive wooden house in the Czech Republic. *Energy and Buildings* 2015, 107, 68–75. https://doi.org/10.1016/j.enbuild.2015.07.068

O'Connor, J., Kozak, R., Gaston, Ch., Fell, D. Wood use in nonresidential buildings: Opportunities and barriers. *Forest Products Journal* 2004, 54(3), 19–28.

Park, J. H., Kim, Y. J., Han, D. S. Application of and changes in construction principles and joint methods in the wooden architecture of the Joseon Era: A case study on the sungnyemun gate in seoul. *Journal of Asian Architecture and Building Engineering* 2018, 17(2), 191–198. https://doi.org/10.3130/jaabe.17.191

Pomianowski, M., Heiselberg, P., Zhang, Y. Review of thermal energy storage technologies based on PCM application in buildings. *Energy and Buildings* 2013, 67, 56–69. http://dx.doi.org/10.1016/j.enbuild.2013.08.006

Ritchie, L., Stephan, A. Engineered timber for apartment buildings in Melbourne, Australia: A construction cost comparison with traditional concrete systems. In *Engaging Architectural Science: Meeting the Challenges of Higher Density: 52nd International Conference of the Architectural Science Association* 2018, 161–168.

Ruxton, G. D. The unequal variance t-test is an underused alternative to Student's t-test and the Mann–Whitney U test. *Behavioral Ecology* 2006, 17(4), 688–690. doi:10.1093/beheco/ark016

Siva, V., Hoppe, T., Jain, M. Green buildings in Singapore: Analyzing a frontrunner's sectoral innovation system. *Sustainability* 2017, 9, 919. doi:10.3390/su9060919

Štefko J., *Wooden Buildings from the Perspective of Sustainable Construction*. Zvolen, Slovak republic: Prolignum 2014.

Štefko, J., Balent V. *Multi-Comfortable Timber Structures*. Zvolen, Slovak republic: Isover, 2014.

Štefko, J., Reinprecht, L., Kuklík, P. *Timber structures: Construction, protection and maintenance*. 2nd edition. Bratislava, Slovakia: JAGA, 2009, 204s.

Štefko, J. et al., *Modern Wooden Buildings*. Slovakia, Bratislava: ANTAR, 2010.

Tang, C.-W. Properties of fired bricks incorporating TFT-LCD waste glass powder with reservoir sediments. *Sustainability* 2018, 10(7), 2503. doi:10.3390/su10072503

Think wood, (cit.03-07-2019) online: www.thinkwood.com/products-and-systems/light-frame-construction

Toppinen, A., Sauru, M., Pätäri, S., Lähtinen, K., Tuppura, A. Internal and external factors of competitiveness shaping the future of wooden multistory construction in Finland and Sweden. *Construction Management and Economics* 2018, 1–16. https://doi.org/10.1080/01446193.2018.1513162

Tsai, C.-Y., A.-S. Chang. Framework for developing construction sustainability items: The example of highway design. *Journal of Cleaner Production* 2012, 20, 127–136. doi:10.1016/j.jclepro.2011.08.009

Vaverka, J. *Wooden buildings for housing*. 1st edition. Prague, Czechia: Grada, 2008, pp. 376.

Vinodh, S., Jayakrishna, K., Kumar, V., Dutta, R. Development of decision support system for sustainability evaluation: A case study. *Clean Technologies and Environmental Policy* 2014, 16, 163–174. doi:10.1007/s10098-013-0613-7

Wagner, K. Generation of a tropically adapted energy performance certificate for residential buildings. *Sustainability* 2014, 6, 8415–8431. doi:10.3390/su6128415

Watts, A. *Modern construction handbook*. Basel, Switzerland: Birkhäuser, 2016.

Wood constructions from prefabricated panels, (cit.03-07-2019) online: https://sk.pinterest.com/pin/541769030157372253/

Xie, X., Lu, Y., Gou, Z. Green building pro-environment behaviors: Are green users also green buyers? *Sustainability* 2017, 9, 1703. doi:10.3390/su9101703

Zhang, Y., Zhang, J., Lü, M., Wang, J., Gao, Y. Considering uncertainty in life-cycle carbon dioxide emissions of fly ash concrete. *Proceedings of the Institution of Civil Engineers-Engineering Sustainability*. London, UK: Thomas Telford Ltd, 2018, 1–9. https://doi.org/10.1680/jensu.17.00058

Zhao, J., Xie, X., Liu, R., Sun, Y., Wu, M., Gu, J. Water and energy saving potential by adopting pressure-reducing measures in high-rise building: A case analysis. *Building Services Engineering Research and Technology* 2018, 39(5). https://doi.org/10.1177/0143624417751056

17 Deployment of the Low Carbon Energy Supply Technologies for Sustainable Development

Karunesh Kant, Amritanshu Shukla,
Atul Sharma, and Camilo C. M. Rindt

CONTENTS

17.1 INTRODUCTION

The global CO_2 emission from fuel combustion was 32.31 Gt CO_2 in 2016, broadly similar to 2015 (32.28 Gt CO_2) and increased by around 40% since 2000. Figure 17.1 represents the global CO_2 emission from 1971 to 2017. During the meeting COP21 (21st Conference of Parties) in Paris, the decision had been taken by 195 countries to limit global temperature rise to well below 2°C

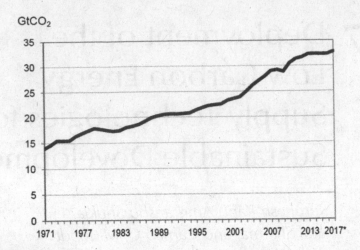

FIGURE 17.1 Global trend for CO_2 emissions from fuel combustion. (From IEA-International Energy Agency, *Technology Roadmap: Hydrogen and Fuel Cells*, 2018.)

above preindustrial levels by the end of this century while aiming at 1.5°C [1]. The Paris Agreement is founded on nationally determined contributions (NDCs) made by countries, which are intended to outline their "highest possible ambition" to address climate change including reducing greenhouse gas (GHG) emissions. NDCs are updated every five years, and each new NDC is to represent a progression from the previous one. Current NDCs cover the period from 2020 to 2030 or 2025, and most include quantitative emissions reductions targets, which are summarized in Table 17.1, Figures 17.2a and 2b for the top CO_2 emitting countries [2]. Countries that have submitted an NDC represent over 96% of global CO_2 emissions. The European Union (EU), having agreed on its post-2020 framework of climate and energy policy [3,4], committed to a 40% domestic GHG emission reduction by 2030 compared to 1990 as the combined contribution of its member states to global climate change mitigation goals [4]. The Paris Agreement has set an ambitious goal of limiting the global mean temperature increase to below 2°C (2DS), and, if possible, under 1.5°C. Achieving this goal requires an unprecedented transformation of the way energy is supplied and used throughout the world, including rapid deployment of low-carbon electricity generation technologies on the supply side and acceleration of energy efficiency improvements on the demand side.

TABLE 17.1

Greenhouse Gas Reduction Targets of the Ten Largest Emitters (based on 2016), Emissions, and IEA Member Countries

Top CO₂ Emitting Parties	1990	2005	2016	2020 GHG Target	Base Year Level	2016 Level	Change to 2016 (%)	NDC GHG Target
	Mt CO₂							
China (incl. Hong Kong)	2122	5448	9102	Emissions/GDP 40–45% below 2005	0.72 kgCO₂/ 2010 USD PPP	0.46 kgCO₂/ 2010 USD PPP	−36%	Reduce CO₂ per unit of GDP by 60%–65% below 2005
United States	4803	5703	4833	17% below 2005	5703Mt	4883 Mt	−15%	26%–28% reduction by 2025 below 2005 levels
European Union	4027	3922	3192	20% below 1990	4027Mt	3192Mt	−21%	40% reduction compared to 1990 levels
India	529	1072	2077	Emissions/GDP 20%–25% below 2005	0.30 kgCO₂/ 2010 USD PPP	0.26 kgCO₂/ 2010 USD PPP	−13%	Emissions/GDP 33%–35% below 2005 levels
Russian Federation	2164	1482	1439	15%–25% below 1990	2164 Mt	1439 Mt	−34%	25%–30% below 1990 levels
Japan	1037	1164	1147	3.8% below 2005	1164 Mt	1147 Mt	−6%	26% below 2013 levels
Republic of Korea (Korea)	232	458	589	x	x	589 Mt	x	37% below BAU emissions of 850.6 MtCO₂ in 2030
Islamic Republic of Iran (Iran)	171	418	563	x	x	x	x	4% below BAU of 1540 Mt CO₂ in 2030; 12% with international support
Canada	420	540	541	17% below 2005	540 Mt	541 Mt	+0%	30% below 2005 levels

(Continued)

TABLE 17.1 (Continued)

Greenhouse Gas Reduction Targets of the Ten Largest Emitters (based on 2016), Emissions, and IEA Member Countries

Top CO$_2$ Emitting Parties	1990	2005	2016	2020 GHG Target	Base Year Level	2016 Level	Change to 2016 (%)	NDC GHG Target
	Mt CO$_2$							
Saudi Arabia	151	298	527	x	x	x	x	Annual GHG emission reduction of up to 130 MtCO$_2$
Mexico	257	412	412	30% below BAU scenario	906 MtCO$_2$ (2020 BAU)	445 Mt	51% below BAU 2020 level	22% below BAU
Australia	260	372	392	5% below 2000 levels	335 Mt	292 Mt	+17	26%–28% below 2005 levels
Turkey	129	216	216	x	x	x	x	21% emission reduction below BAU of 1175 MtCO$_2$
Switzerland	41	44	44	20% below 1990	41 Mt	38 Mt	−7%	50% below 1990 levels. 35% anticipated reduction by 2025
Norway	27	35	36	40% below 1990	27Mt	36 Mt	+29%	40% below 1990 levels
New Zealand	22	34	30	5% below 1990 levels	22 Mt	30 Mt	+40%	30% below 2005 levels

Source: IEA, CO$_2$ *Emissions from Fuel Combustion: Overview*, International Energy Agency, 14, 2018.

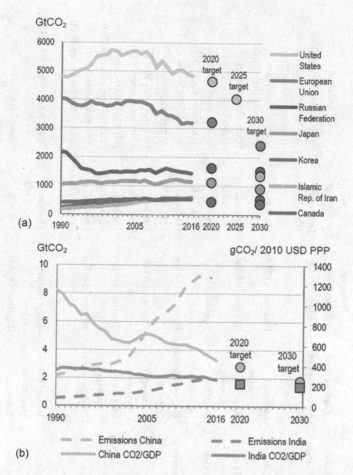

FIGURE 17.2 CO_2 emissions (1990–2016) and emissions reduction targets (2020, 2025, 2030) for: (a) United States, European Union, Russian Federation, Japan, Korea, Islamic Republic of Iran, Canada, (b) China and India.

17.2 CARBON CAPTURE UTILIZATION AND STORAGE

Carbon capture utilization and storage (CCUS) is one of the only technology solutions that can considerably reduce CO_2 emissions from coal and gas power generation. In this technology, the CO_2 is captured and transported from one place to another for its utilization as a resource to create valuable products or services or its permanent storage deep underground in geological formations. Globally, more than 30 million tons of CO_2 is captured from large-scale CCUS facilities for use or storage. More than 70% of this occurs in North America. Several large-scale projects are running to capture, utilize, and store CO_2, and details of these projects are given in Table 17.2.

TABLE 17.2
Worldwide Large-Scale CO_2 Capture Utilization and Storage Projects

Name of the Plant	Capture Rate	Feedstock	Capture Type	Transport Medium	Storage	Location
Great Plains Synfuel Plant and Weyburn-Midale	3 Mtpa	Lignite/Brown Coal	Pre-combustion	Pipeline, 329 km	Onshore, CO_2-EOR	North Dakota, United States and Saskatchewan, Canada (Operating since 2000)
Century Plant	8.4 Mtpa	Natural gas	Industrial separation	Pipeline, > 255 km	Onshore, CO_2-EOR	Texas, United States (Operating since 2010)
Terrell Natural Gas Processing Plant (formerly Val Verde)	0.45 Mtpa	Natural Gas	Industrial separation	Pipeline, 316 km	Onshore, CO_2-EOR	Texas, United States (Operating since 1972)
Air Products Steam Methane Reformer	1 Mtpa	Natural Gas	Industrial separation	Pipeline, 158 km	Onshore, CO_2-EOR	Texas, United States (Operating since 2013)
Petra Nova Carbon Capture	1.4 Mtpa	Sub-bituminous Coal	Post-combustion capture	Pipeline, 132 km	Onshore, CO_2-EOR	Texas, United States (Operating since 2017)
Illinois Industrial Carbon Capture and Storage	1 Mtpa	Corn	Industrial separation	Pipeline, 1.6 km	Onshore, Saline aquifer	Illinois, United States (Operating since 2017)
Alberta Carbon Trunk Line (ACTL) with Agrium CO_2 Stream	0.45 Mtpa	Natural Gas	Industrial separation	Pipeline, 240 km	Onshore, CO_2-EOR	Alberta, Canada (Under construction, expected 2019)
Alberta Carbon Trunk Line (ACTL) with North West Sturgeon Refinery CO_2 Stream	1.3 Mtpa	Bitumen	Pre-combustion	Pipeline, 240 km	Onshore, CO_2-EOR	Alberta, Canada (Under construction, expected 2019)
Petrobras Santos Basin Pre-Salt Oil Field CCS	1 Mtpa	Natural Gas	Industrial separation	Direct injection, 0 km	Onshore, CO_2-EOR	Brazil (Operating since 2013)
Snøhvit CO_2 Storage	0.7 Mtpa	Natural Gas	Industrial separation	Pipeline, 153 km	Offshore, Saline aquifer	Norway (Operating since 2008)
Sleipner CO_2 Storage	1 Mtpa	Natural Gas	Industrial separation	Direct injection, 3 km	Offshore, Saline aquifer	Norway (Operating since 1996)

(Continued)

TABLE 17.2 (Continued)
Worldwide Large-Scale CO_2 Capture Utilization and Storage Projects

Name of the Plant	Capture Rate	Feedstock	Capture Type	Transport Medium	Storage	Location
Abu Dhabi CCS Project	0.8 Mtpa	Natural Gas	Industrial separation	Pipeline, 43 km	Onshore, CO_2-EOR	Abu Dhabi, UAE (Operating since 2016)
Uthmaniyah CO2-EOR Demonstration	0.8 Mtpa	Natural Gas	Industrial separation	Pipeline, 85 km	Onshore, CO_2-EOR	Saudi Arabia (Operating since 2015)
Yanchang Integrated Carbon Capture and Storage	0.41 Mtpa	Lignite/Brown Coal	Industrial separation	Pipeline, 150 km	Onshore, CO_2-EOR	China (expected 2020)
Sinopec Qilu Petrochemical CCS	0.4 Mtpa	Bituminous Coal	Industrial separation	Pipeline, 75 km	Onshore, CO_2-EOR	China (expected 2019)
CNPC Jilin Oil Field CO2 EOR	0.6 Mtpa	Natural Gas	Industrial separation	Pipeline, 53 km	Onshore, CO_2-EOR	China (Operating since 2016)
Gorgon Carbon Dioxide Injection	3.7 Mtpa	Natural Gas	Industrial separation	Pipeline, 7 km	Onshore, Saline aquifer	Australia (expected 2019)
Quest	1 Mtpa	Natural Gas	Industrial separation	Pipeline, 64 km	Onshore, Saline aquifer	Alberta, Canada (Operating since 2015)
Boundary Dam Carbon Capture and Storage	1 Mtpa	Lignite/Brown Coal	Post-combustion capture	Pipeline, 66 km	Onshore, CO_2-EOR	Saskatchewan, Canada (Operating since 2014)
Shute Creek Gas Processing Plant	7 Mtpa	Natural Gas	Industrial separation	Pipeline, up to 460 km	Onshore, CO_2-EOR	Wyoming, United States (Operating since 1986)
Lost Cabin Gas Plant	0.9 Mtpa	Natural Gas	Industrial separation	Pipeline, 374 km	Onshore, CO_2-EOR	Wyoming, United States (Operating since 2013)
Enid Fertilizer	0.7 Mtpa	Natural Gas	Industrial separation	Pipeline, 225 km	Onshore, CO_2-EOR	Oklahoma, United States (Operating since 1982)
Coffeyville Gasification Plant	1 Mtpa	Petroleum coke	Industrial separation	Pipeline, 112 km	Onshore, CO_2-EOR	Kansas, United States (Operating since 2013)

17.2.1 CO_2 CAPTURE

CO_2 capture comprises the separation of CO_2 from industrial processes and energy-related point sources such as power plants. Separating the CO_2 requires energy and often involves modifications to existing processes by adding extra process steps. After separation, the CO_2 stream can be further purified and compressed to make it ready for transport. CO_2 capture is typically divided into four main categories. In certain cases, these categories can be combined to create hybrid routes to capture.

17.2.2 CO_2 UTILIZATION

There is a growing interest in innovative ways of utilizing CO_2 as a feedstock for products that have a market value in addition to their value in moderating climate change. CO_2 use may deliver several other services to society, such as the substitution of fossil fuels as a feedstock for fuels and materials and the conversion of renewable electricity to hydrocarbons that are compatible with existing infrastructure. Figure 17.3 represents the possible use of CO_2 and its classification.

17.2.3 CO_2 STORAGE

The storage of CO_2 involves the injection of captured CO_2 from various sources into deep underground geological reservoirs of porous rock (the reservoir) covered by an impermeable layer of rocks (seal), which prevent the upward migration of CO_2 beyond the storage complex. There are several types of geological formations that are suitable for CO_2 storage mainly saline aquifers and depleted oil and gas reservoirs. Depleted oil and gas reservoirs are porous rock formations containing either mainly crude oil or gas that has been physically held in stratigraphic or structural traps for millions of years. Saline aquifers are layers of porous and permeable rocks saturated with brackish water (brine), which are fairly widespread in both onshore and offshore sedimentary basins.

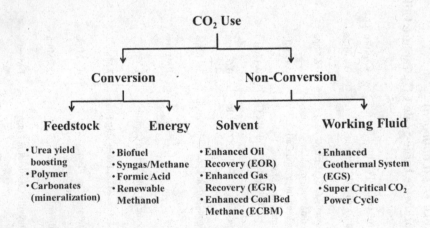

FIGURE 17.3   Classification of CO_2 uses.

17.3 DEPLOYMENT OF LOW-CARBON ENERGY SUPPLY TECHNOLOGIES

17.3.1 SOLAR PHOTOVOLTAIC TECHNOLOGY

In solar photovoltaic (PV) technology, the solar energy directly converts into electricity by means of the solar PV panel. Solar PV technology combines two advantages. On the one hand, module manufacturing can be done in large plants, which permits economies of scale. On the other hand, PV is a very modular technology, and it can be deployed in very small capacities at a time. This quality allows for a wide range of applications of PV technology. Systems can be very small, such as in calculators or off-grid applications, up to utility-scale power generation facilities. In 2017, cumulative solar PV capacity reached almost 398 GW and generated over 434 TWh, representing around 2% of global power output [6]. Utility-scale projects account for just over 60% of total PV installed capacity, with the rest in distributed applications (residential, commercial, and off-grid). The deployment of concentrated solar power (CSP) plants is at a stage of market introduction and expansion. In 2016, the installed capacity of CSP worldwide was 4.8 GW, compared to 398 GW of solar PV capacity. CSP capacity is expected to double by 2022 and reach 10 GW with almost all new capacity integrating storage. The cumulative installed capacity of solar thermal installations reached an estimated 472 GW by the end of 2017. However, the market continued to slow in 2017 for the fourth year in a row, as total annual installations decreased by 9%, owing mainly to a continual slowdown in China. The power generation from solar PV is estimated to have increased by more than 30% in 2018, to over 570 TWh. With this increase, the solar PV share in global electricity generation exceeded 2% for the first time. Figure 17.4 presents the yearwise power generation, forecast, and power generation for Sustainable Development Scenario (SDS) by solar PV and CSP.

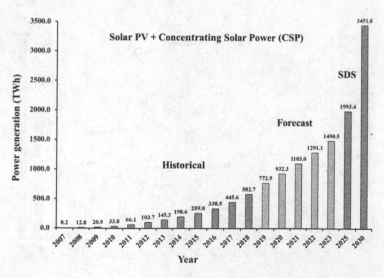

FIGURE 17.4 Yearwise power generation, forecast, and power generation in SDS for solar PV and CSP.

17.3.2 WIND ENERGY

The cumulative installed capacity of grid-connected wind power in 2017 was 515 GW (497 GW onshore wind power and 18 GW offshore wind power) and wind power generation accounted for nearly 4% of global electricity generation. The onshore installed capacity of wind power is expected to grow by 323 GW in the next 5 years and reach almost 839 GW by 2023 in the main case of the IEA's Renewables 2018 prediction. China leads this growth followed by the United States, Europe, and India. As a result, the onshore wind electricity generation would increase by nearly 65% globally over 2018–2023. In 2018, global offshore and onshore wind generation reached an estimated 66 TWh and 1150 TWh, respectively. By 2023, global offshore wind cumulative capacity is expected to reach 52 GW by 2023. The deployment will be led by the EU and China. Enhanced policies and faster deployment of projects in the pipeline could result in a further 8 GW. Figure 17.5 presents historical, prediction, and power generation in SDS for offshore and onshore wind energy. In 2018, onshore and offshore wind energy generation increased by an estimated 12% and 20%, respectively; however, capacity addition grew 7% for onshore wind energy and 15% for offshore wind energy.

17.3.3 BIOENERGY AND BIOFUELS

Bioenergy accounts for roughly 9% of the world's total primary energy supply today. Around 13 EJ of bioenergy was consumed in 2015 to provide heat, representing approximately 6% of global heat consumption. Modern bioenergy is also extensively used for space and water heating, either directly in buildings or in district heating schemes. Additionally, about 500 TWh of electricity was generated from biomass in 2016, accounting for 2% of world electricity generation. In the long-term, bioenergy

FIGURE 17.5 Power generation, prediction, and power generation in SDS for wind.

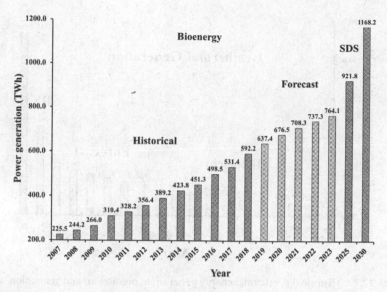

FIGURE 17.6   Historical bioenergy generation prediction and generation in SDS.

has an important role to play in a low-carbon energy system. For instance, modern bioenergy in final global energy consumption increases four-fold by 2060 in the IEA's 2°C scenario (2DS), which seeks to limit global average temperatures from rising more than 2°C by 2100 to avoid some of the worst effects of climate change. Figure 17.6 represents the yearwise bioenergy generation prediction and generation in SDS. In 2018, bioenergy electricity generation increased by over 8%, maintaining average growth rates since 2011.

17.3.4 Geothermal Energy

Geothermal energy can provide heating, cooling, and base-load power generation from high-temperature hydrothermal resources, aquifer systems with low and medium temperatures, and hot rock resources. Geothermal power plants provide stable production output, unaffected by climatic variations, resulting in high-capacity factors (ranging from 60% to 90%) and making the technology suitable for baseload production. In 2017, global geothermal power generation stood at an estimated 87.5 TWh, while the cumulative capacity reached 14 GW. Global geothermal power capacity is expected to rise to just over 17 GW by 2023, with the biggest capacity additions expected in Indonesia, Kenya, Philippines, and Turkey. Only a limited number of countries use geothermal energy directly for heat production, with China and Turkey alone accounting for 80% of consumption in 2017. During 2012–2017, global consumption almost doubled, mostly due to rapid growth in China. Over the outlook period (2018–2023), growth is expected to be lower at 24% but to remain important in a number of countries and sectors. While most geothermal heat is used for bathing (45%) and space heating (34%), agriculture (primarily for heating greenhouses) has long been an important end-use sector in some countries.

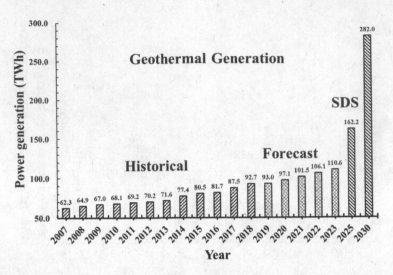

FIGURE 17.7 Historical geothermal energy generation, prediction, and generation in SDS.

Over recent years, the energy-intensive greenhouse sector in the Netherlands has expanded geothermal use due to strong policy support, and the country has become the fourth-largest user of geothermal heat in the agriculture sector after China, Turkey, and Japan. Figure 17.7 represents the historical, predicted, and SDS power generation. The geothermal electricity generation increased by an estimated 6% in 2018. In the next 5 years, average annual geothermal installations are expected to accelerate to 700 MW, driven by strong project development, mainly in Indonesia, the Philippines, Kenya, and Turkey, with smaller-scale deployment in nearly 30 other countries.

17.3.5 HYDROPOWER

Hydropower is the largest source of renewable electricity in the world, producing around 16% of the world's electricity from over 1200 GW of installed capacity. Annual net capacity growth has slowed in recent years, due to fewer large projects being developed in China and Brazil. However, cumulative capacity is still expected to increase by an additional 125 GW by 2023. China is likely to grow at a slower pace than in the past but would still account for over 40% of the net growth, followed by additions from other markets in Asia, Latin America, and Africa [7].

Hydropower is expected to remain the world's largest source of renewable electricity generation by 2023 and will play a critical role in decarbonizing the power system and improving system flexibility. As such, developments can be expected in market segments that can contribute to this flexibility, such as the refurbishment of existing hydroplants and pumped storage projects. Hydropower additions were stable in 2018 (25 GW) with the commissioning of large-scale projects in China and Brazil. Figure 17.8 presents historical, forecast, and SDS power generation. The hydropower generation is estimated to have increased by over 3% in 2018, which is a much

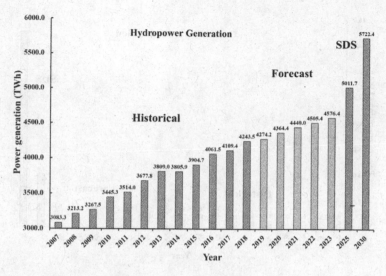

FIGURE 17.8   Historical, forecast, and SDS hydropower generation.

larger increase than the 1.5% increase in 2017; however, it is not fully on track to reach the SDS level. This downward trend is projected to continue owing primarily to less large-project development in China and Brazil, where concerns over social and environmental impacts have restricted projects.

17.3.6   Ocean Energy

Ocean power accounts for the smallest portion of renewable electricity globally, and the majority of projects remain at the demonstration phase. However, with large, well-distributed resources, ocean energy has the potential to scale up over the long term. Five different ocean energy technologies, that is, tidal power, tidal (marine) currents wave power, temperature gradients, and salinity gradients, are under development. Tidal barrages are most advanced because they use conventional technology. However, only two large-scale systems are in operation worldwide: the 240 MW La Rance barrage in France has been generating power since 1966, while the 254 MW Sihwa barrage (South Korea) came into operation in 2011. Other smaller projects have been commissioned in China, Canada, and Russia. Power generation from ocean technologies augmented an estimated 3% in 2018. The technology is not on track with the SDS, which requires a much higher annual growth rate of 24% through 2030 as shown in Figure 17.9. Policies promoting R&D are needed to achieve further cost reductions and large-scale progress.

17.3.7   Hydrogen and Fuel Cells

Hydrogen with a low-carbon footprint has the potential to facilitate significant reductions in energy-related CO_2 emissions and to contribute to limiting global temperature rise to 2°C, as outlined in the high hydrogen variant (2DS high H_2) of the IEA Energy Technology Perspectives (ETP) 2°C Scenario. In addition, hydrogen use

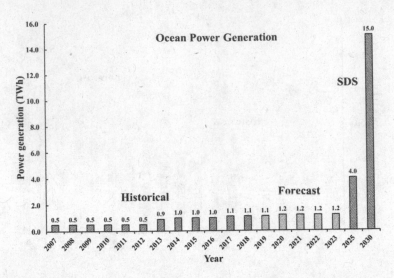

FIGURE 17.9　Historical, projected, and SDS power generation from ocean technology.

can lower local air pollutants and noise emissions compared to direct fossil fuel combustion. By enabling continued use of fossil fuel resources for end-use applications under a 2DS, hydrogen production in combination with CCUS can provide energy security benefits and help maintain a diversified fuel mix.

As an energy carrier, hydrogen can enable new linkages between energy supply and demand, in both a centralized or decentralized manner, potentially enhancing overall energy system flexibility. By connecting different energy transmission and distribution (T&D) networks, sources of low-carbon energy can be connected to end-user applications that are challenging to decarbonize, including transport, industry, and buildings. In remote areas with little access to the power grid, these connections can expand off-grid access to energy services while minimizing emissions.

Fuel cell electric vehicles (FCEVs) can provide the mobility service of today's conventional cars at potentially very low-carbon emissions. Deploying a 25% share of FCEVs in road transport by 2050 can contribute up to 10% of all cumulative transport-related carbon emission reductions essential to move from an ETP 6°C Scenario (6DS) to a 2DS, depending on the region [8]. Assuming a fast ramp-up of FCEVs sales, a self-sustaining market could be achieved within 15–20 years after the introduction of the first 10,000 FCEVs.

17.4　SUMMARY AND CONCLUSIONS

This chapter is mainly focused on the technologies deployed to reduce carbon emission or reuse the emitted CO_2. In the Paris Agreement, several countries participated and agreed to take necessary action to reduce global temperature rise due to CO_2 emission. In this agreement, an ambitious goal had been set up to limit the global mean temperature rise to below 2°C (2DS), and, if possible, under 1.5°C. To achieve this goal, various low-carbon technologies such as solar PV, bioenergy, and wind energy have been installed worldwide. The power generation from these technologies has been predicted for the next 5 years and also for the sustainable development scenario. The decreasing order of various renewable energy supply technologies by capacity and power generation are hydropower (4243.5 TWh) > wind energy (1215.2 TWh) > bioenergy (592.2 TWh) > solar PV (582.7 TWh) > geothermal (92.7) > ocean energy (1.1 TWh). These technologies will be critical to keeping global warming under 2 degrees C, but we need to remain cognizant of their effects on the environment, such as their higher use of metals like steel and copper in manufacturing.

ABBREVIATIONS

2DS	2 degrees Celsius
BAU	Business-as-usual
CCS	Carbon capture and storage
CCUS	Carbon capture utilization and storage
COP	Conference of Parties
CSP	Concentrated solar power
ECBM	Enhanced coal bed methane
EGR	Enhanced gas recovery
EGS	Enhanced geo thermal system
EOR	Enhanced oil recovery
ETP	Energy technology perspectives
EU	European Union
FCEVs	Fuel cell electric vehicles
GDP	Gross domestic product
GHG	Greenhouse gas
IEA	International Energy Agency
NDCs	Nationally Determined Contributions
PPP	Purchasing power parities
PV	Photovoltaic
SDS	Sustainable Development Scenario
T&D	Transmission and distribution

UNITS

Gt	Giga ton
GW	Giga watt
km	Kilometer
kg	Kilogram
Mt	Mega ton
Mtpa	Mega ton per annum
TWh	Terawatt hours

REFERENCES

1. United Nations Framework Convention on Climate Change (UNFCCC). Report of the Conference of the Parties on COP 21, FCCC/CP/2015/10. Conference of the Parties on Its Twenty- First Session (COP 21). Paris, France, 2016;01192:1.

2. IEA-International Energy Agency. *CO2 Emissions from Fuel Combustion*. International Energy Agency. Paris, France, 2018, 13.
3. European Commission. *A policy framework for climate and energy in the period from 2020–2030*. Brussels, 2014.
4. European Council. *2030 Climate and Energy Policy Framework*. vol. SN 79/14. Brussels, 2014.
5. IEA. *CO2 Emissions from Fuel Combustion: Overview*. International Energy Agency. Paris, France, 2018:14.
6. OECD/IEA. Renewables 2018: *Analysis and Forecasts to 2023*. International Energy Agency and Organisation for Economic Co-Operation and Development Publishing. Paris, France, 2018:1–12.
7. IEA-International Energy Agency. *Technology Roadmap: Hydropower*. Paris, France, 2012.
8. IEA-International Energy Agency. *Technology Roadmap: Hydrogen and Fuel Cells*. Paris, France, 2018.

18 Development and Application of Phase Change Materials in the Biomedical Industry

Abhishek Anand, Amritanshu Shukla, and Atul Sharma

CONTENTS

18.1   INTRODUCTION

The phase change material (PCM) is seen as a promising way of thermal energy storage. The material stores heat at a constant temperature by changing its phase. This is accompanied by minimal volume change. The energy density per unit volume is also very high. This caters to its extensive application in the field of engineering and science. Scientists have developed pioneers and brought several modifications to the existing devices using PCM. The PCM is widely available in the form of standing chemicals in the widest temperature range (Shukla 2015). Accordingly, newer PCMs are also developed by extensive thermal testing. The cost-cutting in energy usage is well established and widely accepted. In the field of biological sciences, the handlings, transportation, and storage of the biomedical products viz. blood samples, serums, vaccines, medicines, etc. need a constant temperature. These products are very sensitive. The casing and packaging are required as such that wipe out any temperature variation. PCM can also be a potent candidate for providing hot/cold therapy to the bones and joints. Researchers have effectively utilized PCM for giving thermotherapy to treat various ulcers and lesions.

18.2   BIOMEDICAL PROBLEMS

Some of the biomedical problems can be high/low temperature sensitive and can be removed or heal faster with continuous exposure to the required temperature zone. It is well known to use hot-temperature therapy to an injured body part for healing as well as for the fast recovery of that injured part. Hot-temperature therapy has been a very effective way since the last decade. Now many have tried to develop such types of materials that are capable of giving heat in long-duration without consuming direct energy source. In this context, PCMs play an important role in maintaining required temperature articles in the form of thermal jackets, elbow covers, PCM bandage, etc. The use of PCMs for thermal energy storage is well-known, but if the components are chosen judiciously, one can also generate molecular alloy PCMs, suitable for thermal energy storage and thermal protection. Molecular alloys are made on the basis of organic substances. We all know that vaccines only work within a certain temperature range, so it is required to maintain the desired temperature range during transportation as well as at the end-use when it will be injected. The low-temperature range is also required in the transportation of organs and blood; this required temperature is also achievable with the help of PCMs. Some of the orthopedic problems, like injured ankles, legs, knees, forearms, elbows, and other body limbs that are immobilized or restricted in the movement of the joint, limb, or body part are also heeled at a faster rate with the help of utilizing PCMs jackets. One of the biomedical problems known as a Buruli ulcer (BU) is a skin disease caused by *Mycobacterium ulcerans* and can be cured through incorporating PCMs. More than 30 countries in Africa, America, Asia, and the western Pacific are affected by BUs. This disease occurs most commonly in poor communities in remote rural areas. Around 70% of those infected are children under the age of 15 years. Treatment options for BUs are available today through incorporating PCMs, and the standard treatment over the past decades have been wide surgical excisions and skin grafting, which are hampered by the enormous costs, limited access, and substantial relapse routes. A suitable PCM to achieve the temperatures required to cure a BU infection is sodium acetate trihydrate ($CH_3COONa3H_2O$) (SAT), which has a melting point of 58°C. A significant advantage of SAT is its supercooling behavior, which means that once completely molten, the material will stay liquid, even if the temperature is far below its melting point. Therefore, liquid

SAT can be stored at room temperature without further heat losses. A special starter within the material can initialize the crystallization process if required. This effect is widely used in commercial pocket heat packs.

18.3 PHASE CHANGE MATERIALS: AN INTRODUCTION

PCMs are available in wider temperature ranges that can be classified crudely as organic, inorganic, or mixtures. For suitability, Sharma et al. (2009) and Anand et al. (2019) broadly categorized PCM according to Figure 18.1.

They further describe each of these categorizations and they are as follows:

18.3.1 ORGANIC PCM

It is basically paraffin and non-paraffin compounds. Paraffin compounds are generally alkanes, that is, acyclic saturated hydrocarbons either branched or unbranched. The general formula is $C_nH_{2n}+_2$. Sometimes the acyclic alkanes also come in portrait, which is though saturation, but have the cyclic structure. The general formula is C_nH_{2n}. Paraffin stores a large amount of latent heat during phase change, which intensifies with increasing the chain length. They usually show congruent melting and virtuous nucleating properties. The melting point increasing with increasing chain length caters its applicability to a varied temperature range. The other added advantages are its non-corrosiveness, non-toxicity, and resilience. In terms of economics, it is low cost, safe, non-hazardous. There are also some limitations using these as PCMs, such as low thermal conductivity, non-compatibility with storage cases, particularly plastic, and they sometimes can be a little flammable. Sharma et al. (2009) classified this further into Group I (most promising), and Group II (promising), and Group III (least promising).

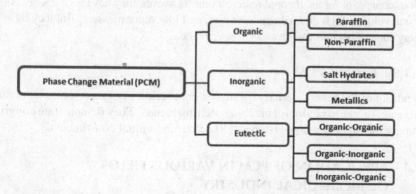

FIGURE 18.1 Classification of PCM based on the application. (From Sharma, A. et al., *Renewable and Sustainable Energy Reviews* 13, 318–345, 2009; Anand, A. et al., *Solar Desalination Technology*, Edited by Anil Kumar and Om Prakash, Green Energy and Technology. Singapore, Springer Singapore, 2019.)

18.3.2 INORGANIC PCM

Inorganic PCMs are classified into two types: salts and metallics. The salts are generally the concoction of two or more salts with water of crystallization. It can be represented as $XY.nH_2O$. The phase conversion is best described as partial or full dehydration. The chemical equation is described as follows:

$$XY.nH_2O \rightarrow XY.mH_2O + (n-m)H_2O$$

$$XY.nH_2O \rightarrow XY + nH_2O$$

The major advantages of using inorganic salts are (1) its high latent heat of fusion per unit volume, (2) comparatively high thermal conductivity when compared to paraffin, (3) minimal volume change during the transition process, (4) little degradation after extended thermal cycles, (5) no supercooling behavior, and (6) non-toxic and non-corrosive and compatible with any packaging materials. The incongruent melting is a major nuisance created by these salts. It creates a supersaturated solution. The solid salt settles at the bottom, which later becomes unavailable for the recombination process with water. This results in an irreversible solid–liquid phase transition. Along with that, the poor nucleating property demonstrates supercooling behavior before the actual crystallization process begins. To avoid this, a nucleating agent can be added, which can act as the initiator, or some crystal can be left at the colder region to arrest these problems. The incongruent melting can be tackled by (1) physical stirring, (2) using surplus water to avoid supersaturation, (3) encapsulation, (4) adding a thickening agent so that salt remains in the suspension, and (5) chemical modification. Some researchers later suggested the use of a rolling cylinder thermal reservoir unit. It delivers (1) effectual chemical equilibrium, (2) arrest nucleation on the wall, (3) augment axial equilibrium, (4) complete phase transition, (5) more than 90 % latent heat release during transition process, and (6) high thermal stability and reliability. Metallics are low-melting metals and their eutectics. They have never been considered a serious contestant for the thermal reservoir unit. However, they have high-energy content per unit volume, high thermal conductivities, and low vapor pressure. But they have low energy content per unit weight and low specific heat.

18.3.3 EUTECTIC PCM

Eutectics are the binary, ternary, or higher concoction of organic-organic, organic-inorganic, or inorganic-inorganic lower melting blends. They demonstrate congruent behavior during phase conversion just like their individual constituents.

18.4 APPLICATION OF PCM IN VARIOUS FIELDS OF BIOMEDICAL INDUSTRY

18.4.1 BURULI ULCERS

A BU is a chronic enervating disease caused by *Mycobacterium ulcernas*. The organism belongs to the class of bacteria that causes tuberculosis and leprosy. It affects the skin but sometimes bones. The lesion caused by BU is shown in Figure 18.2.

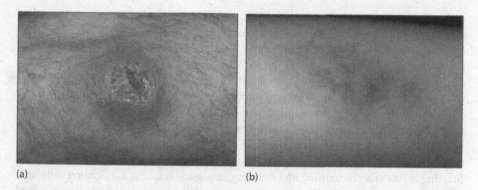

FIGURE 18.2 Lesions caused by Buruli ulcer: (a) knee and (b) arm. (From Combe, M. et al., *Emerg. Microbes. Infec.*, 6, e22, 2017.)

It leads to permanent disfigurement and long-term disability. The mode of spread is largely unknown. The disease is pandemic in tropical and subtropical countries. According to the World Health Organization (WHO), it is reported in more than 33 countries of this region. Out of these, 13 countries regularly report to WHO. It is most prevalent in West and Central Africa, including Ghana, Congo, Benin, Cameroon, and Côte d'Ivoire (Combe et al. 2017). In recent times, the cases of BU have risen in Australia and Nigeria. The cases are reported equally among males and females. The lesions are reported mainly in limbs that consist of 35% on upper limbs, 55% on lower limbs, and remaining from other parts of the body. There are three subcategories of BUs. Category I consists of single small lesions with 35% cases, Category II consists of a non-ulcerative and ulcerative plaque and oedematous forms with 35% cases, and Category III has disseminated and mixed forms such as osteitis, osteomyelitis, and joint involvement with remaining 33% cases. In all countries, 70% of cases are reported in the ulceration stage. The *Mycobacterium ulcernas* grows at a temperature between 30°C–33°C. It further produces a harmful toxin known as mycolactone responsible for damage to the skin tissues and hampering the proper functioning of the immune defense system. The BU initially develops with painless swelling. The mycolactone impedes the local immune response, causing the impairment of skin. It further aggravates, causing gross deformities of bones. The research still has not established its mode of transmission. The BU is most normally diagnosed with an IS2404 polymerase chain reaction. The treatment generally includes a combination of antibiotics, such as rifampicin, streptomycin, and clarithromycin. Bacillus Calmette–Guérin (BCG) vaccine is administered for restricted protection. Drug treatment has several limitations. Antibiotics are not always safer and have their own side effects. Children and pregnant women are most vulnerable and not recommended for these combinations. The alternative treatment conveniently employed is thermotherapy. Because bacteria cannot grow above 37°C, thermotherapy above this temperature is quite effective for the killing and proliferation of germs. The heat treatment with PCM is gaining importance in remote and resource-scare countries. It is largely recommended because of its low cost, reliability, and rechargeability of PCM.

Junghanss et al. (2009) utilized SAT as a PCM for heat application. The PCM was filled with commercially available plastic bags. The dimension of the plastic bag taken was 21 cm × 15 cm × 2 cm having weight about 800 g. The melting point of the SAT is 58°C. Before applying the PCM, an initial cleaning and sterile dressing of BU was done. The heat sensor connected to the data logger was placed on healthy skin, near the site of the BU. To lower the working temperature of the PCM to 40°C, a layer of elastic bandage was applied between the skin layer and PCM. The working temperature will not give much discomfort and can be tolerated for smaller intervals. The thermal insulation layer was also provided to arrest any heat loss to the environment. A general 24-hour protocol was followed daily for the heat treatment. Starting at 08:00 with cleaning and dressing, assessing the heat application progress, photographing, and finally the application of the PCM. At 12:00, there was removal of PCM packs with cleaning and dressing of wounds. A five-hour pause for the heat treatment was provided. At 17:00, there was a renewal of PCM packs with additional cleaning, and finally at 22:00, the removal of the PCM packs. The clinical observations, such as the ulcer's appearance, neighboring heat-exposed skin, and all-around clinical assessment of the patient, were done. The temperature at the skin surface was automatically recorded after an interval of 10 minutes, which was then stored in a data logger. The patients with small ulcers (patients 1, 2, 3) received the heat treatment for 28–31 days, and patients with large ulcers (4, 5, 6) received the heat treatment for 50–55 days. The baseline data, heat treatment schedules, and results are shown in Figure 18.3.

The temperature around the wounds was maintained at somewhat near and above 39°C between 8.4 and 13.2 hours and near and above 40°C between 4.4 and 9.3 hours per day. Epithelialization started developing in all patients between 4 and 11 days after the applications. It was almost completed in Patients 1, 2, and 3 at the conclusion of the treatment. The patients having edematous lesions (Patients 4 and 5), white discharge from ulcers, was only visible during the initial period of treatment. The patients with larger wounds (Patients 5 and 6) had to get skin grafting after the conclusion of the treatment. All the patients were healed and recovered after 18 months of treatment. The healing process has been shown in Figure 18.4.

Braxmeier et al. (2009) developed a mathematical model to assess the thermal behavior of PCM applied for the treatment of BUs. The experiment was mainly conducted to ascertain the skin surface temperature and the amount of PCM with respect to discharge time. The developed PCM bandage was quite effective for both a moderate and hot climate. The diagram for the PCM bandage is shown in Figure 18.5. The temperature of 42°C was quite tolerable. The PCM bandage could keep the temperature above 40°C for about 4 to 5 hours. The SAT was proved effective for heat treatment in this case.

18.4.2 BLOOD STORAGE AND TRANSPORTATION

Blood is a very fragile component to handle. The collection, storage, transportation, and transfusion processes require an adept process. The chain of the management process is known as "cold chain management." Anticoagulants and preservatives are initially added to thwart clotting and increase sustainability. The temperature plays

Patient [no age sex]	Ulcer before start of heat treatment	Description of ulcer	Detection of AFB or M.ulcerans DNA (swab day 0)	Histopathology (punch day 0)	Heat treatment [T above / for hrs/day no of days no of PCM-packs]	Size reduction cm² (%)	
no 1 9 yrs m		Ulcer, left upper arm, no undermined edges, no oedema	Swab neg PCR weak pos	No AFB, some mixed infiltration, no necrosis, some fat cell ghosts, psoriasiform epidermal hyperplasia	≥ 39 °C / 8.4 hrs/day ≥ 40 °C / 5.1 hrs/day ≥ 41 °C / 2.2 hrs/day ≥ 42 °C / 0.4 hrs/day 28 days 1 PCM-pack	start	1 (100)
						1 week	1.1 (110)
						2 weeks	0.3 (30)
						3 weeks	0.1 (10)
						4 weeks	0 (0)
no 2 11 yrs m		Ulcer, right knee, circular undermined edge, cotton wool appearance, little oedema of the margin	Swab neg PCR weak pos	No AFB, acute infiltration (mainly PMNs), slight necrosis, some fat cell ghosts, psoriasiform epidermal hyperplasia	≥ 39 °C / 9.2 hrs/day ≥ 40 °C / 4.7 hrs/day ≥ 41 °C / 1.2 hrs/day ≥ 42 °C / 0.2 hrs/day 31 days 1 PCM-pack	start	0.7 (100)
						1 week	0.9 (129)
						2 weeks	0.6 (86)
						3 weeks	0.4 (57)
						4 weeks	0.3 (43)
						5 weeks	0.2 (29)
no 3 11 yrs f		2 ulcers, left forefoot, oedema between ulcers and around large ulcer, no undermined edges	Swab clear pos PCR clear pos	AFB, necrosis and fat cell ghosts, psoriasiform epidermal hyperplasia, massive mixed infiltration, no granuloma	≥ 39 °C / 10.1 hrs/day ≥ 40 °C / 4.4 hrs/day ≥ 41 °C / 1.3 hrs/day ≥ 42 °C / 0.2 hrs/day 28 days 1 PCM-pack	start	2.8 (100)
						1 week	2.1 (75)
						2 weeks	1.4 (50)
						3 weeks	0.5 (18)
						4 weeks	0.1 (4)
						5 weeks	0 (0)
no 4 6 yrs f		Ulcer, left upper arm, circular undermined edge, cotton whole appearance, oedematous margin up to 4 cm, raised max 1 cm above skin level	Swab clear pos PCR strong pos	No AFB, typical massive necrosis and fat cell ghosts, psoriasiform epidermal hyperplasia	≥ 39 °C / 10.9 hrs/day ≥ 40 °C / 5.9 hrs/day ≥ 41 °C / 1.5 hrs/day ≥ 42 °C / 0.2 hrs/day 53 days 1 PCM-pack	start	1.1 (100)
						1 week	1.3 (118)
						2 weeks	1.1 (100)
						3 weeks	0.7 (64)
						4 weeks	0.6 (55)
						5 weeks	0.5 (45)
no 5 21 yrs f		2 ulcers, right upper arm, separated by skin bridge, undermined edges at 3, 6 and 9 o'clock, oedema of the margin and the skin bridge	Swab clear pos PCR strong pos	No AFB, large necrotic area, minor leukocyte infiltrates, numerous fat cell ghosts, psoriasiform epidermal hyperplasia	≥ 39 °C / 13.2 hrs/day ≥ 40 °C / 9.3 hrs/day ≥ 41 °C / 4.6 hrs/day ≥ 42 °C / 1.0 hrs/day 50 days 2 PCM-pack skin graft	start	12 (100)
						1 week	13 (108)
						2 weeks	13 (108)
						3 weeks	14 (117)
						4 weeks	13 (108)
						5 weeks	12 (100)
No 6 15 yrs f		2 ulcers, right lower leg, separated by skin bridge, undermined edges 9-12 o'clock proximal (x), 1-9 o'clock distal (xx), filled with slough. Induration extending down to the ankle	Swab strong pos PCR strong pos	No AFB, massive necrosis and fat cell ghosts, psoriasiform epidermal hyperplasia	≥ 39 °C / 13.1 hrs/day ≥ 40 °C / 8.0 hrs/day ≥ 41 °C / 4.4 hrs/day ≥ 42 °C / 1.0 hrs/day 55 days 4 PCM-pack skin graft	start	27 (100)
						1 week	43 (159)
						2 weeks	43 (159)
						3 weeks	43 (159)
						5 weeks	38 (141)

FIGURE 18.3 The outcome of the heat treatment of the patient. (From Junghanss, T. et al., *PLoS Negl. Trop. Dis.*, 3, 1–7, 2009.)

a vital role in the overall process from collection to transfusion. The fluctuation of temperature can be detrimental. Hygiene is also essential to prevent soiling. Before we go further, it is essential for us to get acquainted with the various blood components and its shelf life and storage temperature.

18.4.2.1 Whole Blood

Whole blood cells contain red blood cells (RBCs) (45%), white blood cells (WBCs) (~1%), plasma (55%), and platelets. The RBCs, WBCs, and platelets are the whole remains suspended in plasma. The whole blood is stored at the temperature of around +4°C ± 2°C immediately after collection. It can also be stored at a temperature of +22°C ± 2°C after collection, which is essential for the production of the platelet concentrates. After that, blood needs to be stored at a temperature of +4°C ± 2°C. The whole blood has a shelf life of about 21–35 days subject to the type of anticoagulant used.

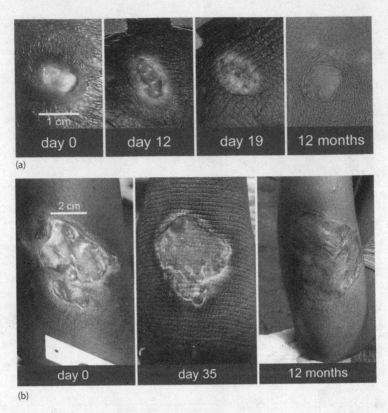

FIGURE 18.4 Healing of the Buruli ulcers for (a) Patient 2 and (b) Patient 5. (From Junghanss, T. et al., *PLoS Negl. Trop. Dis.*, 3, 1–7, 2009.)

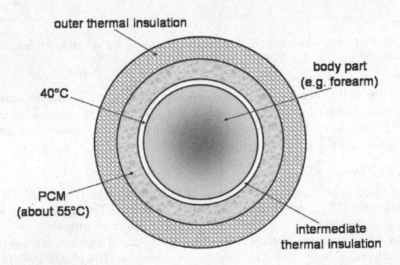

FIGURE 18.5 The diagram for the PCM bandage. (From Braxmeier, S. et al., *J. Med. Eng. Technol.*, 33, 559–566, 2009.)

18.4.2.2 Red Blood Cells

RBCs are the most obvious type of cell found in the blood. Each cubic millimeter (mm³) of blood has about 4–6 million RBCs. It has a diameter of about 6 µm. In humans and particularly in mammals, the RBC is devoid of a nucleus. The lacking nucleus provides more space to store hemoglobin, which is an oxygen-binding protein. This increases the oxygen-carrying capacity. The biconcave shape of the RBC also provides ample surface area for oxygen dissemination. RBCs give the blood its distinct color. RBCs are used to treat anemia in which blood looks pale and the person easily gets fatigued and short of breath. RBCs are stored at a temperature around +4°C ± 2°C. Alternatively, when the RBCs are prepared from whole blood cells, it can be stored at +22°C ± 2°C for a maximum period of 22 hours. This will ensure the production of sufficient platelet concentrates. RBCs have a shelf life of about 45 days, depending on the preservative and the external condition. The shelf life attained can be 21 days for acid citrate dextrose, 28 days for citrate phosphate dextrose (CPD), 35 days for CPD added with adenine, that is, CPDA-1, and 42 days for CPD replaced with a suitable additive solution.

18.4.2.3 White Blood Cells (WBCs)

The WBCs play an important role in the immune response. WBCs come in various shapes and sizes. Some cells have a nucleus with lobes, and others have one single round nucleus. WBCs can be classified into three types: granulocytes, monocytes, lymphocytes. Granulocytes have granules in the cytoplasm. They account for 60% of our WBCs. They engulf and destroy invading bacteria and viruses. Granulocytes are further divided into neutrophils, eosinophils, and basophils. Neutrophils are the main phagocytes. They are the first responders to the site of any inflammation. Eosinophils are involved in allergic reactions and asthma. They kill multicellular organisms, particularly worms. Basophils are about 0.5%–1% of the total WBCs. They are concerned with allergic reactions. They release histamine and serotonin that augments inflammation. They also release heparin, which prevents blood from clotting. Monocytes are divided into dendritic cells and macrophages. Dendritic cells (DCs) are the antigen-presenting cells. DCs can point out foreign cells that have to be destroyed by the lymphocytes. Macrophages destroy and engulf foreign cells in the process called phagocytosis. Lymphocytes are also the body's immune cells, which are categorized into B lymphocytes (B cells) and T lymphocytes (T cells). Both B cells and T cells originate from stem cells in the bone marrow. Some travel to the thymus, where they convert to T cells, and others stay in the bone marrow, where they become B cells. B cells release antibodies that are Y shaped and bind to the infected cells or microbes. It offsets the target microbes or smears it to be attacked by T cells. T cells are further categorized into helper T cells, cytotoxic T cells, memory T cells, and regulatory T cells. Helper T cells release cytokines that facilitate other WBCs. Cytotoxic T cells kill viruses and other cancerous cells. The memory T cell is an experienced cell because of a previous infection or vaccination. During a later encounter, it can generate a superior immune response. Regulatory T cells stop other T cells from earmarking body cells.

18.4.2.4 Platelets

Platelets are known as thrombocytes and are the colorless cell fragments whose main function is to instigate blood clots and stop excessive bleeding. Platelet concentrates must be removed from the whole blood within 8 hours after phlebotomy. It can be then stored at a temperature between +20°C and +24°C with constant agitation to prevent aggregation. It has a shelf life of 5 days with the current plastic bags.

18.4.2.5 Plasma

It is the liquid part of the blood. The RBCs, WBCs, and platelets remain suspended in plasma. Plasma contains mostly water and dissolved proteins, glucose, electrolytes, hormones, O_2, CO_2, etc. Blood plasma is separated from the whole blood through centrifugation where the blood cells are settled at the bottom of the centrifugation tube and plasma is collected at the top. The plasma is then separated. The blood from which plasma is separated within 24 hours of collection is at a temperature below −25°C. It has a storage life of 3 years when stored at this temperature. It can be also stored between the temperature range of −18°C to −25°C but with a reduced shelf life of 3 years.

18.5 STORAGE AND TRANSPORTATION OF VACCINES

Vaccines are temperature sensitive. They lose their potency when they are stored below or above the recommended temperature range. The handlers have to follow complex cold-chain compliance to ensure their effectiveness. Most of the vaccines are stored at a temperature between 2°C and 8°C. The vaccines should be carefully monitored during storage and transportation so that they can't fall outside this temperature range. To handle this, a four-hour and a weekly temperature monitoring has to be ensured. When any vaccines cross this temperature, it should be reported. Table 18.1 gives the maximum and minimum temperature range for vaccines that have to be ensured.

TABLE 18.1

Optimal Temperature Requirement for Different Vaccines

S. No.	Vaccines	Stages	Maximum Temperature (°C)	Minimum Temperature (°C)
1	Oral poliovirus vaccine, BCG, measles, yellow fever	All	+8	−20
2	Hepatitis B, DTP	All	+8	0
3	Diphtheria-tetanus vaccine (DT), Tetanus Toxoid (TT)	Transport	+40	0
4	DT, TT	Storage	+8	0
5	Diluent	Transport	Ambient	0
6	Diluent	Storage	Ambient	0
7	Diluent	Point of use	+8	0

Source: WHO, Temperature Monitors for Vaccines and the Cold Chain, *World Health Organization,* 1999; WHO, Cool Innovations for Vaccine Transportation and Storage, *World Health Organization,* 1–2, 2012.

18.6 PROTECTION OF BIOMEDICAL-SENSITIVE PRODUCTS

In the biomedical industry, sometimes we need constant temperature for biomedical products during casing and transportation. PCMs can be used for these applications. PCMs can maintain a constant temperature, absorbing the additional heat. One such model is shown in Figure 18.6. The working principle of this device is simple. A double-layered pouch can be used that contains PCM surrounding the medical product. The pouch is first placed in a chiller to ensure its solidification. After the proper solidification, the product is placed inside the pouch during transportation. The PCM absorbs additional atmospheric heat, thus protecting the product. During melting, the PCM remains at a constant temperature, thus eliminating any fluctuations in the temperature inside the pouch containing the product.

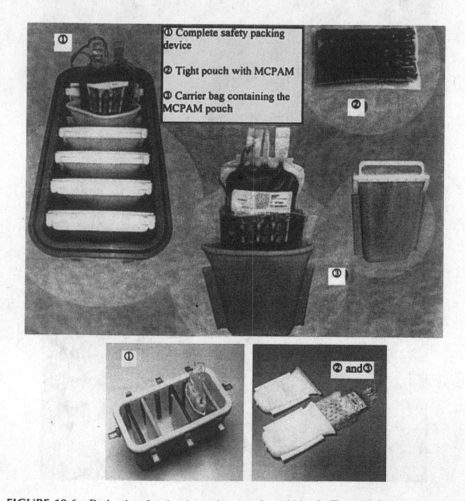

FIGURE 18.6 Packaging for the thermal protection of blood. (From Mondieig, D. et al., *Transfus. Apher. Sci.*, 28, 143–148, 2003.)

18.7   COOLING VEST AND SURGICAL DRESS

PCMs can be used to make surgical gowns, clinical dress, patient uniforms, and clinical beds. This will ensure thermal comfort for doctors, medical staff, and patients in the operation theater. A prototype of one such vest is shown in Figure 18.7. The result shows that the vest has octadecane and hexadecane and has an outstanding cooling effect (Bendkowska et al. 2010). This kind of vest can also be developed for the biomedical industry.

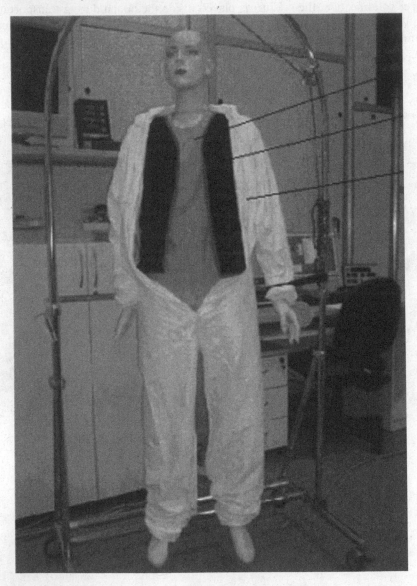

FIGURE 18.7   Cooling vest. (From Bendkowska, W. et al., *Fibres Text East Eur.*, 78, 70–74, 2010.)

18.8 INNOVATIVE PHASE CHANGE MATERIALS

Shukla et al. (2015) have extensively worked on the development of PCMs for the biomedical industry. A list of PCMs developed and suggested by them are presented in Table 18.2.

TABLE 18.2
Innovative PCMs That Can Be Used for Different Applications

S. No.	Application	Optimal Temperature Range (°C)	PCM	Melting Temperature (°C)	Latent Heat of Fusion (KJ/Kg)
1	Buruli ulcers, Pads for orthopedic applications, Pads for back pain (heating)	42–46	$CH_3COONa.3H_2O$	58	264
			Lauric acid	46.13	190.21
			$Na_2S_2O_3.5H_2O$	45–47	NA
			4-Chlorobenzaldehyde	44–48	264–289
			Camphene	48.27	150.41
			LAMA (20–80 wt. %)	46.82	176.73
			MASA (50–50 wt. %)	48.29	183.69
			LAPA (50–50 wt %)	41.25	193.10
			Veratraldehyde	48.50	176.65
2	Pads for back pain (cooling)	10–20	Paraffin C16–C18	20–22	152
			PEG 600	20–25	127.20
			CALA (80–20 wt. %)	21.24	136.91
			CAMA (90–10 wt. %)	22.63	135.84
			LACAMASA (20–30–30–20 wt. %)	15.15	125.08
3	Transportation and storage of vaccines	2–8	Paraffin C14	4.50	165
			Paraffin C15–C16	8	153
			Polyglycol E400	8	99.60
4	Keeping clinical samples	35–40	Paraffin wax	36–38	222–246
			Lauric acid	46.13	190.21
			$Na_2HPO_4.12H_2O$	35.50	264
			$Zn(NO_3)_2.6H_2O$	36.10	146.90
			LASA (60–40 wt. %)	34.59	218.54
			LAMA (60–40 wt. %)	37.58	172.80
			LAMA (70–30 wt. %)	36.22	171.96
5	Surgical dress and clinical beds	30–35	$CaCl_2.6H_2O$	29	190.80
			$Na_2SO_4.10H_2O$	32.40	264
			CAAC (60–40 wt. %)	32.86	153.46
			CAAC (70–30 wt. %)	30.49	160.53
			CASA (90–10 wt. %)	29.54	160.39

Source: Shukla, A. et al., *J. Med. Eng. Technol.*, 39, 363–68, 2015.

18.9 CONCLUSION

PCM is seen as an indispensable material for thermal energy storage because of its high latent heat of fusion and availability in a wide temperature range. PCMs to the biomedical industry are not new. PCM has been effectively utilized for the treatment of BUs, packaging and transportation of blood and vaccines, and the design of clinical beds and garments. PCM is utilized for warming and cooling body parts. When the ice is used for the same purposes, there is a chance of hypothermia. The PCMs are applied to kill germs and used as disinfectants. Hospitals use PCMs for the treatment of rheumatic conditions. It is employed for the treatment of trauma of the musculoskeletal system and circulatory system disorders. It is widely used as cooling care after surgery or accidents. Sleeping bags and blankets are made using PCMs that can be used to prevent hypothermia. There is a need for more PCMs for biomedical applications. The development of new PCMs, thermal testing, field testing, and applicability are some of the challenges ahead.

ACKNOWLEDGMENTS

The author (Abhishek Anand) is highly obliged to the University Grants Commission (UGC) & Ministry of Human Resource Development (MHRD), Government of India, New Delhi for providing the Junior Research Fellowship (JRF). Further, authors are also thankful to the Council of Science and Technology, UP (Reference No. CST 3012-dt.26-12-2016) for providing research grants to carry out the work at the institute. Sincere thanks are also due to the referees for their valuable suggestions.

REFERENCES

Anand, Abhishek, Karunesh Kant, A Shukla, and Atul Sharma. 2019. *Solar Desalination Technology.* Edited by Anil Kumar and Om Prakash. Green Energy and Technology. Singapore: Springer Singapore. https://doi.org/10.1007/978-981-13-6887-5.

Bendkowska, Wiesława, Magdalena Kłonowska, Kazimierz Kopias, and Anna Bogdan. 2010. "Thermal Manikin Evaluation of PCM Cooling Vests." *Fibres and Textiles in Eastern Europe* 78 (1): 70–74.

Braxmeier, S, M Hellmann, A Beck A Umboock, G Pluschke, T Junghanss, and H Weinlaeder. 2009. "Phase Change Material for Thermotherapy of Buruli Ulcer: Modelling as an Aid to Implementation". *Journal of Medical Engineering & Technology* 33 (7): 559–566. https://doi.org/10.1080/03091900903067457.

Combe, Marine, Camilla Jensen Velvin, Aaron Morris, Andres Garchitorena, Kevin Carolan, Daniel Sanhueza, Benjamin Roche, Pierre Couppié, Jean-françois Guégan, and Rodolphe Elie Gozlan. 2017. "Global and Local Environmental Changes as Drivers of Buruli Ulcer Emergence." *Emerging Microbes & Infections* 6 (January): e22. https://doi.org/10.1038/emi.2017.7.

Junghanss, Thomas, Alphonse Um Boock, Moritz Vogel, Daniela Schuette, and Helmut Weinlaeder. 2009. "Phase Change Material for Thermotherapy of Buruli Ulcer: A Prospective Observational Single Centre Proof-." *PLoS Neglected Tropical Diseases* 3 (2): 1–7. https://doi.org/10.1371/journal.pntd.0000380.

Mondieig, Denise, Fazil Rajabalee, Alain Laprie, Harry A.J. Oonk, Thereza Calvet, and Miguel Angel Cuevas-Diarte. 2003. "Protection of Temperature Sensitive Biomedical Products Using Molecular Alloys as Phase Change Material." *Transfusion and Apheresis Science* 28 (2): 143–48. https://doi.org/10.1016/S1473-0502(03)00016-8.

Sharma, Atul, V. V. Tyagi, C. R. Chen, and D. Buddhi. 2009. "Review on Thermal Energy Storage with Phase Change Materials and Applications." *Renewable and Sustainable Energy Reviews* 13 (2): 318–345. https://doi.org/10.1016/j.rser.2007.10.005.

Shukla, A. 2015. "Latent Heat Storage through Phase Change Materials." *Resonance* 20 (6): 532–541. https://doi.org/10.1007/s12045-015-0212-5.

Shukla, A., Atul Sharma, Manjari Shukla, and C. R. Chen. 2015. "Development of Thermal Energy Storage Materials for Biomedical Applications." *Journal of Medical Engineering and Technology* 39 (6): 363–368. https://doi.org/10.3109/03091902.2015.1054523.

WHO. 1999. "Temperature Monitors for Vaccines and the Cold Chain." *World Health Organization.*

WHO. 2012. "Cool Innovations for Vaccine Transportation and Storage." *World Health Organization*, 1–2. www.who.int/immunization_delivery/optimize/.

Index

Printed in the United States
by Baker & Taylor Publisher Services